EVOLUTION UNRAVELED

by Randall Harris

Edited by Dr. Scott Zarcinas

Dedicated to
Lorna,

In 2000, under a Moreton Bay fig tree on the front oval after a Parents and Friends meeting, I offered to research a few facts about how the universe operated, for you to give to David. I never imagined discovering the overwhelming array of intricate complexities that are in place so life can exist on earth, or that so many things I had been taught were known to be scientifically impossible. It has been as much a journey of wonderment for me as it will be for others who read this book. I hope David can enjoy the research I have done and see the facts that exist point to a conclusion that is both liberating and comforting for all of us.

ISBN: 978-0-9942483-6-7 (ebook)

ISBN: 978-0-9942483-5-0 (Print book soft cover)

Published by
Starmonics Pty Ltd
P.O. Box 483 Glenside
South Australia 5065

Cover design and some illustrations by Greg Grace

Illustrations by Ben Peterson

Edited by Dr. Scott Zarcinas

Layout design by Anna Dimasi

Proofread by Dianne Wadsworth, Gumhill Proofreading

CONTENTS

PART THREE – EARLY LIFE

AUTHOR'S NOTE

The world I see doesn't even remotely look like the world I was taught about in school or read about in science books. I was taught that life came from stardust and that creatures reinvented themselves in an accidental process called evolution, and that's how apes eventually turned into humans. Science also told me that matter comes from nothing; that is, the universe had to self-create from nothing.

However, when I examined the layout and operation of the universe, it became apparent that the current theories surrounding its explosive emergence from the big bang are contradictory and even, in some circumstances, scientifically impossible. The conflicting theories bothered me so much that I decided to find out what science actually knows and what the latest scientific minds had to say on life, the universe and everything. That was fifteen years ago; and I was completely dumbfounded by what I learned. At best, the scientific theories on evolution are speculation, not facts. Worse, science itself has disproven every Theory of Evolution to date. Yet, these theories are still taught to our children as fundamental truths and the media still presents them as if they are facts.

Subsequently, there was only one course of action for me to follow. I set about collating all the scientific information that's available in order for you to know and understand what science has put on the table and what they really think behind closed laboratory doors. You have the right to know the scientific truth and, like me, while it may be disturbing to learn that all of us—you, me and the rest of the world—have been misinformed; the truth is actually very comforting.

Randall Harris

PROLOGUE

A New Perspective

The information in this book will make you think hard, very hard, about a lot of things you may never have previously considered. Why? Because, like me, you probably assumed that you were being told the true facts about life, your existence and the universe—who you are, why you are living and how everything came to exist. This book collates the masses of information that have not been brought to the wider attention of society, yet are well known by science. Information that reveals what is being taught about the self-creation of the universe, the solar system and the earth is just not true or even possible—at all. Information that debunks these claims because it confirms that science knows for certain there can only be one way that life and the universe originally started.

Rarely exposed evidence, as revealed in this book, suggests that the universe, galaxies, stars and planets were not assembled as modern scientific thought proclaims and teaches. This scientific evidence also suggests that the universe is a specifically structured, laid out and tightly controlled environment that is fine-tuned so that the earth can support life. In order for life to continue, the earth requires a huge number of strict and unique structures and processes operating in tandem both here and in the wider universe. Science has not been able to find any known method for life to self-create from stardust or chemicals, and in order for life to evolve from one form to another (for example, ape to human), requires the most complicated and specific biological programs far, far more complex than any computer program that the human mind has yet invented, programs that so far we can't even pinpoint where they exist or whether there is any potential for them to be changed.

This book, not only unmasking scientific evidence, also provides a fresh new look at the known facts from science, complemented by real world observations. In many ways, it is a book of scientific facts and processes delivered at a very personal level and presented in great detail with a new and simple perspective.

In the Western world, we have been bombarded with a huge variety of teachings, theories, concepts, and ideas that for the most part we have gradually taken as facts when, surprising as it may seem, they are not necessarily so. We have ideas embedded in our minds about how life, the universe and everything were meant to have happened, but most of these ideas it seems would not even pass the simplest test of evidence. A lot of common scientific information is made up of theories or possibilities that don't work in real life.

But before we embark on this mind-opening journey, we need to remember that each and every one of us is conditioned by our upbringing and environment. The influence of parents, teachers, friends, studies, reading, culture, and media all affect our views and knowledge. As you go through this book, it will be a challenge to accept every single conclusion on every topic because there are some ideas and concepts that you may find difficult to let go. This is normal and natural, even with evidence proving otherwise. Like me, one of the main stumbling blocks to seeing things as they really are is trust: we have trusted our education, we have trusted science, and we have trusted the media. Naturally, as you read, your faith in these three pillars of society will be challenged and tested when it's discovered that a lot of the information they've been supplying isn't proven, logical, reasonable, or even possible. Whilst this may not be deliberate on their part, a lot of things you may have assumed were facts, because of the way they were presented as generalizations, are actually just ideas or concepts that cannot, could not and did not ever happen. When we examine the scientific detail of what is known for sure and how these things actually work in practice, the proven facts will be shown to differ enormously from what you and I have learned.

Accepting that we have been misinformed about the most fundamental issues in life may be discomforting. So be prepared to have many things you believe contradicted by the realities of science that are revealed to you. For the first time, scientific facts about life and the universe are laid out in a simple, step-by-step look at how everything happened from the very beginning. It is my sincere hope that this book will expose the real knowledge that mankind currently possesses and bring a fresh way of analyzing the scientific data that humankind has collated over thousands of years.

Finally, once we have been exposed to the true and verifiable facts about how things actually work, we will at last be able to make sensible and credible conclusions about the origins of the universe and the existence of life. This can only be good news.

How Will We Undertake This Journey?

The basic issue is very simple. To explain how everything exists, there are only two options: either there is a Creator who made everything or there isn't and everything made itself. So, what does the scientific evidence tell us? As most of what we have been taught comes from science and has resulted in the development of evolutionary theories, we are going to work our way through the timelines of history from the very beginning comparing the most common theories with the facts known by science. This method involves digging deep into the mechanics of processes and discovering the reality of what is actually possible of its own accord rather than accepting theories which are unsupported by detail. Evolutionary Theory appears to avoid any real level of scrutiny by using generalizations, and when challenged, often falls back on the unprovable defense that things weren't always as they are.

During this analysis we discover perhaps the most startling fact of all:

There is no official Theory of Evolution proposed by science.

The only place on earth where a Theory of Evolution is laid out in step-by-step detail from beginning to end is here, in this book. It is a complicated and disjointed journey but it is filled with amazing moments of discovery, clarity and awe.

This process will reveal many interesting truths, including:

1. The Evolution of the universe and the solar system

- There is only one way that the universe could possibly have come into existence, according to what science knows.

- The official position of science concerning the big bang theory.

- The unusual, incredible and structured layout of the universe.

- Why many different theories on evolution are proposed.

- Why there is no official Theory of Evolution.

- Why there are many differing opinions about the timing of the universe's beginnings.

- The truth about how our solar system was formed and how many of its bodies were actually made at different times.

- Why the planets and their moons don't collide with each other or the sun in defiance of their massive gravitational attraction.

- What components exist to make all the things in the universe.

2. The Evolution of Biology and Life

- The science of how life actually started.

- How the atmosphere, water, oxygen, and microbes appeared on earth.

- Whether any natural biological process can actually invent itself.

- How the sudden arrival of complex life forms suited to water, land and air appeared complete, without any trace of origin.

- The basics of how life operates as a conglomeration of tiny cells, what cell mutation actually is and how it happens.

- The indisputable biological structures and processes that prove whether one life form can change into another.

- The actual steps necessary for creatures to change from living in the oceans to living on land and flying in the air.

- How the first male and female genders appeared from nowhere at exactly the same time and were perfectly matched from the beginning.

- How the human body, and all creatures' bodies, operate under a vast series of automatic, interrelated, control programs so complex that humankind cannot determine how they work, let alone find them.

- How science determines facts, theories, hypotheses, and laws and why they can change in an instant.

- Whether instincts can change from one creature to another in an evolutionary process.

- What the world would actually look like if mutations were responsible for changes to living beings.

- The truth about the fossil records and how they undoubtedly prove whether or not biological evolution ever actually happened.

- An analysis of how cells form then change into different cells types.

- The amazing process of forming a living creature from an embryo and how it operates under the strict control of internally timed, sequential, instruction programs embedded in its make-up from conception.

- The first ever exposure of the process that is the KEYSTONE of biological evolution. An intricate revelation about how progressive biological mutations would actually have to happen at a detailed level never previously examined. A seemingly unassailable barrier for any theory that a life form can change itself into something else.

3. The Evolution of Humanity

- The human body, its basic design, operation, and features.

- Revelations about the origins of human traits such as reason and intuition.

- The amazing insights of Albert Einstein.

- What the astrophysicist Stephen Hawking really thinks.

- What Charles Darwin actually said.

- Whether or not humans came from the apes.

- Why the scientific community doesn't always tell us what it knows.

And as a bonus, you will be presented with scientific facts that will let you answer that most famous of questions once and for all:

"Which came first, the chicken or the egg?"

PART ONE

THE THEORY OF EVOLUTION

1

EVOLUTION AND CREATION — THE BASICS

As generally agreed, there are only two options to explain the origins of the universe and our existence: evolution or creation. First, though, to get an understanding of the underlying principles of evolution and creation, we will start with the basics.

What Is Evolution?

Evolution is a theory or concept that everything in the universe created itself from nothing without any specific purpose, design, meaning or plan. The actual meaning of the word evolution is *change over time*.

What Is Creation?

Creation is the belief that there is a God or Higher Being who is First Cause and who designed and made everything that exists.

Evolution

As it is fundamental to understand how the concept of evolution arose in our thinking, we will now examine this topic in some detail.

Over the course of human history, there have been a few noteworthy people who have raised the specter of evolution in living creatures. A Sicilian called Empedocles, circa 492 BC, was thought to be the first person to propose that living creatures somehow evolved from each other. Aristotle, a Greek philosopher born in 384 BC, put more definition to the idea as being a gradual development over time where single forms eventually changed to become humans. Then, as religious ideas spread across the world, most notably the belief of creation as told in the Judeo-Christian-Muslim traditions, the idea of evolving life took somewhat of a backseat. It wasn't until Sir Francis Bacon (1561-1626) known as the 'Father of Empiricism' (knowledge that comes from evidence via sensory experience) and the French philosopher Rene Descartes (1596-1650), who was best known for the statement, "I think, therefore I am" (Cogito ergo sum), that the idea of evolution was revitalized. Charles Darwin (1809-1882), an English naturalist, made various general observations about self-development and the process of natural selection. Darwin was given the title of 'Father of Evolution' for his work, most notably published in his book *On the Origin of Species* in 1859.

Since its humble beginnings in Sicily, the concept of evolution has been fuelled by a surge of scientific data to the point that evolution now encompasses both the origins of the universe and the development of living creatures. It is also important to clarify that, in this book at least, the adaption of a species and the evolution of a species are not the same process. For example, the adaption of Darwin's finches into different groups (by definition species) that did not interbreed is not the same as the assumed evolutionary development of reptiles changing completely into birds. Humans, too, have adapted to their environment. For instance, some anthropologists claim that the strength of the human thumb was a developmental adaption to our use of stone tools. This is to say that our thumbs did not evolve from our use of tools, but that what already existed adapted and became better.

It is also surprising to discover that important scientific information on evolution does not exist neatly compiled in one place. This is predominantly due to the division of science into many disciplines—Astronomy, Geology, Physics, Chemistry, Biology, Botany, Mathematics and so forth. Individuals in each of these disciplines present their own fairly detailed analyses to substantiate the evolutionary theories regarding their particular branch of science, but they tend to stop short of linking their theories to the next chronological step because they defer to another area of expertise. An example of this is in the appearance of life on earth where chemists maintain most of the ingredients to make life existed in their basic forms, and biologists tell us how these forms operate, but there is a gap where neither party has the responsibility to explain how these chemicals combined together to invent life which is, so far, considered scientifically impossible of its own accord.

Consequently, there are many gaps in the theories of evolution that require filling, loose ends that need tidying up, if we are to ever have a full and complete, scientifically validated theory. Another example is that no natural scientific explanation exists for the initial appearance of matter, leaving a chasm between nothing existing and something existing. Complicating matters further, there are also many scientists within each individual discipline of science who provide differing opinions supporting alternative theories, resulting in literally millions of documents arguing for and against evolution. This does not even include religious, philosophical and personal beliefs.

Therefore, in order to completely understand the totality of evolution as such, it is necessary to link all the individual and separate scientific theories into a chain of chronological events, whereby the order of events in evolution is determined by their historical occurrence. However, within most of the scientific disciplines, there are numbers of differing theories about the same issues and, further, the chronological sequence of events is regularly in dispute, often arising because of an overwhelming lack of scientific evidence. The end result is a lot of theories that are based on many individual's specialized research, their own perspectives, ideas and thoughts, which leads to theories contradicting each other and even contradicting the way things are observed to work today. The most obvious of these is that evolution claims all types of creatures are the result of mutational changes by other creatures, and yet this is not observed. Of the trillions of creatures on earth, not one is known to be changing by mutation into a different creature and we will learn in Chapter 8 why the fossil records do not change this observation.

As it stands, there is no step-by-step universally accepted theory covering the entire Theory of Evolution. In other words, most of what we know and understand of evolution is just a blend of different theories that most scientists disagree about because most of it can't be scientifically proven. Analyzing the available scientific information on evolution that is scattered around the sciences, books, media, and the internet reveals something remarkable:

> *The Theory of Evolution is predominantly assumed by inference and supported by statements that are presumed to be fact.*

As examples to elaborate these two points:

1. Many articles, books, scientific writings, and documentaries will regularly refer to previous life forms as our ancestors, even though there is no evidence that such an amazing transformation could ever occur. Simply because it is stated regularly, it is *assumed by inference* that all creatures developed from other life forms, despite the complete lack of evidence.

2. The big bang is regularly referred to as a known event. It is *presumed to be fact* by the majority of the population. However, the overwhelming evidence, now widely recognized by physicists and cosmologists, is that the layout and movement of the universe could not have been

formed from this type of large explosion. Despite Hubble's discovery a century ago that space is stretching apart and not static, as Einstein had assumed, the big bang is still only a theory, one that receives constant mention; it is not a fact. (This topic will be examined in greater depth in Chapter 3).

Unfortunately, in the Western world at least, what we are taught on evolution at school or university is limited and lacking in detail, especially the amount of detail required to understand the myriad of processes involved in evolution. That is, how the changes proposed by evolution actually occurred in practice, bit by bit, little by little, over the course of time. For example, we have been told that land life came from the oceans but when we look at how a fish's body actually operates, cell by cell, we discover that this cannot actually happen. This process is examined fully in Chapter 16.

Contrary to popular opinion, there are no text books with substantiated evidence for the processes that happened from the instantaneous arrival of matter to the formation of the life on earth. There don't even appear to be any government-based subjects, qualifications or formal acknowledgement of this theory as a specific topic. Parts of various evolutionary theories are scattered in general teachings in science, biology and geology, but no single, all-encompassing theory with supporting evidence exists in our schools and universities. In fact, because these described processes contradict real world scientific observations, all we really have in regards to evolution are generalizations without scientific explanation or validation.

Moreover, there is considerable confusion surrounding the actual definition of evolution. The general consensus is that evolution only refers to the process of living things changing into other forms (i.e. biological evolution). But an actual, all-encompassing Theory of Evolution should encapsulate the complete story, such as:

- Why does anything exist? How, why and when did matter and energy self-create out of nothing? Is this process continual or has it stopped?

- How did the universe form itself into such a specific layout that defies the known physical forces of nature?

- How did earth happen to be in the perfect location and have all the components required, and in the correct ratios, for life?

- Can life self-create from non-living matter? If so, how?

• What is the process by which living entities physically change from one form (species or class) to another? Can this be scientifically proven?

Defining The Theory Of Evolution

One of the problems with discussing evolution is that, surprisingly, there is no official definition for it. The explanation of this aberration appears to be due to evolution being merely theory and not fact. The key word here is theory. However, there are many descriptions from science and dictionaries. The following quotation is from *Evolution and the Myth of Creationism,* Stanford University Press 1990, Dr. Tim M. Berra, (Professor Emeritus of Evolution, Ecology and Organismal Biology at The Ohio State University). It has been chosen above others because it offers a broad definition:

> "Common usage of the word 'evolution' is the idea that living things in our world have come into being through unguided naturalistic processes starting from a primeval mass of subatomic particles and radiation, over approximately 20 billion years. A more precise understanding of the above statement divides the 'atoms to people' transition into four realms:

1. *Cosmology* is the branch of astronomy which deals with the origin and formation of the general structure of the universe.

2. *Abiogenesis* refers to first life—the production of living organisms from inanimate matter.

3. *Micro-evolution* or speciation refers to populational and species change through time. There are many published examples of speciation, if by the development of a new "species" we mean the development of a new population of individuals which will not breed with the original population to produce fertile offspring. Micro-evolution is a scientific fact which no one, including creationists, dispute.

4. *Macro-evolution* or *general evolution* refers the progression to more complex forms of life. The mechanisms of macro-evolution, including whether or not micro-evolution over a long enough time leads to macro-evolution, can be regarded as a "research topic".

The first and most obvious point to make about this summary is that the description starts by stating that evolution is an idea, not a fact. Also noted is the manner in which item 2 (*Abiogenesis*) is described: *the production of living organisms from inanimate matter.* This is just a description of the scientific field of study but it is written in a way that can be taken as though it is an actual fact—as something that has happened. Well, it's not a fact and, as far as we know, it hasn't and can't happen. Abiogenesis is the study of the theory that life made itself from non-living matter. There isn't any evidence or proof; it is just a theory to explain where life came from if a Creator didn't make it. Abiogenesis is simply a field of study to determine how such a thing could possibly have occurred, but has still yet to deliver any substantiated factual proof. We will examine this subject in greater detail later, but the point to keep in mind is this: if we are constantly subjected to this type of wording, we can often subconsciously accept it as true when it is not.

Thus, what is made available regarding the Theory of Evolution is a misshapen construct of numerous independent theories from different areas of science that are only connected by chronological dating. Furthermore, within each field of science, there are often many, many theories for a given event. If we choose just nine major steps in the history of the universe, each having just three differing but reasonable theories for each step it would look something like this:

	Matter & energy appear	Universe forms	Solar System forms	Earth's atmosphere forms	Microbial life starts	A living body forms	Land life	Plants appear	Flight starts
Nothing exists									
No. of theories	3	3	3	3	3	3	3	3	3
Cumulative Total	3	9	27	81	243	729	2187	6561	19683

As we can see, there are almost twenty thousand different ways to make up a complete Theory of Evolution from ideas and concepts that people have come up with using just these nine topics. In reality, there are thousands of topics that evolution needs to explain. What this highlights is that there is no obvious, single, clear path to follow; rather, there are many, many thousands of different possibilities because of the lack of clear evidence to support or give credence to any one particular Evolutionary Theory.

This raises a serious issue for evolution as a theory requiring substantive scientific proof to validate it as a fact of life, not a theory of life. The issue is this: for any Evolutionary Theory to work, it is reliant on one step happening after another, sequentially, as time goes on.

Evolutionary Theory is therefore dependent on time being the link between events. If suddenly the dates that things occurred were found to be different, for example, if there was new, undisputed evidence of humans appearing on earth before apes, then the theory would change that humans evolved into apes, not vice versa, as is currently the case. This seems ridiculous at best, but this outcome is actually possible because most of the current theories are only based on the chronological sequence in which life appeared. As such, because science has identified that apes appeared on earth before humans, evolutionary theorists claim that apes have changed into humans over the course of time.

Although this seems simple and logical enough, this approach doesn't match the evidence of available scientific data, as we will see. What has come to light is that evolutionary theorists have only used the chronological order of consecutive events as the base for evolutionary theories because *there is no actual evidence of evolutionary events ever happening.* What makes this approach worse is that many of the processes involved are known to be impossible. Chronological sequencing as the foundation for Evolutionary Theory is like building a house on sand—unstable, unsound, and not very good science.

Herein lays the enigma that is the Theory of Evolution and why it is probably not an official subject taught anywhere in our educational institutions: *Every part of it is subject to an opposing view and it is open to change at any time* and, most likely, no credible scientist wants to put their name to a full theory where parts of it can and do regularly change.

When chasing down these theories we regularly bump into events or circumstances that elicit a response of "unexplained" from science. This doesn't mean there is no explanation possible; it means there is no way to explain those events within the confines of the theories proposed. For example, several moons in our solar system display all the characteristics of relatively newly formed bodies, something that cannot be explained if the solar system was formed in one process 4-5 billion years ago. Remember solar system formation of itself is just a theory, it is not fact. There is, however, a logical explanation put forward in our analysis which complies with the scientific data and makes sense. We will encounter many examples throughout the book that are assessed purely on the simplicity of the evidence and it soon becomes clear why evolution's proposals do not reflect the reality of the way things are.

The Principles Of Evolution

As there is no complete, definitive reference document to tell us the step-by-step details of what is theorized to have happened in the process of evolution, the most up-to-date current theory can therefore only be taken as an assumed likelihood of possible events, based on a collection of current available information. Taking this into account, below is a list of evolution's underlying principles from which it has been created and delivered to us, the general public:

- There is no specific order, plan, meaning, or design to the universe, life or anything.

- Life and the universe exist purely by chance.

- Natural things that happen are essentially random, unintentional and possibly coincidental.

- Because there is no external control over things, chaos is the only legitimate determining factor. [1]

- There is no end point to which 'evolution' is heading; it is a continual process without end.

These principles leave us with a picture of evolution as a boulder barreling downhill out of control—banging, bumping, crashing, and bouncing from one spot to another. It doesn't matter where it goes, what it hits or what it does. The evolutionary boulder is on a course of its own—unstoppable, unpredictable and unplanned.

This concept of uncontrolled chaos is important because the universe and the world we live in:

1. Should show clear signs of the destructive path of these uncontrolled processes. We should also be able to pinpoint exactly what has happened by the impressions and debris evolution has created; and

2. Should comply with the subsequent mathematical predictions that there should be all sorts of crazy, weird things happening beyond the realms of quantum dynamics, and these should be evident in macroscopic (visible) form everywhere. As we will see throughout this book, there really shouldn't be controlled structures and systems in place, nor should there be clear, defined living entities.

---

1. Chaos theory is a mathematical field of study applied to various scientific endeavors, which essentially says the outcome of an operating system will be affected by the initial conditions. For example, meteorology is a science predicting weather based on physical variables. In the 1960s, as a result of changes in the outcomes of experiments by meteorologist Edward Lorenz, it was recognized that there were unidentified events that changed weather conditions. A concept called the 'butterfly effect' was put forward where it was supposed that if a butterfly flapped its wings in Beijing in March then hurricane patterns in the Atlantic will be different by August that year. Effectively, minor changes in variables resulted in major changes to outcomes.

Evolutionary principles predict unstructured layouts in most things as well as masses of partly developed life forms. As everything is meant to be randomly changing, rather than staying consistent, current living creatures would be expected to be developing new and different bodily parts, functions and processes. Yet this is not observed to be happening, without abnormal contamination that is (such as radiation or chemical poisoning).

As we progress through the Theory of Evolution, we will examine these issues in depth to see if the contents, structure and layout of the universe do actually represent such a wild journey.

Creation

The opposing view to Evolutionary Theory is Creation. Creation is a concept based in culture, religion and spirituality and has existed well before the idea of evolution was even considered a possibility. There are many accounts of Creation that have developed from oral traditions in many different cultures around the world—Australian Aboriginals, Mayan, the ancient Greeks, Norse, Hindu, Shilluk, Tahitian, Judeo-Christian, almost every single culture that has existed. The general consensus is that God, as a Supreme Being, created life and the universe with a plan, with meaning, with purpose, and with love. Since Darwin, however, science has gradually taken the view that the Western world (Judeo-Christian) account of Creation is a mythical or fictional account of how the universe and life exist, but as yet has not accumulated the evidence required to irrefutably disprove its validity as a reliable and trustworthy account.

However, the basis for Creation, especially the Judeo-Christian adaption, is not only a major barrier for many scientifically minded people but a major barrier for everyday people in general. The story of Creation is told in the opening chapter of the Bible, and is generally interpreted as taking less than a week to create the universe, solar system, plants, animals, and humans. Naturally, this conflicts with the scientific validation that the universe is billions of years old and herein lay many of the problems associated with the Judeo-Christian explanation of the universe. Many people cannot accept either the one week creation timetable or the literal teachings in the Bible. Accordingly, they dismiss both the likelihood of God existing and of Creation being possible.

What is not readily recognized, however, is that there is a Theory of Creation and a process of Creation. The Theory of Creation is that God made everything. The main documented process of Creation is that physical matter and life were made in the manner described in the Bible.

However, the aim of this book is not to present information that can determine or validate any religious beliefs, rather it is to examine whether or not the scientific evidence that exists to date can actually support the Theory of Evolution. The aim is to present and analyze the truth of the information we are reviewing to determine a valid conclusion.

Keep It Simple

Before we start the detailed review on evolution, we are going to do one simple but revealing thing that will dramatically change the way we are able to assess the validity of everything that is presented from herein. This concept will be a constant reference point to keep our reasoning on track. In the past, the Creation versus Evolution debate has primarily been focused on the time after the formation of matter and energy, such as the development of the universe, galaxies and planets. This is because scientific research has been unable to find data that can explain the origins of matter and energy.

There are too many variables that lie beyond scientific explanation and observation. In fact, this time before the big bang, this 'time-before-time' eternity, is actually inaccessible to science and its observing instruments because it lies outside the universe—worldly instrumentation cannot observe events before the universe and world existed. This leads to something of great importance that cannot be explained even by theory, that the existence of matter and energy require a non-scientific explanation. In other words:

Matter and energy cannot have appeared from nowhere without the input of an external source as science, using the First Law of Thermodynamics, states that it is impossible (as explained in Chapter 2).

The act of going back one step to before the known existence of anything—to the 'time-before-time' eternity when there was/is no physical matter and no physical time—enables the deconstruction of evolution and creation theories and processes to their basic level or foundation, thus allowing discussion and consideration of these conflicting arguments from an entirely different perspective.

To start our analysis, we will now examine the Theory of Evolution and how it describes the processes with which the universe came into existence completely by accident, how it arranged itself, and how atoms, elements, stars, planets, and life forms came into being. We will look at the important steps in the development from nothing to life as we know it, to determine how this amalgam of theories is supported by evidence and logical reasoning.

Chapter 1 Unraveled

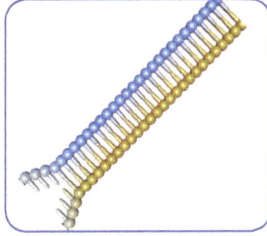

- *There is no official, definitive Theory of Evolution. There are many possible theories that can be amalgamated from a range of chronological datings from various branches of science. They are open to change and do change regularly. Not all branches of science agree on a theory for their discipline—in fact, most disciplines of science propose multiple theories.*

- *Much of the information that we receive from many sources—articles, books, scientific writings, and documentaries—is based on false presumptions and assumptions, and on theories that have not been proven.*

- *Creation has long been a well-established concept in human thinking across the world.*

- *There is considerable debate among creationists about the timing of creation, which distracts from the underlying principles involved.*

- *The existence of matter and energy is either accidental and chaotic or planned and created.*

2

ACCIDENTAL EVOLUTION

To recap, Evolutionary Theory assumes that life, the universe and everything happened accidentally, by chance, without any design or purpose. This theory is based on philosophical reasoning dating back as far as the 5th Century BC and is allegedly supported by the sequence of events found in the fossil records and from scientific observations on earth and from space. The key words for evolution are: accidental, chance, chaos, and time.

To start our deconstructive analysis of this theory, we will define the major steps that are widely accepted as being integral to the process of evolution and then we will work our way through the available scientific data to see what is fact and what is fiction.

Here is a summary of the major steps, as far as we know, in the evolution of life and the universe:

- Before anything physical appeared, there is nothing anywhere. There is nothing but blackness—no space, no time, no-thing.

- For no reason at all, matter and energy made themselves from nothing and then stopped making themselves. In other words, something appeared from nothing: a one-off event.

- At some unknown time later, this material collected itself together until it became so dense and hot that it underwent a massive explosion, called the big bang, and eventually formed itself into the universe.

- Our solar system, the sun, planets, moons, asteroids, and comets formed themselves from localized clouds of matter.

- Our planet, earth, coincidentally finished up in the perfect position to support life—before life existed.

- All of the elements on earth, without which life would not exist—chemicals, gases, water, and atmosphere—self-created without reason.

- Life started itself on earth completely by accident and for no reason.

- The first form of life, being single (bacterial) cells, made themselves from chemicals that were floating around in water. These new life forms continually divided into halves and mutated to perform the many different, complex and fundamental processes necessary for future life forms to exist, before those life forms existed.

- Single cells changed themselves into highly complex multiple (nucleated) cells.

- Nucleated cells continually divided into halves, mutated into other different cell varieties and then joined themselves together to make living things, such as plants and creatures, initially in water and then moving onto land.

- Most of these new life forms self-created perfectly matched male and female genders, splitting the information required to procreate and then recombining to make a new miniature form of one gender or the other.

- Descendants of the first living creature changed every component in their bodies to mutate into all of the different creatures known on earth—in water, on land and in the air—culminating with humankind as the last known new life form.

- This process of random biological change will continue indefinitely, if not until the death of the earth.

We know that matter and energy exist. Matter can be seen, felt and measured with scientific instruments. It can exist on the macroscopic level in solid, plasma, liquid or gas forms. In the microscopic, quantum level, matter exists as energy, as confirmed by Einstein's famous equation $E=MC^2$, where E is energy, M is mass (matter) and C is the constant speed of light. On the macroscopic level of existence, matter is the substance whilst energy is that force which moves the substance. There are two basic types of energy: kinetic energy, which is the power doing the work (e.g. superheated steam driving the pistons of a train engine), and potential energy, which is stored power waiting to do the work (e.g. the body of water at the face of a dam waiting to propel an electric turbine). There are various forms of electromagnetic energy and there is debate among scientists as to whether gravity is a form of energy or results from curved space.

Let us start by finding out how evolution explains that out of nothing, matter and energy came to exist.

We will begin our journey at the 'time-before-time', the nothingness prior to physical space and time, when there was nothing at all, anywhere— no time, no space, no-thing. This can be visually represented as blackness, as shown below.

Figure 1: No Time, No Space, No-Thing

The current size and contents of the universe are massive. There are millions of billions of billions of stars that we know about. Possibly, there are further innumerable numbers of galaxies and stars and unknown numbers of planets. The amount of material that we have been able to identify in space makes our planet so miniscule by comparison, even smaller than a grain of sand on the beach, as to be completely insignificant in the greater scheme of the universe.

Evolution has to account for the sheer quantity of matter in the universe, but even trying to imagine it is a difficult task. Nonetheless, let us try to visualize the arrival of this material when it first formed by doing two things:

First, imagine filling a five-pack copy-paper carton with regular twelve millimeter diameter marbles. There would be approximately ten thousand marbles in the carton. Now imagine making a stack that is eighty cartons wide, fifty cartons deep and twenty-five cartons high. In other words, there are about one hundred thousand cartons, roughly sixteen meters square and seven meters tall. The total number of marbles is about one billion, standing at about the size of an average house.

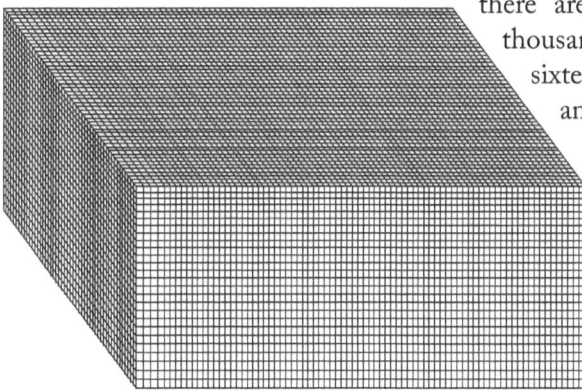

Figure 2: One Billion Marbles

Next, *for each* of these billion marbles in the stack, make another identical stack of one hundred thousand cartons next to each other. That is, a billion stacks of a billion marbles. The wall of packets would reach over sixteen million kilometers (ten million miles) long. This wall would reach the moon over forty times.

Figure 3: One Billion Billion Marbles

The total number of marbles in all of these cartons is less than the number of stars that we know about. Whilst this is extremely difficult to imagine, it does give us some idea of the staggeringly enormous amount of matter in the universe.

Second, we will now attempt to visualize the arrival of all this material that makes up the universe. At this point, it is important to understand that science identifies the universe as a place that formed after matter already existed. In fact, the time frame over which matter first appeared is hotly disputed as this is a time before the universe came into being. In reality, Evolutionary Theory has no explanation regarding how this matter could self-create or when it arrived or where it came from, only than it just existed.

Imagine then that you are you are floating in total darkness. There is no light anywhere. You cannot even see your hand. It is pitch black and deathly quiet. You can take a moment to visualize this by closing your eyes. You cannot see anything in any direction, as there is nothing to see. Looking ahead in the distance far, far away you are about to witness the appearance of matter from nothing—not the start of the universe or the big bang, as that is claimed to have happened after matter first existed.

What you are about to see is the arrival of enough matter, gases and energy to make up a huge star for each of the marbles that we have just seen in Figure 3: One Billion Billion Marbles. To put this in some perspective, try to imagine each marble being as big as our sun, which is over one million times the size of earth by volume. It is estimated that our sun converts twenty thousand trillion tons of hydrogen to helium each year and that it will last for billions of years. The amount of material in just this one star is incomprehensibly massive, and if we multiplied this by the known number of stars it would amount to trillions of trillions of trillions of tons.

Then, a truly daunting event occurs. The space in front of you fills up to the full width and height of your vision, as shown below in Figure 4: The Primal Universe. All you can see is a wall of gas and bits of matter. Perhaps the energy is so powerful you can feel it. The amount of material is so inconceivably big there is no room for anything else.

Figure 4: The Primal Universe

This is how the matter and energy in the universe is theorized as having appeared, accidentally, out of nothing, and for no reason. But, as there is no formal theory of the period of time that it took, it is generally thought it happened instantly. Ironically, you probably couldn't see anything at this stage, as light had not yet started. Matter therefore appeared in total darkness.

Evolutionary Theory, however, does not tell us what the matter was made of (for example, atoms or gases). Nor does it tell us what type of energy appeared, where and how it was stored. Furthermore, and just as importantly, it does not explain why the amount of matter being made in the primal universe suddenly stopped.

What we have then is an unexplained situation wherein masses of matter and energy arose instantaneously:

- From nowhere.
- For no reason.
- In an unknown form.
- Then suddenly stopped.

This concept of 'instantaneous matter' is how the universe was able to exist, according to the evolutionary protagonists, who assert that the beginning of the universe was just an accident. In other words: this is the start of the Theory of Evolution.

If this amount of matter formed all at once, and was an accident, it raises several issues, the most important being: Where did physical matter and energy come from if there was nothing physical beforehand?

It is a tricky question, and evolutionists have as yet to work out how something can appear from nothing. Some even avoid dealing with the issue altogether. The inconsistency with an accidental universe is that, whichever way it's examined, the formation of matter and energy is unexplained or, as previously noted, unable to be explained within the framework of Evolutionary Theory.

Even at this early stage, this scientific theory doesn't appear too scientific. There is one enormous obstacle that this theory needs to overcome if it is to prove that matter was formed by accident from nothing. In particular, it needs to contend with one entrenched scientific observation that always applies. This is called The First Law of Thermodynamics.

The First Law Of Thermodynamics

A broad definition of this law is:

Energy is maintained within a system and that energy can be transferred into or out of the system by either heat or work interactions.

In essence, what this principle says is:

- In the physical world (the universe), the levels of both mass (physical things) and energy (light, heat etc.) are fixed at a total that cannot be increased or decreased.

- It is possible to change between the two: one state of matter can be converted into another and one form of energy can be converted into another. It is possible to convert matter into energy and vice versa. However, it is not possible to reduce or increase the total of both.

- Therefore, the total amount of energy and matter in the universe remains constant.

For example, if we light a fire, the total level of heat increases (infinitesimally) in the same proportion as the wood decreases (again infinitesimally).

In other words, there is no way of making more matter or energy: we can only change what already exists. Nor can we get rid of any matter or energy. As such, in terms of the matter that made up the universe, how can we explain where it came from? It cannot have made itself, as matter cannot do that. Did it form itself from some type of energy? If so, where did the energy come from, as energy cannot self-create either.

Where does this leave us? What can we say about the matter that appeared at the beginning of the universe?

According to scientific principles, the matter making up the universe could not spring into existence by itself. It is impossible.

Ask any scientist or physicist espousing the Theory of Evolution to explain the origin of matter by itself from nothing and the response will probably be that it cannot be explained by science. That is, the accidental appearance of matter from nothing cannot be explained by scientific observation or scientific reasoning.

This, therefore, is a summary of what Evolutionary Theory is saying in regards to the birth of the primal universe:

All of the material that makes up the universe made itself out of nothing, in some unknown form, for no reason, even though science cannot explain how it happened and there is even an immutable law of thermodynamics that says it is impossible.

Chapter 2 Unraveled

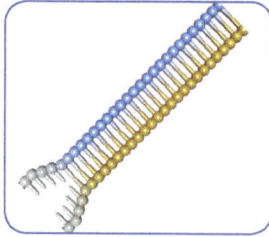

- *The amount of matter and energy in the universe is massive and perhaps even beyond our comprehension.*

- *Before the primal universe there was nothing: no space, no time, no matter, no energy, no-thing.*

- *Science cannot explain how, why or when matter and energy made themselves from nothing.*

- *It is assumed that the primal universe sprang into being in total darkness, before light existed, but science cannot explain the make-up of its primal matter or energy.*

- *The First Law of Thermodynamics tells us that matter and energy cannot make themselves from nothing nor can they be destroyed.*

PART TWO

ATOMS TO GALAXIES

3

AN EXPANDING EVOLUTION

The Big Bang Theory

The Theory of Evolution, even though science cannot explain how matter came into existence by itself, still seeks to describe what happened after that point. This part of the theory was developed by working backwards from the scientific information that we have today and theorizing what were the most likely happenings. Generally, science has had a familiar theme regarding the most likely explanation for the self-formation of the universe. Most of us were taught this at school.

This part of the theory tells us that pre-existing primal 'matter', most likely comprised of gases and solid materials and probably some form of energy, gravitated together into a very dense mass over an unknown period of time.

Figure 5: Contraction Of Primal Matter

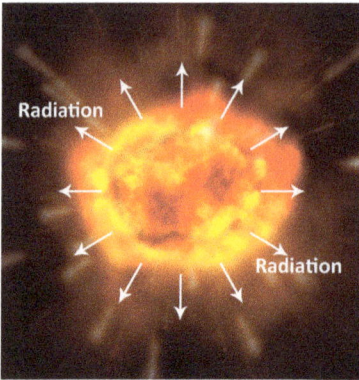

Figure 6: The Big Bang

This mass compacted itself into a tiny space, no bigger than a pinhead, which became so overheated it caused a massive explosion that propelled untold billions of fragments of matter outward with enormous force at ultra-high speed and temperature. This explosion is now commonly referred to as the big bang.

At this point, we have to realize something else quite important. If primal matter did first appear by itself, it mustn't have exploded into existence; otherwise it would have shot out in all directions and been unable to regather itself to then create the big bang. It must have been a relatively ordered and peaceful event. Essentially, what Evolutionary Theory is proposing as the likely scenario, is that masses of matter and energy calmly popped up out of nowhere, in the same location, then hung around in the same general locality so gravity, a generally weak force, could draw it all together to ultimately create the stars and planets at some later time.

In 1965, two astronomers, Arno Penzias and Robert Wilson, working at Bels Labs in New Jersey, discovered cosmic microwave background (CMB) radiation, which to the scientific community confirmed that the observable universe had started as one large explosion, approximately 15 billion years ago. At this time, it has been estimated that the temperature inside the explosion was billions or even trillions of degrees Celsius.

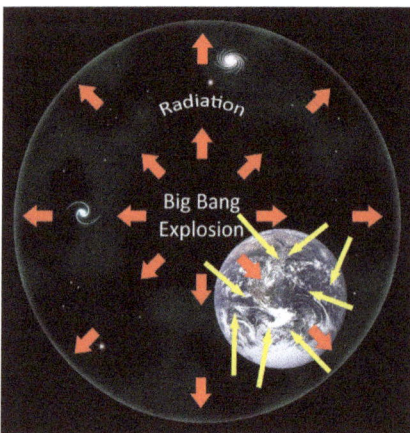

Figure 7: Cosmic Microwave Background (CMB) Radiation

One of the problems with this part of the theory is that this background radiation would only travel in one direction—away from the initial explosion—because there was nothing else to deflect it in another course or direction. But the radiation that Penzias and Wilson discovered attacks the earth from many different directions, as shown in Figure 7: Cosmic Microwave Background (CMB) Radiation.

This CMB radiation is also at a very low level that some scientists have calculated should be up to one thousand times stronger. Furthermore, the CMB radiation is apparently not hot enough if it happened 15 billion years ago.

CMB radiation therefore appears to be an unreliable way to measure the chronological history of the universe. The big bang theory could probably be dismissed on this evidence alone. Furthermore, after the big bang explosion, as countless pieces of matter rocketed outward and flew away from each other, most of them then gravitated together into smaller groups to eventually form stars by a process of fusion (which we will discuss later).

The big bang scenario actually leaves more questions unanswered than it does answered. Some physicists believe that not even the basic structures of atoms and their nuclei could have existed at such enormous levels of heat. Which leads to another issue that requires closer examination.

Early Matter

The basic building blocks of matter are atoms (which we will review in more detail in Chapter 4 and later in this chapter when we talk about quarks and other subatomic particles). All physical matter is comprised of atoms. Different combinations of atoms form different types of matter. However, and this is self-evident, atoms can't form any type of physical matter until they first exist. That is, only once atoms existed can they combine to create physical matter.

Accordingly, the first thing that ever appeared would therefore have to have been the subatomic components that make up atoms. To get around this point, some evolutionary proposals have suggested that during this early upheaval in the universe there was some other, alternate, unspecified form of matter. But this is unsubstantiated and lies on the outer fringe of accepted cosmological thinking, especially as it is without explanation as to what or where this material might be now. From any accepted scientific viewpoint, physical matter is only made from atoms (excluding dark matter, which isn't physical matter, as we will discuss later in this chapter). The only possibility therefore is that the original forms of physical material were the components of atoms appearing instantaneously from nowhere.

Again, Evolutionary Theory cannot explain where atoms' components came from and how they could invent themselves, particularly with such massive levels of energy trapped in their nuclei. In essence, what Evolutionary Theory is actually proposing, in reference to the big bang, is this:

- The components of atoms appeared from nothing and compiled themselves into atoms.

- Then they compacted together and became so superheated that they destroyed themselves.

- Their released energy exploded in a big bang, only to remake themselves again and evolve into the universe as we know it today.

This big bang explosion, at maximum possible velocity, is estimated to have overcome the force of gravity that was originally holding all the matter together. This makes sense, otherwise its own density would have pulled it back on itself like a piece of stretched elastic. Billions of years later, this outward velocity of the universe is still believed to be happening, and while it is assumed there is nothing to hold it back, for many years (until 1999) its speed had thought to have been decreasing. This slowing down of momentum was theorized to be the effect of gravitational forces within the universe decelerating its forward thrust at a rate slightly greater than its burst speed, which is thought to be maximized at the speed of light. The eventual effect of this would be that all matter would re-congregate into one huge ball again, a process sometimes referred to as the Big Crunch. The velocity of the universe's expansion is assumed to have decreased considerably since the big bang, as the speed of movement of the stars and galaxies is now measured at less than ten percent of the speed of light.

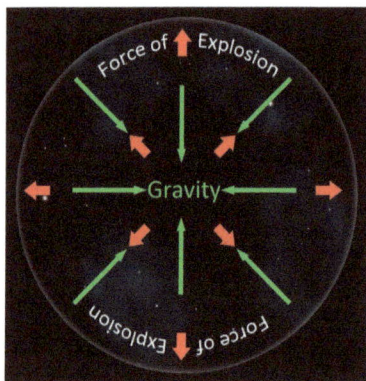

Figure 8: The Big Crunch

At this point, the atoms required to make the stars, planets, moons, asteroids (and the living bodies we know of) were not yet formed. The question, therefore, is how did the force of gravity exist in the early universe without atoms and matter?

This is just one gap in Evolutionary Theory that requires explanation. Science does have some theories regarding the emergence of atoms from nothing, but it still can't explain what type of atom-less matter existed before atoms emerged into the universe and created the matter we are familiar with today.

Here's an example of one such theory: Some time after the big bang explosion, huge amounts of the hot scattered matter came together in separate groups or clusters under gravitational pull and repeated the big bang process. These groups or clusters are, it's claimed, superstars or supernovas. Within these supernovas, new atoms were made by thermonuclear fusion reactions. This is despite thermonuclear explosions requiring atoms to fuse and make the explosion in the first place. Nonetheless, these big bang Mark II explosions then flung matter into space again and are considered as the most likely building blocks from which our life forms and most of the elements we know originated. Over a period of billions of years, as

Figure 9: A Spiral Galaxy

they spread apart, these huge, hot, burning pieces of matter formed into stars. These stars then formed groups with a gravitational center, and we call these star groups galaxies.

To summarize, the Theory of Evolution assumes that the universe, as we see it, occurred as the result of five, separate events:

- The relatively calm appearance of matter and energy from nothing.

- A contraction then a super explosion of this matter (the big bang).

- Billions of smaller explosions from separate groups of this matter (Supernovas).

- The ensuing formation of basic atoms, then the full range of atoms that we currently know of.

- Atoms forming themselves into gases, solids, liquids, stars, planets, and galaxies.

According to evolution, these events are unrelated, coincidental and accidental.

Cosmic Inflation Theory

The Cosmic Inflation Theory deserves mention at this point. This theory claims that in the immediate moment after the big bang (something like a billion, trillion, trillion, trillionth of a second), the universe expanded one hundred trillion, trillion times at a speed considerably greater than the speed of light (see below sub-chapter: Dark Matter and Einstein for more discussion on this theoretical event).

Albert Einstein calculated that such an expansion would create gravitational waves (ripples) in space-time where it would compress then expand space itself and this violent expansion would have stretched these initial waves so much that the distortions would be imperceptibly small.

In March 2014, a group of scientists from the Harvard-Smithsonian Centre for Astrophysics USA announced they had detected swirls (called curl or B mode pattern) in CMB radiation via the BICEP2 telescope in Antarctica. They proposed these swirls are evidence of the effect gravitational waves have on the CMB radiation. They then claimed these confirm the rapid Cosmic Inflation Theory and, by default, proved the big bang.

Several issues to consider are:

• The radio telescope used does not actually detect gravitational waves; it measures fluctuations in CMB radiation (electromagnetic waves). In this event, the CMB radiation was not moving up and down in random directions, rather it showed some swirling patterns. It is therefore an hypothesis, not a proven fact, that the swirls are caused by the effect of gravitational waves.

• CMB radiation is considered by cosmologists only to have appeared about three hundred and eighty thousand years after the big bang, which would have placed any gravitational waves far out in space by that time and their effect on the CMB radiation, which is essentially behind its outward movement, is at best speculative.

• There are other signals from space which create foregrounds that can result in effects similar to gravity waves. These are considered to exist in the Milky Way galaxy and it is well known by scientists that BICEP2 has limited ability to distinguish these signals.

• The Cosmic Inflation Theory purports that the initial rapid expansion of the universe had to suddenly slow down to the current rate of expansion. To do this would require an inward force just slightly less than the big bang's outward force to rein in the speed of the components moving away from the centre. If it were a greater force, the universe would have collapsed again. If not strong enough, the universe would be expanding much quicker. There is no actual evidence of any force that might have caused this slowdown, making the Cosmic Inflation Theory a highly speculative concept, thereby categorising the observed swirls as likely to be caused by some other non-associated anomaly.

• New data has since rendered this claim as invalid with proponents appearing to have jumped the gun based on unsubstantiated evidence. In January 2015 it was confirmed that the BICEP2 signals were not gravitational waves but were actually dust also observed in September 2014 by the Planck satellite. This claim is now emphatically dismissed leaving Comic Inflation Theory as it started - simply another concept without evidence.

Location, Location, Location

Regardless of its rate of expansion, the time involved or the current size of the universe, we know that the universe does exist and it has a significant level of order to it. We also know that gravitational forces are a major influence in the behavior and workings of the universe. Furthermore, the universe is full of CMB radiation that scientists use as a measuring stick to gauge its age.

• This is how CMB radiation is used to estimate the birth of the universe: bodies such as stars, planets, comets, and other physical entities in space give off thermal radiation at different wavelengths relative to their temperature; for example, as microwaves, infrared, ultra-violet, and so forth. Scientists use this information to determine the original level of heat of the universe. By measuring the current heat levels of stars and other materials, and tracking their movement, scientists theorize they can locate and determine the original point of the big bang explosion. However, when looking at the swirls in CMB noted above, the fluctuations appear to be concentric from earth's perspective. If so, they would point to the original centre of the universe possibly being earth.

Nevertheless, scientists and astronomers have as yet not been able to do this. One problem is that the universe is so large that we have not measured all of the galaxies, let alone found them. Compounding the problem is the expansion of space itself, both at the edges and from within, which may mean locating the theoretical point of origin of the big bang remains just that, theoretical.

Another factor to consider is the nature of explosions: an explosion blows out from its middle, forcing everything outward and leaving nothing in the center at ground-zero. Theoretically, there should be a corresponding leftover 'hole' in the middle of the universe where the big bang took place—a Ground-Zero to dwarf all ground-zeroes (refer Figure 10: The Hollow Center of the Universe). To date, scientists have not observed such a hole, and they probably won't because there isn't one, as we shall see shortly.

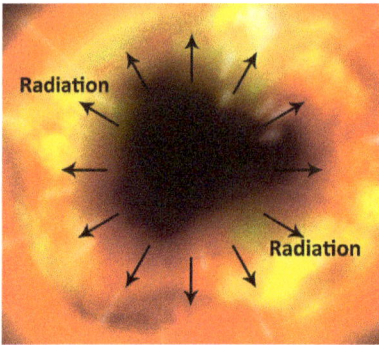

Figure 10: The Hollow Center Of The Universe

The general scientific view today, however, is that the concept of the universe beginning with a large explosion is implausible. There is growing scientific support for the theory that there was not an explosion into space but that space itself is growing in all directions with an unidentifiable starting point in much the same manner as a loaf of bread or a scone will rise in a heated oven. To get a clearer picture of how this works, we will need to examine the structure and movement of the universe in finer detail.

Dark Matter And Einstein

Let us consider another inconsistency raised by the big bang theory: the evenness of the universe. This was originally postulated in the 1920s by Russian mathematician, Alexander Friedman, whose work also confirmed the expansion of the universe with moving matter. Although Friedman's work in dynamic cosmology would eventually form the standard for the big bang theory, scientists have discovered an inconsistency between the theoretical models and the natural laws of the universe. The inconsistency is this:

Following the big bang explosion, mathematical calculations show that the universe must have balanced both its temperature and density in less time than it would have taken light to cross from one side to the other.

To explain what this means, imagine the early universe as a loaf of bread or scone (see Figure 11: Heat and Expansion of the Early Universe) expanding at or around the speed of light. The universe at this stage of its expansion is thrust outwards in all directions, scattered with numerous hot spots, like raisins in the loaf of bread or scone, produced from the early supernovas (as discussed in the previous sub-chapter Early Matter). If the raisins were superheated, in order for the temperature to be evenly spread within all areas of the ballooning universe, including the edges of the expanding surface, the overall temperature would have to even itself out quicker than the balloon was expanding.

In other words, the primordial speed of temperature balancing itself into evenness had to be greater than the speed of light, which implies that the early universe defied natural laws of physics because it suggests that energy travelled faster than light, and this is considered impossible by Einstein's famous equation, $E=MC^2$, which strictly forbids anything to travel faster than the speed of light.

White light comprises a vast range of slightly different colors in a spectrum starting from red and changing to violet at the end of the spectrum. If the light coming from a star is measured on earth as more towards the red end of the spectrum, then that star is increasing its distance from earth—it is moving away.

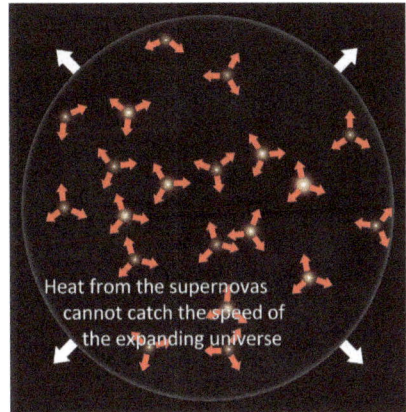

Heat from the supernovas cannot catch the speed of the expanding universe

Figure 11: Heat And Expansion Of The Early Universe

This contradiction means one of two things:

1. Science should reconsider the big bang hypothesis, and even reject it, as the data does not match the theory; or

2. Science should reject Einstein's theory, as something has been shown to travel faster than light (without going backwards in time).

Until recently, the reason the current theory hadn't been rejected is because no other model existed that was superior to the big bang theory. Until now. New scientific evidence has been announced that contradicts the big bang theory. Two groups of astronomers—one from the Lawrence Berkley National Laboratory in California and the other at the Mount Stromlo Observatory in Australia—were attempting to calculate the expansion rate of the universe by studying supernovas. Comparing the standard brightness of a specific supernova, in this case, Type 1a (which are also known as 'candlestick' supernovas due to their consistent brightness throughout the universe), and measuring its brightness as it appears on earth, allows scientists to determine the distance of the supernova from our planet. Combined with the redshift[2] in the light spectrum, this then

---

2. Redshift is also known as Doppler shift. It is the change in wavelength of, in this instance, light from a stellar object. Light is in fact a spectrum of different colors starting from red and changing to violet at the end of the spectrum. When a distant star is moving away from the earth, its wavelength is increased or stretched, shifting it to the red end of the spectrum. If a stellar object is moving toward the earth, its wavelength is decreased or scrunched, shifting it to the blue end of the spectrum. In fact, almost all stellar objects outside our own Milky Way have a redshift pattern and are thus moving away from the earth.

allows scientists to calculate the speed they are moving away from us. By measuring a range of 1a supernovas, scientists were able to determine how fast they are travelling now and how fast they were moving at different times in the past.

In 1999, both groups of astronomers arrived at the same conclusion: the expansion of the universe is not slowing down; it is accelerating, as if some force is pushing it outwards at an ever-increasing velocity. The most likely cause for this acceleration is considered to be the creation of new matter.

That is, scientists are now coming to the conclusion that the universe is not expanding because of a big bang; instead, the most likely cause is an ongoing act of creation.

Albert Einstein originally introduced the concept of a cosmological constant, which he called a form of anti-gravity, because his general relativity theory seemed to contradict the details then known about the cosmos. He admitted to disliking the insertion of a cosmological constant into his equations, but begrudgingly accepted it as a necessary evil because it enabled his equations to explain cosmological observation. But recent evidence suggests that his cosmological constant is more than a necessary evil to maintain the validity of mathematical equations: it is a necessary part of the functioning universe.

Einstein's cosmological constant is currently thought to be something called 'dark energy', and its force is greater than the entire force of gravity within the universe. Astronomers also believe that there is an invisible substance called 'dark matter'. It's called dark because we just can't see it with the naked eye or with scientific instruments, only the effect that it has on cosmological bodies such as galaxies, stars and planets. For example, according to physics, spiral galaxies (see Figure 9: A Spiral Galaxy) should not spin in the orderly way that is observed through our telescopes. The outer arms of a galaxy, by law, should lag behind and spin slower than its center. But they don't. Galaxies spin at exactly the same velocity throughout their bodies, whether at the center or at the tips of its arms. Mathematically, this is only possible if the mass of the outer galaxy is significantly greater than the mass that is currently observable. Which is why scientists have postulated the existence of matter that is dark and cannot be seen.

Dark matter acts as an opposite to the substance of stars and planets in a ratio of somewhere between six and ten to one, although it isn't anti-matter[3]. This means that galaxies, our solar system, earth and all physical matter are made from matter that is in a significant minority to the majority of matter in the universe. Explaining where this newly found material came from, and continues to come from, only adds to the complications for Evolutionary Theory as there is no scientific explanation for the existence of any forms of matter regardless of how many are identified.

The evidence suggests that more matter (dark matter and probably atomic matter) is being made on a progressive basis, even today, and that is what is pushing the universe outwards, even accelerating it.

Consequently, now that the rate of the universe's expansion has been measured as increasing, not decreasing, and that it is possible that dark matter fills the void of space, there is extraordinary pressure for the Theory of Evolution to change, especially in regards to the big bang. Just how the theory will or can change is not known, because, as we have already discussed, an official Theory of Evolution doesn't actually exist. One thing is now certain: scientists can no longer reasonably argue that the available scientific evidence supports the notion that a big bang occurred at the origin of the universe.

Two obvious points stand out. Because matter is now assumed to be made on a continual basis, it stands to reason that:

1. Matter was made on a continual basis, in the form of operating stars, dark matter and so forth, from the very beginning, rather than as a one-off explosive occurrence in some unexplained form, which then required an inexplicable reversing force to slow down the rate of expansion claimed in the Cosmic Inflation theory (see above).

2. The universe expanded gradually over an extended period of time, which explains how density and temperature are balanced across the universe.

3. It confirms recent observations, noted above, that at any point in the universe everything is expanding away from that point.

---

3. Anti-matter, also referred to as negative matter, comprises particles which are the opposite to normal particles having the same mass but with the polar opposite charge. For example, a positron is the antiparticle of an electron. When these opposites collide they destroy each other, releasing vast amounts of energy in the form of gamma rays as well as joined pairs of lower mass particles. There is very little known anti-matter in the universe. Anti-hydrogen and anti-helium atoms are known to exist but have not as yet been able to be manufactured artificially.

In the face of this information, the scientific community cannot be expected to make a public announcement that the big bang did not occur because there is no official agreed position that it did; there are only individuals and groups that hold differing theories. There is no consensus. It is therefore up to us to keep reminding ourselves and the general public about the frailties of Evolutionary Theory:

- It is just a conglomeration of unproven theories, based on a series of chronological measurements, within which there are many areas of conflicting theories.
- There is no defined, fixed theory.
- It is not fact.
- It is not provable.
- It cannot be accepted as truth because it is constantly changeable.

Still, despite the glut of contradictory information, many scientists—and of course the general public—stick with the current theory in the absence of a better one; one that might substantiate a viewpoint of self-creation. Ironically, the big bang concept is now so widely accepted as an actual event that it is known as the 'standard cosmology'. It is however, an excellent example of the manner in which society accepts theories as truth because they are constantly referred to as real events. The regular repetition in everyday writings, textbooks and language has been misleading, to say the least, across a wide range of false assumptions.

Not all scientists are alike, however. There are some who have been prepared to contest the big bang concept. The first to postulate an alternate theory was Sir Fred Hoyle (1915-2001) who, with Hermann Bondi and Thomas Gold, authored a work on a steady state universe—one in which matter is continuously being created as the universe expands. Their conclusion agrees with the latest data and dismisses the big bang concept of a huge explosion as the origin of the universe. Ironically, it was actually Fred Hoyle who termed the phrase "big bang" as a sarcastic shot at the theory.

Another critic of the big bang was French astrophysicist Evry Schatzman (1920-2010) quoted below: (*Our Expanding Universe*, 1992, McGraw Hill Horizons of Science Series; English language edition):

"Everyone talks of the big bang and thinks of it as a 'Historic' event that science 'discovered'. And this supposed event is described as a gigantic explosion that flung material into space. But this is mere talk and says nothing because the big bang, if big bang there was, occurred at a moment when there was neither matter nor space as understood by modern physics. Indeed this is why modern physics has nothing to say about this big bang; this is why it is only a hypothesis".

The analysis that we have just been through is a good example of how the story of evolution is *assumed by inference* and accepted by *presumption of facts* despite significant contradictory evidence and non-acceptance by modern scientific reasoning. But the real problem is in the communication of new information. It is probably no exaggeration to say that most of the general public would not be aware that the big bang theory was redundant and no longer in vogue with the scientific community, if ever it actually was. Yet, there are still many new documentaries, writings, books, correspondence and media articles that include the theory as an assumed fact.

Unfortunately, we have accepted the big bang theory most of our lives as fact and now we learn that it is not true, it was just an idea. It was never agreed as a scientific fact, yet that is what we have been taught to believe.

What is left for us to do now is to replace it with something that makes more sense and is more in tune with scientific observation.

The Large Hadron Collider

It might seem incredible, considering the failings of the big bang theory we have been discussing, but the international scientific community has managed to convince governments across the globe to invest between six and ten billion US dollars to make the largest and most complex scientific machine ever built, for the main purpose of "duplicating events just moments after the big bang started". This machine is called the Large Hadron Collider (LHC) and was built over a period of twenty years in a twenty seven kilometer underground tunnel on the border of Switzerland and France. Running through a vacuum at minus two hundred and seventy one degrees Centigrade, beams of protons travelling close to the speed of light are crashed into each other in order to duplicate the (assumed) original moment of the big bang. The main areas of interest are:

- To study quark-gluon plasma, which is a form of matter thought to have appeared within thirty seconds of the big bang.

- To search for a subatomic particle, named Higgs Boson, that is thought to give mass to all types of matter.

- To search for new unknown particles by applying magnetic fields one hundred thousand times stronger than earth's.

- To study the differences between matter and anti-matter through B-meson particles.

There are literally thousands of scientists involved in this project at enormous cost. The first operations started in 2008, but various problems hindered progress over several years. To quote from the home page of the LHC website in October 2011:

"How Did Our Universe Come To Be The Way It Is?

The universe started with a big bang—but we don't fully understand how or why it developed the way it did. The LHC will let us see how matter behaved a tiny fraction of a second after the big bang. Researchers have some ideas of what to expect—but also expect the unexpected!"

The most obvious anomaly that stands out is the first sentence: "The universe started with a big bang". Yet science and scientists themselves tell us that it didn't. They also tell us that the big bang theory is not and never has been an official position of science.

It seems that world governments have accepted the big bang theory as a given and unwittingly spent billions of tax-payers' dollars on investigating an event that has been proven to contradict the known scientific facts. What this highlights is the depth to which the big bang theory has embedded itself into our modern thinking.

Nonetheless, the LHC is not a complete waste of tax-payers' money. The experiments being undertaken are really quite fantastic and there has always been an expectation that some interesting discoveries would arise that may be far more important than the reasons for building the machine in the first place, such as the existence of the Higgs field or Boson.

Here is a brief background to help explain what scientists are trying to achieve in the discovery of the Higgs Boson. In essence, the smallest theoretical particles identified so far are called quarks. These are believed to have no mass yet they are the particles that make up everything else—which has mass[4]. In theory, quarks are the building blocks that make up the components of atoms' nuclei (neutrons and protons); electrons are technically leptons having no smaller components. There are, however, various dilemmas that quarks and other particles create for physicists because of their lack of individual mass. The most basic problem is that

---

4. The mass that a body possesses is the amount of its resistance to a force of acceleration, for example, gravity, and the combined attraction it has with another body. The higher the mass, the less resistance to the combined gravity. In the observable world that we normally see, mass is generally considered to be the amount of material (matter) that something possesses. But mass is really a measure of an object's inertia (its ability to resist attraction). Objects also have weight, which is their mass times the force of gravity.

atoms, life or even the universe couldn't exist without something or some force that can hold atoms together and make them stick.

This is where the Higgs Boson gets involved. In 1964 Peter Higgs, a theoretical physicist (and later an Emeritus Professor at the University of Edinburgh), among others, proposed that, at the moment of their construction, a field was generated around the massless quarks that gave rise to the acquisition of mass. This field, or particle, created the mass when it interacted with an unknown, theoretical medium or field that exists everywhere. This would mean that space everywhere is replete with this unknown medium—which is not visible or currently detectable. Effectively, as particles move through this medium or field they push against it and the interaction (or friction) produces mass and marginally slows their movement.

One of the main purposes of the LHC is to see if this particle is identifiable as a breakaway when protons (comprised of quarks) are smashed into each other. As it happens, the scientists involved believe that they have already done this. In early July 2012, scientists at the European Organization for Nuclear Research (CERN) announced that two separate groups of scientists had recorded the Higgs Boson breaking away from the quarks. Whilst the benefits of such a discovery may not come into play for some time, this is considered to be a watershed discovery and alone probably justifies the building of the collider.

The discovery of the Higgs Boson is fascinating news for several reasons.

1. Although not proving that the big bang explosion occurred, the actual discovery is that, at the moment of their making, quarks had the Higgs Boson inserted around them in order to give mass to everything that otherwise wouldn't have it.

2. Importantly, the evidence reveals what happened to quarks at the moment they were made and not the process that made them. Whilst the LHC was built on the premise that quarks were made in a big bang explosion, this is not what the experiment results confirmed.

Any two objects in a vacuum will be attracted to each other by their mass. The smaller object will move further towards the larger one than vice versa depending on the relative difference in mass. If two smaller, but different sized objects, are attracted to a third large object, the smallest object will be the slowest in its approach as it has the smallest combined mass with the larger object. Objects also possess density, which is the amount of material in them. A foam ball will have less density than the same sized lead ball, as lead has a higher density, which can be measured by the gravitation force (acceleration) on it.

3. The evidence suggests that atoms are comprised of components that are highly ordered, structured and balanced so that they can operate in a specific manner. In other words, their components are assembled in order to function in a deliberate and predefined way.

4. The evidence also suggests that space is crammed full of a yet to be determined medium that can help explain the functional anomalies of the universe outside the normal behaviors expected by physics.

5. The Higgs Boson and the unknown medium/field are required to exist together and to exist complementary to one another, thus ensuring that the universe and life can exist as mass and form.

The results obtained from the LHC confirm that everything in the universe is made up of infinitely small particles that are constructed in a clever manner with brilliant components. These have a specific function and they interact in a particular manner with a specific (although as yet unidentified) medium, in order for the universe and life to exist and function effectively.

Interestingly, in an obscure kind of way, the Higgs Boson was dubbed the 'God Particle', even before it was confirmed as existing.

There is however another conundrum that science has created for itself here. It is claimed that in the immediate moment after the big bang, Higgs particles attached themselves to quarks (and, assumedly, other particles). This requires that Higgs particles and quarks were all created prior to that moment, that is, during the super-heated contraction of energy. But science tells us this fusing conglomeration of compacted, uncontrolled energy prior to the big bang was so hot that none of these components could have existed. Further, there is no explanation about when the matching Higgs medium/field appeared or the materials used in its making.

Chapter 3 Unraveled

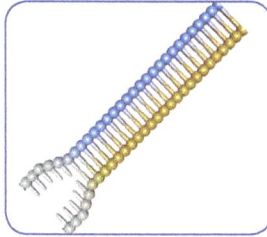

- *The movement and growth of the universe is not, and cannot be, the result of an initial big bang explosion.*

- *Matter is probably still being made today on a continual basis.*

- *The big bang is just a theory. It is full of contradictions and the evidence indicates that it never occurred.*

- *Science has not ever officially proposed the big bang as an actual event.*

- *The discovery of the Higgs Boson reveals the universe is constructed in a clever manner with brilliant components.*

- *Science offers contradictory theories regarding the existence of the Higgs Boson particles and protons.*

4

ATOMIC AND STELLAR EVOLUTION

Our next step is to examine how evolution proposes that atoms actually formed and then how they accidentally made so many billions of stars. The proposed steps put forward by evolution include:

- The formation of supernovas from atom-less material.
- The self-creation of hydrogen atoms.
- Star self-creation from self-made hydrogen atoms.

Like the big bang, there are numerous theories regarding the appearance of atoms. Some theorists claim hydrogen and helium atoms were present in these early supernovas, while others claim atoms could not have existed at that time. What we do know, though, is that stars are made mostly of hydrogen, which is the first, simplest and most abundant element in the universe (disregarding dark matter, which is not considered an element).

The theory of atoms has existed even longer than any Theory of Evolution. Historical evidence suggests that in the 6th Century BC, Hindus in India were discussing the possibility of minute, unseen building blocks making up the visible matter they saw around them. The ancient Greeks definitely had theories on atoms (the word atom comes from the Greek word "uncuttable"). In the 5th Century BC, Leucippus and his pupil, Democritus, put forward the notion that matter was made of atoms and surrounding void. Plato rejected it, and even Aristotle thought the elements of Fire, Wind, Earth, and Water better explained the makings of the universe than a construct of tiny, unseen atoms.

However, the scientific proof of atoms was only established in 1802 when John Dalton, an English chemist, showed that matter was made of lumpy particles, evidence confirmed by independent scientific research that has continued ever since the presentation of Dalton's famous papers, *Experimental Essays*. Today, there are new theories about even smaller particles that constitute atoms and these are covered under the topic of String Theory below.

String Theory

String Theory is a relatively new concept about how the universe is constructed. Einstein's General Relativity theory details the universe at a macroscopic level where things are smooth and ordered. At the infinitesimally minute scale of quantum physics, things appear chaotic and out of control, unbalanced, uncontrollable and unpredictable. There is an imbalance between these two levels in how they allow the universe to operate. Developed within the last forty to fifty years, String Theory, also referred to as "The Theory of Everything", attempts to explain how microscopic and macroscopic forces can operate predictably in the universe.

Essentially, String Theory is a mathematical equation that identifies infinitesimally minute structures, or strings, that make up quarks, which then make up the components of atoms' nuclei—protons and neutrons. Strings are believed to be trillions of times smaller than atoms and the source of gravity at a quantum level. To do this, strings are calculated to operate in a minimum of six dimensions[5], with each dimension being different for each type of string. According to some theories, these numerous spatial dimensions disperse the universe's gravity throughout each dimension, thus accounting for why gravity is such a weak force in the observable four dimensions of space-time. Figures 12 to 14 are diagrammatic representations of how String Theory is thought to operate.

Figure 12: Three Dimensional View is the macroscopic view of the universe we normally perceive in our three dimensional world (height, width and length).

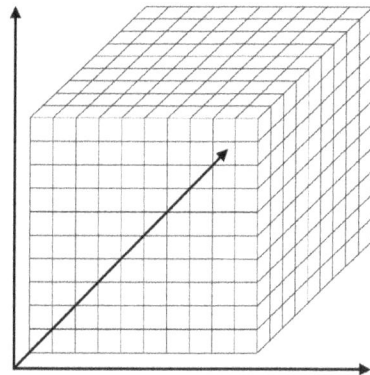

Figure 12: Three Dimensional View

---

5. Variations on String Theory range between six and eleven dimensions, including the three dimensions of visible space.

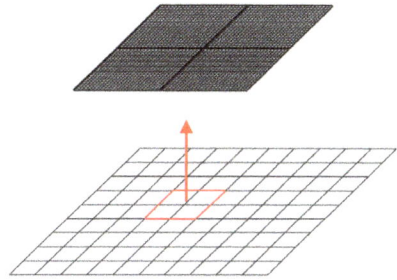

Figure 13: Smaller Flat Surfaces with Grids delves deeper into the two dimensional surfaces to expose further grids.

Figure 13: Smaller Flat Surfaces With Grids

Figure 14: Six Dimensional Balls is the final two-dimensional surface showing further dimensions at the smallest grid junctions, inside which are the theoretical strings.

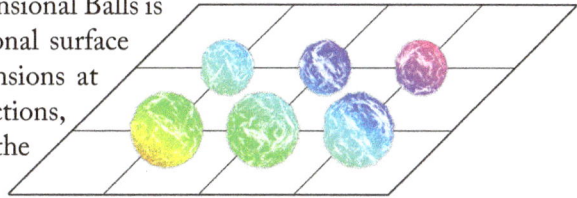

Figure 14: Six Dimensional Balls

Any pictorial representation of what a multi-dimensional string might look like is speculative as our previous understanding of dimensions is limited to four—height, width, length, and time. It is theorised that strings vibrate at different harmonies within their six dimensional framework like a violin string, and these differences determine the type and functionality of each atom just as a violin string determines the pitch and quality of the sound we hear. In String Theory, all atoms are vibrating and have a specific harmonic level. The hypothesis is that these vibrations smooth out the universe and overcome the imbalance between the Theory of Relativity and quantum physics.

Possibly this concept may explain the massive energy levels that exist in atoms. When splitting hydrogen atoms, the release of energy may actually be masses of vibrating strings escaping from their highly condensed environment. Similarly, when hydrogen fuses into helium in stars, the radiation may be related to strings.

Whilst all of this is fascinating, challenging, mysterious and currently unprovable, it does highlight one important thing: the universe is a place where we exist but science is only scratching the surface of the level of its complexity.

The Basics Of An Atom

There are three basic components to an atom:

1. Proton   — a positively charged particle.

2. Neutron   — a particle without any electrical charge (the proton and neutron are held together in the middle and are called the nucleus)

3. Electron   — a negatively charged particle that is believed to orbit the nucleus.

If we could physically see the way that electrons operate around the nucleus of an atom, we may find that they fan out and decrease with distance. The following diagram represents how many scientists imagine atoms operate. It must be said that this idea is relatively recent.

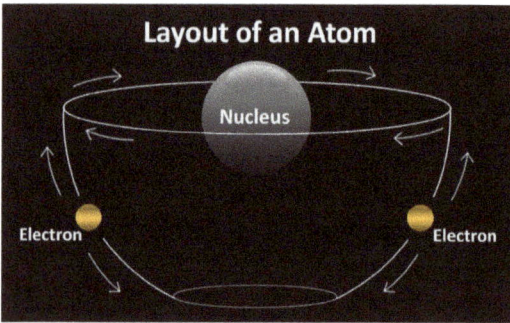

Figure 15: Layout Of An Atom

This recent discovery shows a three dimensional curved shape with the smaller parts rotating around the larger centerpiece. Most of the mass of the atom is in the nucleus while the rest is vacant space. The electrons are very small and scientists cannot yet precisely pinpoint their positions. Currently, electrons are considered to vibrate rather than circle around the nucleus.

Any substance that is made from just one type of atom is called an element (for example, hydrogen, helium, boron, lithium). The number of protons remains equal within a particular element, but the number of neutrons can change, and this difference is referred to as its isotope. This can be complicated and is not relevant to our discussion. What is relevant is the number of electrons. Each atom is unique to its assigned element: hydrogen has only one electron, helium has two electrons, and lithium has three electrons. As an example of a more complex atom, lead has eighty two electrons around the nucleus, as shown in Figure 16: Hydrogen, Helium, Lithium and Lead, which illustrates how the type of atom changes as the number of circling electrons increases. There is the same number of protons as electrons in atoms but the number of neutrons changes. Using these four elements the numbers are:

Element	Neutrons	Protons	Electrons
Hydrogen	0	1	1
Helium	2	2	2
Lithium	4	3	3
Lead	125	82	82

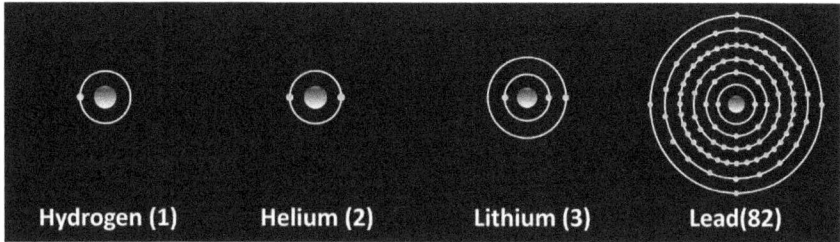

Figure 16: Hydrogen, Helium, Lithium And Lead

There are ninety two naturally occurring elements. Scientists have been able to make a further twenty elements artificially in particle accelerators in laboratories, but these require setting up extreme conditions that may not occur naturally. In theory, there could be any number of elements but those with larger atomic numbers are radioactive and unstable to the point where the nuclei will decompose before forming a solid mass. (For more information, there is a chart showing all of the elements on the website www.evolutionunraveled.com).

It is not necessary to go into an explanation of the elements in any depth. We know that elements can exist by themselves or mix with others making compounds. For example, two hydrogen atoms joined with one oxygen atom make water (H_2O); one sodium, one hydrogen, one carbon, and three oxygen atoms make sodium hydroxide ($NaHCO_3$). The issue for us to consider is not what they can do now but where they came from in the first place and how they constructed themselves.

To recap, Evolutionary Theory claims that the big bang forced matter to spread out in every direction, then supernovas formed out of an unknown, atom-less matter, then atoms appeared as the primal universe cooled down. This proposes that atoms made themselves after other, unspecified matter existed, which then disappeared.

Atomic Evolution

Most evolutionary theories assume atoms only came into existence about three hundred thousand years after the big bang explosion when things had cooled down significantly. According to the theories, massive amounts of energy gathered into the tiny, tiny, tiny space that makes up the nucleus of an atom. If you consider that one atom holds all the energy that explodes from an atomic bomb, then the amount of energy stored in that pinhead area just before the big bang is mind-boggling. Where this energy came from is also not addressed in any satisfactory detail by Evolutionary Theory.

We know that at the time of an explosion there is a compressed amount of energy that is released at a rapid rate. The energy is in fact trying to even itself out to match the same level of energy as its surroundings. If there is a lot of energy, it will release in a hurry, making a large explosion. If there is little energy, the explosion will be smaller. But either way, the energy balances itself fairly quickly, effectively decompressing. For example, if we look at a bomb, it is only a few seconds later that the compression effects are gone, or with a volcanic explosion, perhaps a few minutes. This means that the energy is quickly balancing itself. It also means that the released energy effectively immerses itself into the surrounding environment such that it can't be identified separately from the surrounding energy. In other words, energy, like heat, gases and liquids, dissipates to the lowest level of intensity—it does not gather itself together in multiple tiny places and cover itself in some sort of casing or shell.

Because Evolutionary Theory claims that atoms formed three hundred thousand years after the big bang explosion, it infers the energy from that explosion must have been spread over an incomprehensible area, thinning by the moment and still moving outwards, as there was nothing to stop it. There are several obvious questions that require explanation:

1. How did this release of energy from the big bang compact itself into atoms when it was so thinly dispersed?

2. How did this energy compress when there is no known scientific explanation for it to compress itself?

3. Even though there is no explanation for the origin of electrical charges that atoms possess, how did atoms overcome the natural force of repulsion and draw themselves together?

Elemental Evolution

Some evolutionary theories claim that the first two elements, hydrogen and then helium, may have accidentally formed during the big bang, but there is no universal agreement with this. In effect, this means that the first atoms ever to exist had to be built from some components that science has not yet defined.

This leads to a fundamental issue that Evolutionary Theory must explain. As far as science is concerned, anything that exists and is made up of smaller pieces has to be assembled by those pieces being put into place, either deliberately or randomly. Everything is built from its components. By inference, the smallest components have to exist before they are combined to make something else, something bigger. This means that electrons, protons and neutrons had to exist before atoms.

To date, there is no explanation from evolutionists regarding the formation of electrons, protons and neutrons. There are only generalized statements that atoms accidentally formed themselves. We will, therefore, have to follow the most logical assumption that each of these components somehow appeared in groups of their own kind and were probably lingering in the same place before the first atoms were assembled.

Science, as we have discussed, claims that the first element to form was hydrogen and that all other elements were subsequently made from the assemblage of this primal element. In order to make one hydrogen atom, therefore, one positively charged proton and one negatively charged electron had to meet and unite. This atomic bonding sounds sensible enough, but regardless of the method, for there to be trillions of hydrogen atoms, the process would through necessity have to be easily enacted. While these protons and electrons were coupling in a flurry of hydrogen making, there must have been a corresponding group of neutrons hanging around that weren't getting into any action at all. At least, that is, until helium was made and they were invited to join the party. You will remember that helium has two protons, two neutrons and two electrons. This means that two neutrons, which have no charge, had to locate and fuse with two hydrogen atoms so that helium could be made.

To counter the implausibility of this, science now tells us that protons and neutrons are made from quarks (whereas electrons have no smaller components). This, though, only pushes the problem upstream. Instead of explaining where protons and neutrons came from, Evolutionary Theory must now explain where quarks came from and why electrons are different having made themselves. We briefly touched upon this in the previous chapter in our discussion on the Large Hadron Collider, but this only raises more questions on the validity of Evolutionary Theory.

The next elements, lithium and beryllium, throw Evolutionary Theory into yet another conundrum. These elements are now considered by some to be made by cosmic rays striking carbon atoms produced in stars and breaking them apart—which is the opposite of how the first two atoms are theorized to have formed. That is, the first two elements were made by congregating atoms and the subsequent two by separating atoms. The process gets even trickier with heavier elements.

Even if we put these issues aside and accept that hydrogen atoms created themselves from nothing, despite there being no scientific explanation, and that they then fuse together to form helium, we still need to explain where the other ninety elements came from. Most theories assume that hydrogen formed first then these hydrogen atoms fused together into all other atoms through the enormous power of exploding stars and supernova. But as we will see, there are atomic barriers or gaps, known as mass 5 and mass 8, where there are no stable atoms and where it appears impossible for hydrogen to fuse into any element heavier than helium.

These atomic barriers or gaps in atomic mass are evident when reviewing atomic weights. The atomic weight of hydrogen is 1.008 (effectively the amount of material in it). This is followed by deuterium (an isotope of hydrogen called heavy hydrogen) with double the weight at 2.016. Helium follows at 4.003, then there is a gap at mass 5, called 'the helium mass gap 4', where there is nothing. Next is lithium at 6.939, then another gap at mass 8, then beryllium at 9.012 and boron at 10.811 and so on.

The gap at mass 5 is because protons and neutrons cannot bind to a helium nucleus (at mass 4) due to atomic instability. A similar instability occurs at mass 8, which means no element can form at this atomic mass. Effectively, because the process of stellar element formation has missing links, the later elements cannot be made by the former; they require some other method to occur. (*refer Universal Order #4, Page 61 regarding the formation of carbon atoms*).

However, these gaps appear to offer a specific outcome relevant to life on earth. Were it not for the gap at mass 5, the processes of stellar atomic fusion would cause hydrogen to eventually convert into all heavier elements of the periodic table. This continuing chain of events would ultimately result in uranium, which would be emitted from our star, the sun. Thankfully, this does not happen, as the consequences of radioactive uranium bombardment would have dire consequences for life on earth.

What's more, neither combining nor splitting hydrogen can bridge these gaps either. A hydrogen bomb fuses hydrogen atoms together into deuterium (helium 2), then into helium 4 at which point it can go no further. A hydrogen explosion (fusion), even in a star, does not provide the necessary environment for hydrogen to cross the mass 5 gap. In a nuclear explosion using uranium (element #92) where atoms are broken apart they separate into barium (element #56) and technetium (element #43).

The only process for which Evolutionary Theory can explain the existence of atoms and elements, other than hydrogen, is that they are made from hydrogen fusing itself into heavier atoms, and the most likely place for this to occur is within stars. In fact, spectral spectroscopy (discussed later in this chapter, *Stellar Spectra*) indicates that higher elements do exist in stars and galaxies, but, as detailed above, science tells us this can't happen because of the atomic barriers—in other words, the theory contradicts scientific reality.

The Nature Of Stars

At their cores, like our own sun, stars are assumed to be very dense. It is here that scientists propose that hydrogen is converted into helium in a process of nuclear fusion. This process emits radiation called gamma rays which, it is theorized, collide with subatomic particles and change into X-rays, which themselves are carried to the surface, possibly by gases. Gamma rays, X-rays, ultraviolet light and sunlight are all part of the spectrum of electromagnetic (EM) radiation.

This process is finely tuned. The number of atoms fusing together is delicately controlled through massive gravitational pressures in order to prevent the process escalating out of control in a massive stellar explosion. As there are vast quantities of hydrogen within stars, this fusion process allows stars to burn without shrinking as they would do if they were simply burning fuel such as wood or coal. The sun is actually vaporizing into space, whereby particles such as helium and hydrogen are emitted as solar wind, particles that are actually deadly to earthly life.

Our sun is in the top ten percent of stars by mass and is called a yellow dwarf. It is comprised of approximately seventy five percent hydrogen and twenty five percent helium. The sun rotates around its own axis, but there is no agreed reason why the sun spins or why it would have an axis. It is theorized that all bodies in the universe spin as a result of gravitational forces. The only body in the universe strong enough to spin the sun is a galaxy. The sun's outer layers rotate at differing speeds. The equator rotates about every twenty five days whereas the poles rotate every thirty six days. Calculations reveal the sun's age at 4.5 billion years and that seven hundred million tons of hydrogen are converted to helium every second. At some stage, the hydrogen will start to run out and the core will fill up with extremely hot and dense helium. Scientists believe this should occur in around two to three billion years from now. The helium will contract and get hotter until helium atoms fuse together. Burning helium gives off carbon and makes the star hotter and brighter, causing it to expand to possibly one hundred times its current size, swallowing the earth and all its life. At this stage of its lifecycle it will be called a red giant. Theoretically, our sun will have a life of 8–10 billion years.

Consider for a moment what is happening inside our sun, and indeed any star, remembering that Evolutionary Theory states that this process is occurring by accident:

- All stars, regardless of their size, have a precisely balanced ratio between pressure and fusion; otherwise we would have no heat, sunlight or life on earth.

- Many billions of stars are able to control this balance, yet the original mass of matter that caused the big bang was unable to find this balance. Instead of making just one super-huge star, or even imploding in on itself to make a black hole, as large stars are claimed to do (which we will discuss later), this matter simply exploded.

Stellar Evolution

Following our stepwise retrospection of Evolutionary Theory, the next thing to consider is the evolution of an early star. Evolutionary Theory suggests that after the emergence from nothing of trillions of trillions of trillions of trillions of hydrogen gas molecules, now floating about in space, they suddenly gathered together under their own attraction in super-dense gas pockets scattered around the early expanding universe. As these gas pockets compacted together, their temperature began to rise. This primal stellar gas is then theorized to have produced EM radiation that overcame the force of gravity contracting the stellar mass any further. The new star then reached an unknown critical point where the energy trapped inside the nuclei of individual atoms exploded outward, without bursting the group of atoms away from its combined gravity. These same atoms then started fusing into helium in a controlled manner and created a light producing star.

According to Evolutionary Theory, this stellar process occurred independently, accidentally, and often simultaneously on billions and billions and billions of separate occasions, forming all of the stars that we have identified to date. There are, however, some fairly important things that Evolutionary Theory needs to answer in order to be valid and reliable.

Firstly, because there is no atmosphere in space, hydrogen gas in space is less dense than it is on earth. More so, as the primal hydrogen molecules were theorized to be travelling outward from the big bang explosion, the gas cloud would be getting progressively less dense with each passing moment as it spreads itself thinner across an increasing surface area (see Figure 17: Expanding Hydrogen Gas).

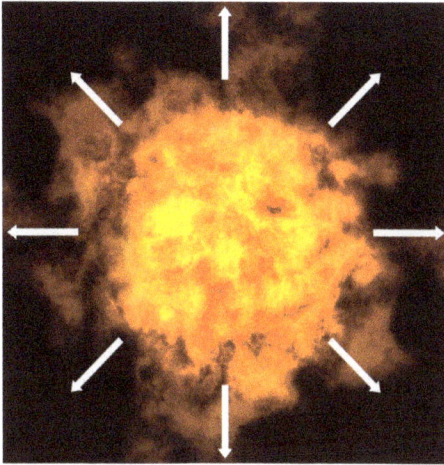

Figure 17: Expanding Hydrogen Gas

The laws of physics state that gases don't gather themselves together but naturally move to the lowest density possible by spreading out into the available area. That is, gas molecules don't contract together; they dissipate. As noted in Chapter 3 Footnote 4, objects in a vacuum will be attracted to each other. However, they have to be in range of each other's gravity. In the case of hydrogen atoms at the beginning of the universe, it is likely these would have been spread out over such vast distances it is questionable they could have been infinitely close to each other to utilise their weak electrostatic forces and group together. As this proposal, which is called 'competitive accretion', is not a scientifically observable phenomenon, evolution has invented a new theory to explain the formation of stars.

The new theory is called 'gravitational collapse'. It states that clumps of heavy molecules break up into gaseous cores, which then collapse to make a star. Those proposing this theory have indicated that the original accretion process is flawed because the modelling is based on clumps of gas that have never actually been observed (for more information, please refer to www.physicsworld.com). However, the new theory of gravitational collapse does not yet explain where these heavy molecules come from and why they are gathered together in billions of scattered locations around the universe. In other words, both proposals are still unsubstantiated theories.

Secondly, hydrogen atoms are limited in their attraction with each other. When two hydrogen atoms are close enough, each of their electrons is attracted to the proton from the other atom, bonding the two atoms together; effectively sharing the mutual attraction in becoming a hydrogen molecule (H_2). This is called covalent bonding. At this point, each hydrogen atom has two electrons in its first (outer) energy level which becomes full (that is, no more room exists) and no more covalently bonding is possible. This means that H_2 molecules have no attraction to each other. The evolutionary concept of the self-attraction of hydrogen molecules as the basis for forming stars is unsupported by reality. Hydrogen atoms are not found as separate, individual entities, they are always bonded with other atoms such as in water where two hydrogen atoms bond with one oxygen atom (H_2O).

Thirdly, to further complicate matters, is the issue of Coulomb repulsion, which basically states that one of the fundamental laws of nature is that like charges repel each other. Hydrogen nuclei have a positive charge and therefore repel all other hydrogen nuclei. This in itself makes fusion extremely difficult to achieve, and only then at high temperatures and pressures, as is proposed to occur in stellar cores like our own sun.

But here is the problem: science claims there are actually two different levels of temperature for the process of hydrogen fusion. There is a temperature level required at the core of stars, and there is also a much lower temperature level, calculated by science, with which to overcome Coulomb repulsion. That is, in order to explain nuclear fusion, Evolutionary Theory claims that two different levels of temperature have been set by nature, which is scientifically nonsensical and of course means that this particular rule of science is not a rule at all.

In summary, there is no scientific explanation how stars appeared—self-made and by accident—because there is no corroborating evidence for:

• Competitive accretion.

• The natural temperature with which to overcome Coulomb repulsion.

The most obvious explanation to account for the anomalies of competitive accretion and Coulomb repulsion is that there is another force at work, one that science either ignores, can't calculate or just can't measure.

Stellar Spectra

Stars vary in luminosity (brightness) depending mostly on age, composition and size. In the late nineteenth and early twentieth centuries, stars were classified into seven main categories by the Harvard College Observatory, Massachusetts U.S.A. with much of the detailed catalogue work attributed to Annie Jump Cannon, an American astronomer (1863 - 1941). These classifications go from hot to cool with O being the hottest and M the coolest: O, B, A, F, G K, M. Most stellar temperatures vary from two thousand to fifty thousand degrees Kelvin (K) with limited numbers being at the lower end (less than two thousand K) and at the top end (over one hundred thousand K). Later modifications to these classifications include additional codes for luminosity, density and composition (that is, which elements are present).

There are various categories of stars including brown, red, green, blue, yellow, and white to giants and supergiants such as VY CMa, which is over two thousand times the size of our sun (being about one million, four hundred thousand kilometers in diameter). Brightness and heat are not directly proportional to each other as stars with different temperatures can have the same luminosity.

Generally, brighter stars are less common, such as O and B stars. Our sun is classified as a 'main sequence star' coded as G2V, being a yellow dwarf with a temperature around five thousand eight hundred K and predominantly in the phase of converting hydrogen to helium (thought to last for a period of about five billion years). Other dwarf stars can be twenty thousand times brighter than our sun.

Light energy is emitted from stars in the form of photons, which transfer directly into a photon quantum field, which itself exists everywhere. The 'light/photon' energy spreads itself evenly into the field, allowing stars to be observed from great distances. If individual photons were dispatched without a controlling field, at some (enormous) distance from the star, the spread of photons would be so sparse that they might be undetectable, having been broadcast across the vastness of space.

Light is also considered to be part of electromagnetic radiation and to travel through space and other mediums as a waveform. The light we see consists of a spectrum of wavelengths that we interpret as colors. We cannot see all the spectrum of light, as our eyes are limited to the visible spectrum: red, orange, yellow, green, blue, indigo, and violet, the colors of a rainbow. Red has the longest wavelength and violet the shortest wavelength of the visible spectrum. The combination of all the colors in the visible spectrum creates an appearance of whiteness. As wavelengths become longer, beyond our visible range, the colors extend to 'infra-red' (longer wavelengths than red) and beyond 'ultraviolet' (shorter wavelengths than violet).

The color spectrum can be seen when splitting a beam of sunlight through a glass prism. Likewise, a stellar spectrum can also be seen when a distant star's light is also viewed through a glass prism, and even the light of far distant galaxies. Through spectroscopy, scientists have been able to derive various stellar and galaxial properties, such as temperature, luminosity, density, distance from earth, and even their relative motion to earth using red-shift or Doppler shift[6]. Scientists have also been able to determine the chemical composition of far and distant stars and galaxies.

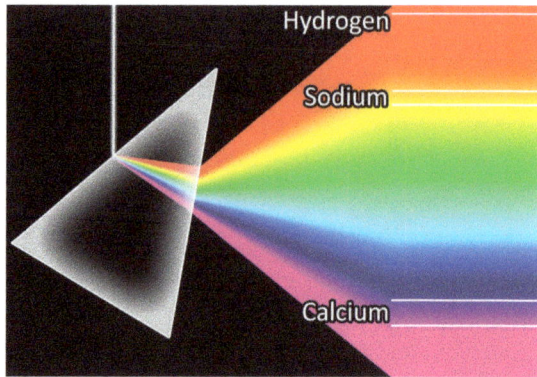

Figure 18: Stellar Spectroscopy

Based on stellar surface temperatures (determined by Stefan's Law), different stars exhibit different variations in the widths of the color bands in their stellar spectrum. Furthermore, there are also other dark lines evident between the color bands, as shown in Figure 18: Stellar Spectroscopy.

The black lines are known as absorption lines and indicate the chemical composition of a star or galaxy. They are the result of a photon energy unit passing through a medium such as a gas. Specific atoms (or molecules or nuclei) in those gases interact with the photon, giving different results (absorption lines in the spectrum) depending on the gas through which the photon is passing. For instance, when a photon passes through a cold gas, it is effectively absorbed and then re-emitted in different directions, thereby reducing the light intensity and creating dark (absorption) lines in the stellar spectrum observed on earth.

Scientists use this information to compare the dark lines with known emission spectra of gases. Essentially, different gases emit light at different frequencies and appear as specific lines in the visible spectrum. When the dark lines of a star's spectrum are compared with the known emission lines of gases, scientists can match and identify which elements are contained in the star's gases. Effectively, the spectral lines become identifiable codes for elements within stars (and are also used for other celestial bodies). Some of the elements identified in stars include hydrogen, helium, carbon, lithium, barium, and zirconium, which are generally considered to be in gas form in the outer atmospheres of stars.

However, the identifiable existence of various elements in stars' atmospheres, other than hydrogen and helium, appears to create a conundrum. This is because, as previously discussed, science has determined

---

6. Refer to sub-chapter *Dark Matter And Einstein*, Chapter 3.

that mass gaps 5 and 8 in atomic weights do not allow any element other than helium to be made from hydrogen by natural fusion processes. This presence of heavier elements in stars raises the issue of determining how identifiable levels of other elements in gas form became part of stellar atmospheres.

Galaxial Evolution

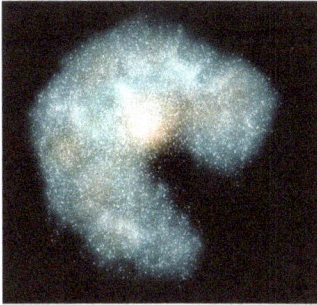

Figure 19: Disc-Shaped Galaxy

Galaxies are neatly arranged, patterned groups of stars estimated to contain on average around one to two hundred billion stars each. There are a staggering three thousand billion galaxies identified with the Hubble Space Telescope to date, making an estimated six hundred billion, trillion stars in the universe that we know of. Galaxies can have a range of forms but are essentially spiral, elliptical or disc shaped. Some have an irregular layout.

Scientists have recently discovered evidence of a black hole near the center of our galaxy, the Milky Way. Whilst it is thought to have the mass of three million of our suns, it appears to have a relatively weak gravitational pull. There is a second smaller black hole nearby with an estimated mass of thirteen hundred suns. It is thought that there may well be black holes near the center of many other galaxies and that, even though black holes account for only one half of one percent of the mass of their galaxy, they are now considered the engine or driving force behind the development of each and every galaxy. Put another way, these theories claim that black holes are the core around which stars congregate and galaxies grow.

The Milky Way galaxy is about one hundred thousand light years wide; whereby a light year is the distance that light travels in one year, approximately nine and a half trillion kilometers. Our sun is one hundred and fifty million kilometers away from earth. Compared with the size of our galaxy, this distance is equivalent to just one millimeter in a length of five thousand kilometers. Our solar system is miniscule in the vastness of the universe. The nearest galaxy to us is Andromeda, which is about two million light years away. Most galaxies are observed as moving apart but Andromeda is measured as approaching us, contradicting recent observations noted above (see sub-chapter: Dark Matter and Einstein, Chapter 3) that everything is moving away from everything else.

Another issue is that a random big bang explosion would result in material flung haphazardly and erratically throughout the universe.

Evolutionary Theory states that, in order to form, galaxies would by necessity overcome this natural haphazardness and rearrange themselves into orderliness. But this defies another law of thermodynamics, the law of entropy, which states that the natural order of the universe is for energy to dissipate from orderliness to chaos.

The Layout Of The Universe

One of the unusual characteristics of galaxies is that they are not evenly distributed in space, as would be expected. Research of the galaxies within about one billion light years from the Milky Way shows that they are strangely arranged, as if clinging to the curved surfaces of huge balls. These balls are almost empty in the middle and measure about thirty million light years in diameter. Where the surfaces of these balls meet are clusters of galaxies, which take on the appearance of galaxy strings. There is no specific theory that accounts for the apparent layout of galaxies, but they do not appear in the way they should if a big bang occurred. There is an apparent force pushing them away from the center of these huge balls, a force strong enough to overcome their gravitational attraction which is theoretically meant to be pulling the galaxies together.

We can visualize this layout with a simple exercise. Imagine a normal size room. Fill the room with blown up party balloons so that they are all touching and are set out in layers. Imagine the galaxies are the size of a pen nib and masses of them are dotted on the outer surfaces of the balloons, like tiny polka dots. Now it becomes easier to imagine how they form three dimensional strings of galaxies. These galaxy strings appear to be typical of how the universe is set up. If we could look at it from afar, we would see a pattern that resembles the image (Figure 20: Layout of the Universe).

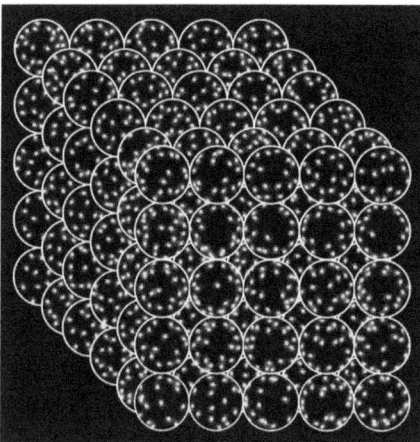

Figure 20: Layout Of The Universe

The result of this layout is that the observable universe is in a type of cubic pattern, much like our image of balloons in a room. It is a specific, consistent layout that is not expected to occur from a random explosion. The reason for universe's unusual shape is not yet understood, but the evidence points to the influence of a significant 'anti-gravity' force within each of these balls that appears to control the consistent layout of the universe.

Furthermore, scientists believe that these balls are expanding from within, pushing the galaxies and universe further outwards, as more matter or dark matter is injected into these balls, inflating them like the balloons in our room example.

As explained early in Chapter 1, the common process used to determine Evolutionary Theory, particularly in the field of astrophysics, is retrograde. That is, scientists work back from the present moment unravelling assumed events until the base components are reached. Then they work forward, proposing a chronological sequence of events that is called evolution. However, if this retrograde process was employed on the actual structure of the universe (Figure 20: Layout Of The Universe) to work back in time, it is highly improbable, if not impossible, that the path would lead to a singular event such as the big bang explosion.

We know that the universe is expanding in all directions such that wherever one is, everything is getting further away. There is a common but relatively new understanding of the universe as having expanded in all directions without any common initial starting point.

Everything Is In A Spin

As previously discussed, cosmologists calculate the age of the universe using a range of measures such as red and blue shifts in the light spectrum. Unfortunately, the accuracy of these measurements is in debate, particularly factors that can affect their measurement, such as distance, passing stars or nebulae clouds, and the angle of measurement, which can cause widely varying results. One reason is that science is really only in its infancy in regards to the equipment and instruments it uses and the range of measurements and observations that can be made with those instruments. This does not necessarily mean that our current knowledge is wrong—just limited to our existing abilities.

We will now examine an interesting observation to highlight this issue, spin.

Electrons are visually described as spinning around or orbiting their nucleus.

(Note: even though we have previously stated that current theories suggest that electrons actually vibrate on and off, we will use the imagery of a circling electron to demonstrate our point).

Electrons orbit
their nucleus.

Moons orbit
their planet.

Planets orbit
their star.

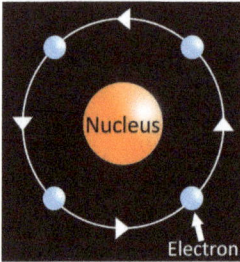

Figure 21: Electron Orbit

Figure 22: Moon Orbit

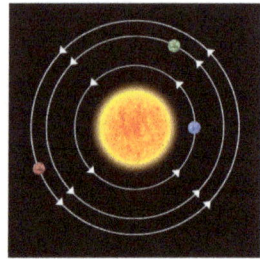

Figure 23: Planetary Orbit

Stars orbit their galaxy
quite probably around
a central black hole.

Galaxies orbit in
a circling motion
through space.

Does the whole
universe rotate?

Figure 24: Stellar Orbit

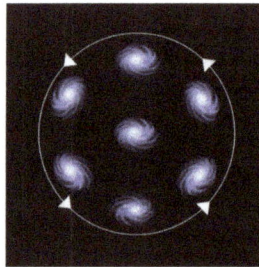

Figure 25: Galaxial Orbit

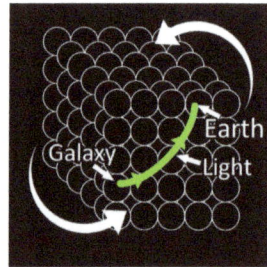

Figure 26: Universal Orbit

The question therefore is: Is it possible that the whole universe is actually circling, spinning or orbiting too?

If the universe is spinning in some way, it may be possible that the light we are measuring is actually coming to us at an angle (see Figure 26: Universal Orbit) rather than in a straight line. This will affect the observation of the red-shift in the color spectrum and give a false reading, such that it will appear the observed star is moving away from us. Which may not necessarily be the case. It could well be circling in the same motion as our planet and solar system but at a different position, which would give the appearance that it was moving away but in fact is only circling further out.

Quasars emit masses of energy with brightness levels over a million times that of a normal star. Recently, quasars have been identified as massive black holes with immense volumes of material swirling around them in a circular manner. This causes friction, the result of which is the emission of light and energy. As it turns out, various quasars have been measured using the red-shift theory, and at least three of them are said to be moving eight times faster than the speed of light. As we already know, this can't be true because nothing travels faster than the speed of light. It does mean, however, that red-shift observation doesn't provide proper, meaningful information and is thus not a reliable measure for determining the movement of stars, and by inference the age of the universe.

Universal Order

The layout of the universe indicates it is ordered, as Einstein suggested. Looking a little deeper at some of the values and structures of its components we find some interesting examples:

1. The strength of gravity has to be exact for stars to stay together, such that, if it were even a billionth of a percent different, higher or lower, the universe would not have stars galaxies or planets. That is as exact as one can imagine.

2. All atoms contain protons and electrons with the mass of a proton being one thousand, eight hundred and thirty six point one five two six (1,836.1526) times the mass of an electron. If this ratio were different by any significant level, it would affect the stability of many chemicals and not allow the formation of DNA molecules, which allow life forms to survive (see Chapter 8 for more details). Because protons and electrons don't change their sizes, this ratio had to be exact from the moment the atomic components first appeared otherwise neither matter nor life could have formed.

3. The Higgs Boson particle and the invisible matching field that it rubs against had to perfectly match each other in size, location, strength, density, resistance, and interaction or nothing would be able to hold itself together—there would be no stars, planets, galaxies, moons, light, or life.

4. The strong interactions within atoms and the strength of electromagnetism are so specifically matched that they allow carbon to be made in a highly efficient manner by fusing together three helium atoms, which would not normally happen without these aligned values. Carbon is the element on which all known life is based and without these particular forces matching perfectly, the universe would be unlikely to support life.

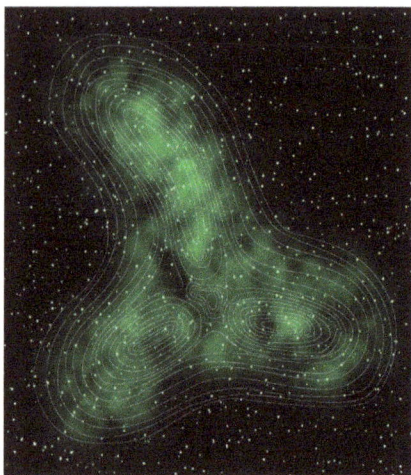

Figure 27: Dark Matter

5. There is a recently identified, enormous invisible structure, thought to be dark matter, that is believed to be holding two galaxy clusters together, Abell 222 and Abell 223. This encircling arrangement looks like a series of thin rubber bands progressively encompassing a wavy circular form (as depicted in Figure 27: Dark Matter). This structure was detected by distorted light caused by gravitational ripples in some form of invisible medium (dark matter)[7].

This type of binding is thought to hold the entire universe together and may well be the first real evidence of curved space as predicted by Albert Einstein.

Evolutionary Theory speculates that these types of particles and ratios formed by chance and accident, values that are assumed to be hit and miss or coincidental. Yet, if these levels, and thousands of others in the universe, were not implemented at these precise and exact determinants, there would be no universe, no atoms, no stars, and no planets. Such specific values are required in order for the universe to operate effectively and support life.

6. The sun is not just a ball of gas; it is a complex multi-layered structure being one million times the size of earth. Nuclear fusion in the core blasts out lethal gamma radiation at about twenty seven million degrees F. This radiation is forced through the surrounding layer of plasma (an ionized gas where both ions and electrons coexist) which is ten times the density of rock. This process is thought to take from 170,000 to 1,000,000 years. When it reaches the chromosphere near the surface, the temperature has reduced to about seven thousand eight hundred degrees F.

The relative combination of the sun being a single star, its size, level of core activity and distance from earth, the density, thickness and composition of plasma, the intensity and temperature of gamma rays, and the earth's atmosphere all combine perfectly to allow life to exist here.

---

7. It cannot occur in a vacuum.

Chapter 4 Unraveled

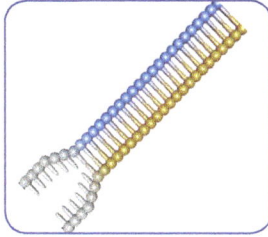

- *There is no known scientific way for atoms to have made themselves.*

- *There is no scientific evidence for the formation of stars by hydrogen congregation (competitive accretion) or the gravitational balancing of the fierce outward force of nuclear fusion reactions.*

- *It is highly improbable that the layout of the universe could have formed from an explosion such as the big bang.*

- *Could the universe be spinning?*

- *The specific values of the components in the universe interact with each other to support life.*

5

OUR SOLAR SYSTEM

Evolutionary theories about the formation of our solar system are widely taught in western education. In this chapter, the broad brush approach of these concepts is analyzed in depth, revealing many processes that not only contradict themselves but are contradictory to science and physics.

In our solar system, the major visible components are:

• One star: the sun

• Eight planets (formerly nine, as Pluto has been downgraded to a dwarf planet)

• Over sixty moons

• Rings of small bits of space matter orbiting several planets, such as Saturn and Uranus

• An asteroid belt (between Mars and Jupiter)

• Various regularly orbiting comets and asteroids

Evolution generally theorizes the formation of our solar system as occurring in a sequence of processes, a sequence we will now explore in detail.

The Early Solar System

According to Evolutionary Theory, there first existed a cloud of dust and gases, roughly the same size as the current solar system, as displayed in Figure 28: The Start Of Our Solar System. This cloud is assumed to have contained hydrogen, helium, nitrogen, carbon, oxygen, sulfur, iron, gold, uranium, silicon, phosphorous, and aluminum. The main components in the Milky Way were hydrogen at seventy four percent and helium at twenty four percent, with the remaining elements comprising just two

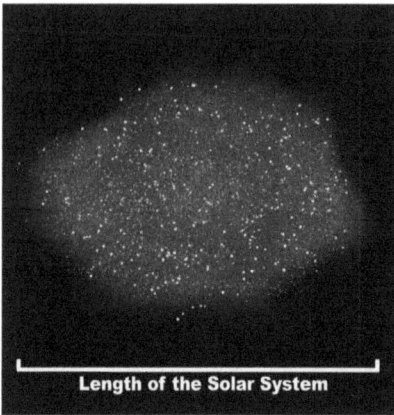

Figure 28: The Start Of Our Solar System

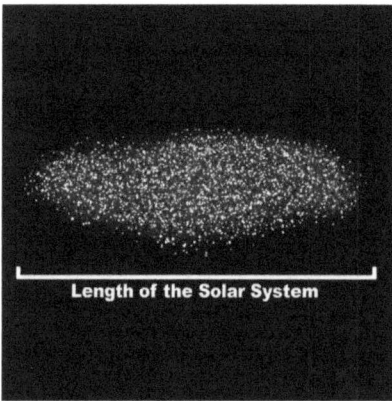

Figure 29: The Compaction Of Our Solar System

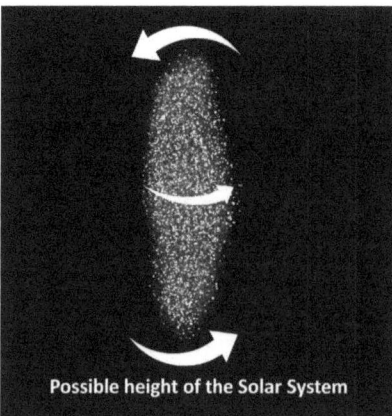

Figure 30: The Alternate Plane Solar System

percent. As previously noted, the nuclear gaps at mass 5 and 8 mean these heavier elements could not have made themselves, making their existence in the early solar system an unexplained mystery. Furthermore, Evolutionary Theory also fails to provide an explanation concerning how these particular elements would, or could, group themselves together in such a high concentration in one remote location in the vastness of space, our early solar system.

Nevertheless, Evolutionary Theory states that over time this matter compacted into a flattish, revolving disc as a result of gravitational forces, as displayed in Figure 29: The Compaction Of Our Solar System.

The shape expected by physical forces, however, would be a more elongated diamond or two-dimensional egg shape, with its axis ninety degrees in the alternate plane to the flat disc we observe today, as shown in Figure 30: The Alternate Plane Solar System. Various theories include suggestions that the cloud disc contained a lot of (unspecified) energy, which assisted in the collapse of material into the current flat shape.

Moreover, there are no scientific specifics regarding the initial cause of the early cloud's revolving movement. What would cause the early solar system to start revolving? Science has no verifiable answer to date.

Even at this early stage, Evolutionary Theory is making assumptions about the early solar system that science is saying cannot happen:

- Science cannot verify the formation of the theoretical cloud of gas, how it aggregated, or how the heavier elements were present.

- Science cannot verify how the solar system became disc-shaped or how it started revolving.

Notwithstanding the absence of scientific evidence, Evolutionary Theory then states that the new disc of gas developed a dense center constituting approximately ninety percent of the total mass of the early solar system. This dense center consisted predominantly of hydrogen and is referred to as a 'protosun'. The inference is that hydrogen atoms were drawn into the center of the early solar system, leaving other heavier atoms behind. Such a process, however, contradicts accepted gravitational physics.

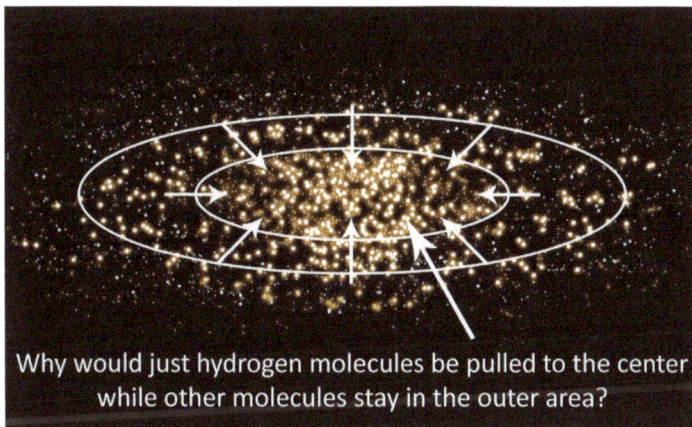

Why would just hydrogen molecules be pulled to the center while other molecules stay in the outer area?

Figure 31: Hydrogen Defies Physics

As seen in Figure 31: Hydrogen Defies Physics, hydrogen, being the simplest element, has less mass and therefore a weaker gravitational force than heavier atoms. As such, it should have been the last element to be pulled into the center, not the first, putting the self-formation of protosuns and stars by this evolutionary method in reasonable doubt.

The theory continues: As the elements were drawn to the central mass, gravity caused its size to reduce and become so dense that a star formed (by the process reviewed in Chapter 4). For our sun to have developed to its current maturity, scientists calculate that the starting explosions would have occurred within ten million years after its initial formation.

This is an event of enormous force, with thousands, if not millions, of nuclear explosions. The force of these explosions would, according to the laws of physics, have expelled the remaining dust and gases surrounding the new formed sun far into the distant realms of space (see Figure 32: The Early Sun's Explosive Beginnings).

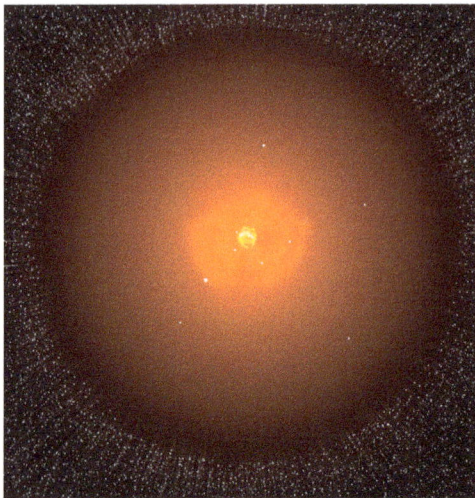

Figure 32: The Early Sun's Explosive Beginnings

Yet Evolutionary Theory seems to ignore, or at best set aside, this natural process. Contradicting physical law, it proposes that a series of rings of matter remained around the sun after the massive nuclear explosions and then formed into the planets. Furthermore, the planets are theorized to have formed in this manner even though there was no material from which they could be made, as it had all been blown away leaving nothing behind. The forming planets were thought to be extremely hot and dense, pulling any heavier elements inwards to the center through gravity, resulting in heavy cores surrounded by lighter gaseous elements.

But there is a stark contradiction here between the early sun's formation and the formation of the early planets. While the sun was forming, Evolutionary Theory proposes that the lighter elements were pulled into its center, whereas with the planets, the heavier elements were pulled in.[8] In one sentence, Evolutionary Theory is claiming two different and opposite methods for the same process.

The Early Planets

The next phase in the formation of the solar system also raises questions about the proposed evolutionary theories.

The theories explaining the early formation of the planets begin not with gravity but with electrostatic attraction, whereby very small particles of dust with positive and negative charges cling to one another and clump

---

8. Hydrogen is by far the most abundant element known and is assumed to have dominated the early universe as the primary form of matter. Relative to iron there is about 66,000 times more hydrogen and 5,100 times more helium.

together. For instance, positively charged particles attract and bind with negatively charged particles. The particles, although initially neutral, gain their polar charges through the frictional force of rubbing together, much like rubbing a woolen sweater to produce a static charge in your hand that can give a mild shock to somebody you touch. The rubbed particles are able to attract other, oppositely charged particles because the vacuum in space offers virtually no resistance to the particles coming together.

This process of electrostatic clumping, however, only goes so far. The electrostatic forces are only strong enough to clump grains less than the size of gravel. For planets to get bigger, it is theorized that another force is required—gravity.

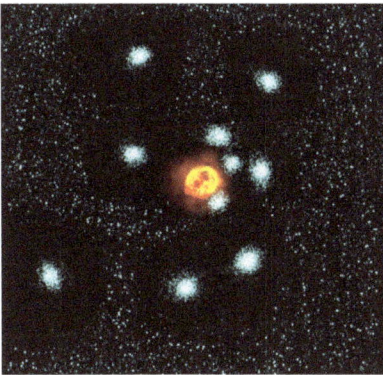

Figure 33: Clearing Zones Around Early Planets

However, there are inconsistencies with the use of gravity. For instance, Figure 33: Clearing Zones Around Early Planets highlights how the area around the early planets would have been cleared of particles through the process of gravity, leaving onion-like zones of near empty(ish) space. Over time, gravity acts to pull in everything within each planet's range, with the result that other particles outside the influence of each planet's gravitation still remain in the general vicinity of the solar system between the planets. Evolutionary Theory proposes that most of this remaining, smallish debris was then pulled into the sun by its strong gravitational attraction, effectively clearing the solar system of most of the remaining cloud particles.

Again, however, Evolutionary Theory in this instance contradicts the known science. Due to its much higher mass, any forming planet or moon will have a significantly stronger combined gravitational pull with the sun than that of smaller dust particles and the sun. This means that the planets would be pulled into the sun, leaving most of the dust particles behind. This effect is recognized in Newton's Law, which is one of the three laws of gravity, and states that the attraction between two bodies is proportional to the mass of each (the amount of material in them) and inversely proportional to the square of the distance between them.

Where:

$$F = \frac{G\ m1\ m2}{r^2}$$

- F is the force between the masses
- G is the gravitational constant (which is a constant of proportionality, in this case between two bodies)
- m1 is the mass of the first body
- m2 is the mass of the second body
- r is the distance between the center of the two bodies

In essence, if the mass of either object doubles, the attraction will double. If the distance between them doubles, the force becomes one quarter rather than a half. This is displayed in Figure 34: Gravitational Physics, where a theoretical unit of gravity (1g) is given as the base unit and M is one unit of mass.

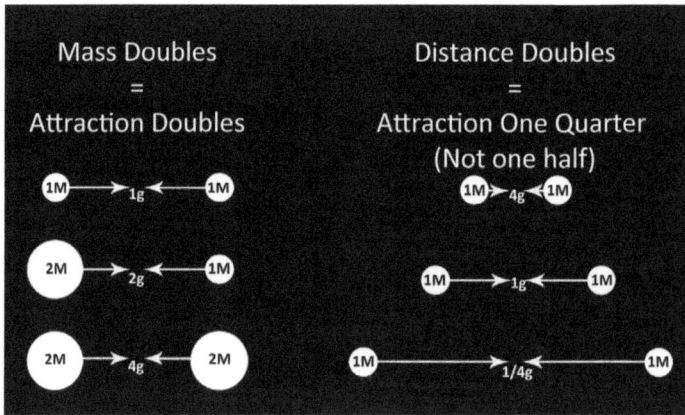

Figure 34: Gravitational Physics

The greater the combined mass of the objects, or the closer objects get to each other, their combined attraction increases sharply as shown in Figure 35: Gravitational Forces In Space. These numbers are representative only.

The massive forming planets (represented with red dots) would be pulled into the sun well before any tiny dust particles, due to their much stronger combined gravity. A planet on the outer circle has an attraction millions of times stronger (1200G) than a dust particle on the inner circle (.0001G). If this physical process was actually the controlling force in forming our solar system, it is likely there would not be any planets or moons, only scattered smallish particles, asteroids and comets.

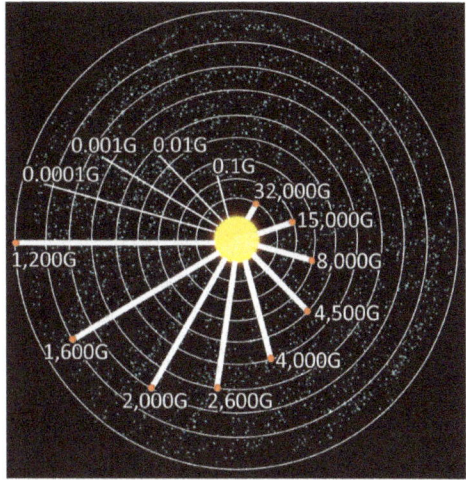

Figure 35: Gravitational Forces In Space

Furthermore, if in the formation of the solar system, the early sun's gravity was such that it could attract small particles, everything would have been pulled into it, rather than around it into orbit, because the planets' initial speeds were as yet insufficient to overcome these massive gravitational forces. There would be no planets, comets, moons, asteroids or anything else, as displayed in Figure 36: The Sun's Strong Attraction.

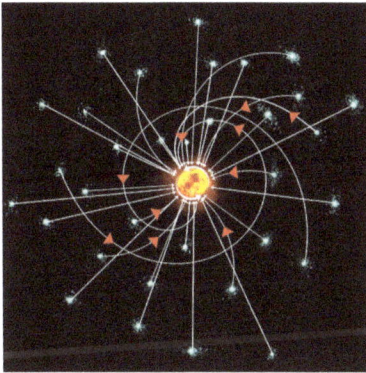

Figure 36: The Sun's Strong Attraction

Evidence from the behavior of our solar system shows that the sun's gravity does not pull everything towards it. The asteroid belt between Mars and Jupiter is made up of millions of small pieces of material sparsely spread in orbit and relatively close, in planetary terms, to the sun. These fragments should be gradually collecting themselves together in a planet building process and/or be drawn toward the sun by its gravity. Both these options are proposed by evolution, yet are not scientifically observable. The asteroid belt does not release most of its occupants, defying the attraction theory, and the belt is showing no signs of creating a new planet based on observable data. To add to this, Mars or Jupiter, being the two closest planets, should theoretically be pulling the asteroid belt apart, yet only relatively few asteroids are moving away from the main pack.

As previously seen with Newton's Law, the level of gravitational attraction or force between the planets and the sun depends on their relative distance and masses. Evolutionists estimate the planets to have taken anywhere up to 4.5 billion years to form after the sun was fully formed. Over this time, their mass and, therefore, gravity will have increased substantially. Physics maintains that the reason the planets don't crash into the sun is because their speed of movement balances out their combined gravity with the sun, thus causing them to stay in orbit. The calculations suggest, however, because the planets have increased their size by hundreds or thousands of times since they first began forming billions of years ago, and have maintained roughly the same speed over a significant percentage of that time, the gravity-to-speed ratio should have increased to the point whereby the planets should have careened into the sun long ago.

Before moving onto the next phase of the solar system's formation, let's summarize the main points in this chapter so far:

- Science cannot verify the formation of the theoretical cloud of gas, how it aggregated, or how the heavier elements were present.

- Science cannot verify how the solar system became disc-shaped or how it started revolving.

- Evolutionary Theory predicts two contradictory gravitational models of solar and planetary formation: in solar formation, hydrogen elements are assumed to have aggregated together in a dense core, whereas in planetary formation heavier elements, including gold and iron, are thought to have aggregated under gravitational forces whereas lighter elements such as hydrogen and nitrogen remained in outer layers.

- Science cannot validate how small minute particles in space were attracted to the newly formed sun, but much larger planetary bodies remained at stable orbits despite their slow initial speeds.

- Scientific observation shows us that the force of gravity is not necessarily the creative force behind the formation of the sun and planets.

The Rotating Solar System

In the next developmental phase of our solar system, the forming sun started rotating around its own axis. To date, the causes of this rotation and the reason for the existence of an axis are unexplained. The rotation of the planets is theorized as occurring as a result of the sun's spin. Immediately contradicting this theory is the fact that Venus rotates in a retrograde or clockwise direction against the sun's spin, as displayed in Figure 37: Venus's Clockwise Rotation, while Uranus spins on its side. Various ideas, mostly based around large objects striking these two planets in a 'big bash' have been proposed to explain these odd rotations. These concepts are reviewed in detail in Chapter 6.

Rather than colliding into it, some asteroids and comets make use of the sun's immense gravity to whip them back into orbit, positioning the sun itself as the innermost point on their elliptical journey. They do not get dragged into the sun; rather, they use its gravity to propel themselves into orbit again as if they are on an invisible track. The core of Albert Einstein's Special Theory of Relativity is the concept that space is curved in a space-time dimension, which helps to explain

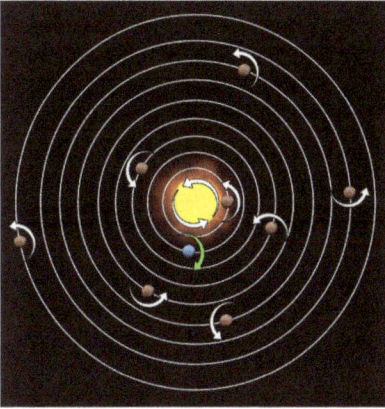

Figure 37: Venus's Clockwise Rotation

the anomalies where planets and comets appear to defy the natural pull of gravity. This concept can be visualized by imagining rowing a small boat between two points in a flat ocean. It is relatively easy to travel in a straight line as depicted in Figure 38: Flat Ocean.

Imagine next that a whirlpool forms between you and the point towards which you are rowing. The water will curve around the center forcing you into a spin beneath the normal water level. Should you manage to get out safely, the journey you have travelled is actually three dimensional—up/down, left/right, and forward/backward—which is displayed in Figure 39: A Whirlpool In Water.

Figure 38: Flat Ocean

Figure 39: A Whirlpool In Water

It is in this manner that Einstein postulated that the planets are attracted into orbit around the sun. The cause of this is unknown, but maybe the sun's sheer mass creates an indent or curve in space toward which the planets 'fall' through the attractive force of gravity. Their sheer speed, however, prevents the planets from colliding with the sun, much as a golf ball can lip around a putting hole without falling into it. Yet nothing physical can be seen in the void of space.

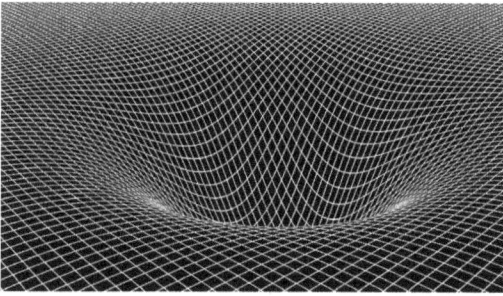

Figure 40: Curved Space

Visually represented, it may appear something like Figure 40: Curved Space.

The perspective of the other planets from earth is a flattish disc shape. If curved space exists, the viewpoint inside the 'whirlpool of space' may result in an appearance of all planets being in the same plane. This may explain how a comet uses the invisible curvature of space as its path around the sun (as with the lipping golf ball). Curved space may also explain why comets do a full orbit and come back on a regular path.

The age of our solar system has been measured at 4.6 billion years from meteorites found on earth. In reality, all that was measured was the theoretical age of the meteorites because the assumptions are that:

• These meteorites were formed in our solar system.

• These meteorites were formed at the same time as everything else.

• Space has no effect upon the meteorites or the measured results.

In 1969, Russian mathematician Viktor Safronov calculated that, if formed according to this Theory of Evolution, the creation of the inner four planets (Mercury, Venus, Earth, and Mars) would have taken roughly 100 million years. These calculations showed that the further from the sun, the greater the time to build up a planet. His calculations were confirmed by American George Wetherill (1925-2006, who was the Director Emeritus, Department of Terrestrial Magnetism, Carnegie Institution of Washington, DC, USA) using computer modelling and are part of the Accretion Theory. This theory assumes microscopic particles of dust and ices are attracted together by their minute electrostatic charges. As the mass builds, the total attraction also strengthens, increasing the collection area. Based on these calculations, Jupiter would have needed at least 100 million years, Saturn longer, Uranus and Neptune billions of years, and newly discovered dwarf planet, Eris, still not totally formed today. These calculations contradict other theories that Jupiter was the first planet to form.

This, however, raises a further inconsistency. When the sun's start-up explosions blew away gases in the solar rings at the 10-million year mark (see Figure 32: The Early Sun's Explosive Beginnings), the planets were only just starting to pull in material to make themselves. This means the matter required to build them was not present for at least ninety percent of the calculated time it would have taken for these planets to form. The question therefore remains: How did the planets form with only ten percent of the material required to build them?

Another inconsistency with the Accretion Theory is consistency. That is, the consistency of the solar system. It stands to reason that all of the elements that are consistent within the Milky Way galaxy should also be consistent throughout every solar system in the Milky Way, including our own. However, it is interesting to note that this is not so. The general make-up of the planets (including the dwarf planets Pluto and Eris) differs greatly from the Milky Way's components of seventy four percent hydrogen and twenty four percent helium.

The makeup of each planet is as follows:

- Mercury: a dense iron core with a silicate mantle and crust

- Venus: iron and nickel interior

- Earth: thirty percent iron, thirty percent oxygen, fourteen percent magnesium

- Mars: iron and sulfur interior with silicate surface

- Jupiter: possibly a rocky core, ninety percent hydrogen ten percent helium

- Saturn: possibly a rocky core, ninety percent hydrogen ten percent helium

- Uranus: rocks and ice made up from methane, ammonia and water

- Neptune: rocks and ice made up from methane, ammonia and water

- Pluto: not known but assumed to be rock

- Eris: assumed to be similar to Pluto, being mostly rock

These differing components bring into question the reliability of current evolutionary theories. The first four planets have very small levels of hydrogen and helium, whereas the next four planets have high levels. Evolutionary theories maintain a precise order of formation in which the outer planets were made later, not earlier, than the inner ones. This, however, brings the Accretion Theory into conflict with the known scientific observations because it predicts the outer four planets had to be made earlier than the inner four, otherwise their hydrogen components would have been drawn into the sun before they could be made.

Furthermore, the Accretion Theory cannot explain the fluctuating proportions of elements on earth, with almost negligible levels of hydrogen and helium, as compared with those in the Milky Way.

So, what does science tell us so far about the formation of the solar system?

Firstly, science cannot validate or prove the current evolutionary theories about its formation. Secondly, science in part refutes the processes described by these theories, which are contradictory and inadequate at best, and which ultimately fail to satisfactorily describe how the early solar system could have self-formed.

Rotations And Revolutions

The rotations and revolutions of the planets highlight some interesting observations. Most evolutionary theories propose nine planets revolve around the sun, each having developed independently from a cloud of dust at different locations and began revolving as an effect of the sun's gravity and rotation. The big bang theory suggests that our newly forming solar system would have had expansive momentum away from the point of the original big bang explosion. Then, at a particular point in time, and for no particular reason, the sun accidentally started spinning in an anticlockwise direction (from the north), simultaneously creating an axis. As the sun and planets were moving in the same direction, there are three possibilities regarding their relative speeds:

Scenario 1: The Sun And Planets Are Travelling At Equal Velocity.

In this most commonly assumed scenario, the sun's spin would be the only mechanism that could start the planets moving. Regardless of their positions, the sun's enormous gravity would draw the planets into it on a fairly direct path of eventual collision rather than around it on a regular orbit, particularly if the sun was in front of the planets. A trailing planet would be pulled in a straight line and crash into the back of the sun or even be flung out into space.

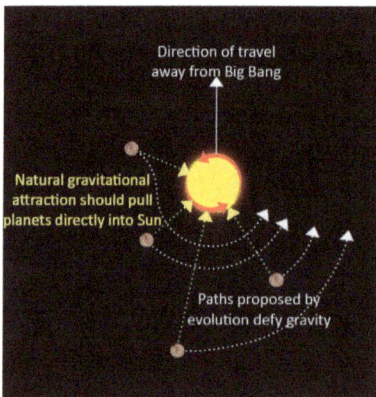

Take a look now at Figure 41: Planets And Sun Equal Velocity. For all planets to the right of the sun, it is easy to imagine them revolving in the same direction as the sun, but this

Figure 41: Planets And Sun Equal Velocity

requires ignoring the gravitational physics. For the planets to the left of the sun, the sun's spin would overcome their natural momentum and propel them in a reverse orbit to their initial direction. As an outside observer, you would see the planet stop and reverse direction in order for it to enter an anticlockwise orbit. At that moment, the lack of speed would see the planet plunged into collision with the sun.

Consequently, this scenario cannot explain how the planets were forced into orbit using only the effect of the sun's gravity and spinning.

Scenario 2: The Planets Are Moving Faster Than The Sun.

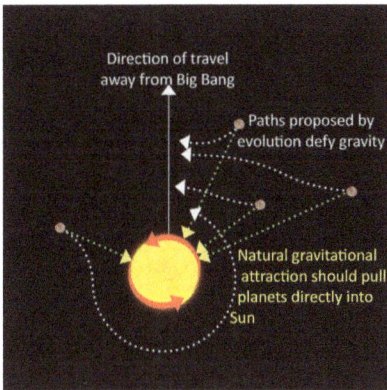

Figure 42: Greater Planetary Velocity

In this scenario, the planets have to be fully formed and attain their orbiting speed by some other method. The only possible way this can happen is if the planets were not formed in our solar system. The sun's gravity would 'catch' the faster moving planets as they neared the slower moving sun and then proceed to hold them in orbit. The planets to the right of the sun would head in anticlockwise orbit. The planets to the left of the sun, however, would head in an opposite or clockwise orbit, until at least the sun's spin took over and changed their direction, resulting in the same scenario as Scenario 1 where they would collide with the sun. Figure 42: Greater Planetary Velocity displays these actions.

Scenario 3: The Sun Is Travelling Faster Than The Planets.

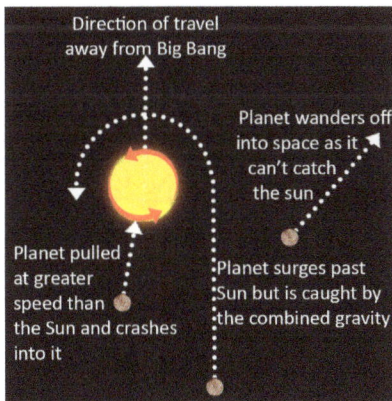

Figure 43: Greater Solar Velocity

In this scenario, the planets could not catch up to the sun unless the sun's gravity pulled them closer at a speed greater than the sun was travelling. They would either crash into the sun or speed past, only to be held up again by the sun's gravity. As they were going past, the sun would only have one chance to hold them in orbit. This would have to be achieved by a delicate balancing act, matching the pull of gravity with

77

the speed and distance of the planet. If a planet escaped, it would most likely still be drifting in space, relatively nearby our solar system. Lacking such evidence, evolution can only assume that the sun corralled them all.

For any of the three scenarios above to work, the planets had to be fully formed before they could move into steady orbits otherwise, as they drew more material into themselves, the balance between their speeds and sizes would disrupt the possibility of orbiting. Realistically, none of these scenarios works in practice leaving Evolutionary Theory about planets' orbits and rotations at odds with physics.

As briefly referred to earlier in this chapter, the planets orbit the sun in an anticlockwise direction, creating a natural physical reaction to 'roll' with the pull. This should result in anticlockwise orbits and rotations of all the planets. However, one planet, Venus, defies these forces, rotating clockwise on an anticlockwise orbit against the pull of the sun. The physics of this situation dictate that it should throw itself out of our solar system, yet it remains intact. (see Figure 44: Venus Should Eject Itself From Our Solar System). This effect can be displayed when hitting a table tennis ball with a bat. Using topspin, the ball will quickly turn toward earth. If underspin is used, the ball will float further until gravity eventually catches it. In the case of the solar system, once Venus has broken out of its orbit because of its reverse spin, its speed and spin would continue to drag it away from the ever decreasing pull from the sun.

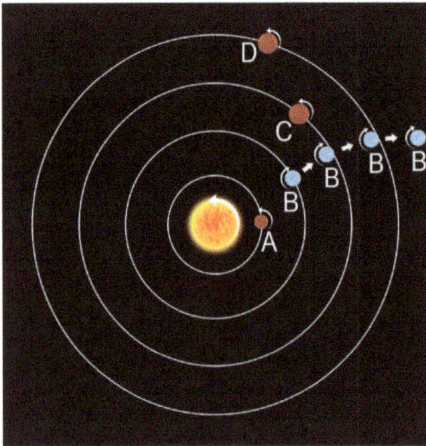

Figure 44: Venus Should Eject Itself From Our Solar System

Nevertheless, Venus spins against the flow and is held in place by some undetectable force that scientists are unable to determine. It seems to be a specific mechanism that controls the positioning of the planets.

Our Solar System: A Summary

Our solar system is unique, in that no other similar solar systems have yet been found. It is believed that there has been a solar system forming for twelve million years around Beta Pictoris, which is a star surrounded by swirling dust and gases. It is hypothesized that it will self-create into a solar system in the same manner as is theorized to have occurred in ours. At this age, this star would have theoretically burst into life relatively recently, pushing away most of the dust and gases when it fired up. Yet, the dust and gases are still there. What's more, there is a massive orbiting planet about nine times the size of Jupiter that astronomers claim has made itself in just a few million years. Yet, this time frame defies science and is completely at odds with Evolutionary Theory, as we will discuss in the coming chapters. Saying that, it is important to note why this planet provides three counter reasons to Evolutionary Theory on solar system formation:

1. According to science, the time frame is too short: It cannot have formed itself in such a short period of time. Either the time frame is miscalculated, or the time frame is correct and other non-evolutionary forces were at work in its construction.

2. Because of its relative nearness to Beta Pictoris, the planet should have been pushed further away when the star exploded into life. It should also still be moving away; but it doesn't appear to have been impacted by any type of repelling force.

3. Beta Pictoris is almost twice the size of our sun, making the gravitational pull between this planet and its star to be very powerful, considering its own size and proximity (see Figure 34: Gravitational Physics). The calculations show that both the star and planet should have already been dragged together by this strong combined gravity.

So what does science tell us so far about the rotations and revolutions of the solar system?

- Science does not validate the explanations of Evolutionary Theory on how the planets entered into orbit around the sun.

- Science cannot explain how Venus rotates in a clockwise direction or explain the unknown force keeping Venus in orbit.

- The evidence from Beta Pictoris contradicts current evolutionary theories on solar system formation: time frames are dubious; and nuclear forces and gravitational forces don't appear to act within known mathematical calculations.

Outside our solar system, there are over a thousand other planets that are claimed to have been identified by astronomers and the number is growing regularly. These have not actually been physically seen, rather, astronomers have identified the wobble effect that they have on their parental stars. This wobble effect, without going into specifics, occurs when a planet passes in front of its star (from our point of view in space). Its gravitational pull makes the star wobble ever so slightly, and it is this infinitesimally small wobble that astronomers detect.

However, many of these newly discovered planets are so big and in orbits so tight that they completely defy the current theories on solar system formation. These discoveries have forced scientists to adjust their theories, as it was assumed that giant planets were only able to form at a much greater distance from their sun, where it is cooler and where gravitational forces wouldn't rip them apart before they could form.

Some scientists have proposed that perhaps these newly discovered giants have been pulled in closer to their stars by gravity. Newton's Law calculates that as the planets get closer to their stars, the gravitational pull, and therefore acceleration, is much stronger (four times as strong at half the distance). Yet protagonists of this theory haven't explained how the planets suddenly stopped in orbit around the star before crashing into it. Our largest planet, Jupiter, is suggested as being of a very similar makeup to these planets, yet it hasn't been pulled into orbit as close to the sun as these newly discovered planets.

However, with many solar systems having large Jupiter-sized planets in close proximity to their stars, orbiting in just days at enormous speeds, the current theories about how our solar system formed are no longer considered valid. It is thought that Jupiter gathered the material to make itself by being closer to the sun and that Saturn, which had built itself further out, coincidentally wandered towards the sun and miraculously dragged both planets away into their current distant orbits. This idea contradicts the existing theories that each planet formed in one place and then started orbiting as a result of the sun's spin. In other words, planets drift about within their solar systems. This new information creates a major conundrum for evolution as the theories put forward over the last century are now contradicted.

Here we have a good example of the regular problems that every evolutionary theory encounters: theories don't work in practice as they contradict known facts. The main reason evolutionary theories are proposed is because science works on the premise that everything made itself, even though this is known to be impossible. Science doesn't look at the facts unencumbered, it looks through a blurred lens of assumed self-development which doesn't allow science to present an unbiased view of the facts.

For more up-to-date information on solar system formation, please visit our website www.evolutionunraveled.com

Chapter 5 Unraveled

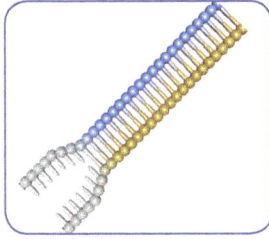

- The proposal that all stars formed from clouds of gas and matter contradicts known physics. Gases dissipate and do not gather together of their own accord. There is no explanation about where these cloud formations came from. Rather, clouds of matter should be more evenly spread following a big bang explosion and there should be no stars.

- It is extremely unlikely that our sun and its planets formed themselves from small bits of matter via gravity because the attraction between the sun and the planets is far greater. Accordingly, the planets would have collided with the sun well before any small bits of matter would have, effectively destroying them.

- If formed by gravity, the solar system should either be full of dust and small rocks, or almost completely barren, with only the sun existing by itself.

- The proposal that the sun started its own controlled thermonuclear fusion process that blew away the remaining dust and gases at the 10-million year mark means that the planets could not have formed their current sizes. The five outer planets, from Jupiter to Pluto, were calculated as forming well after the 10-million year mark, contradicting this theory because there would not have been any material from which to create themselves.

- Of the planets discovered outside our solar system, over ninety nine percent contradict the Evolutionary Theory about how planets were meant to have formed by themselves.

- The materials that make up planet earth are completely different from both the rest of the solar system and the universe. There is no explanation where these materials came from or why they appeared in one isolated spot, by accident.

- Although gravity is the cornerstone of evolutionary modelling for solar system formation, the scientific evidence does not support the theories.

6

A DETAILED VIEW OF OUR SOLAR SYSTEM

In Western society, most people have learned through schooling, media and scientific articles that the entire contents of our solar system—the sun, planets, moons, asteroids, comets, and so forth—formed around the same time some 4 billion or so years ago and from similar materials, primarily as a result of gravitational forces. As we have seen, however, these theories contradict natural forces of attraction and lack sufficient time frames to allow self-building. Current scientific data therefore implies that the solar system could not have formed in the manner described by Evolutionary Theory and must have been made by some other method.

In recent times, science has been able to make far more accurate measurements about the universe regarding size, age, chemical and physical composition, rates of movements, and so forth. Summarized in chart form in Figure 45: Current Planetary and Lunar Data are the measurements of various actions and positions of the planets and their moons in our solar system. The times quoted are relative to earth time. This data shows our solar system to be a very different place from what we have previously believed.

An orbit is the time it takes for a planet to go around the sun once—for example, 1 earth year. Mercury, being closest to the sun, has the shortest length of orbit and orbits the sun in the shortest time. As expected, this pattern follows each successive planet; those further from the sun have a longer orbit and take more time to complete one orbit. Vice versa: the closer a planet is to the sun; the greater it's orbiting speed.

However, the rotations of each planet (where one full spin equals 1 day) produce wild variations. Earth, Mars, Jupiter, Saturn, Uranus, and Neptune take between 10 and 25 hours to complete one full spin on their axes, whereas Pluto's rotation is roughly 6.5 days. Mercury (58 days) and Venus

(243 three days) are even slower. These speed variances are contrary to the scientific calculations made by physicists and astronomers, as planets closer to the sun with faster orbits should theoretically spin faster if the sun is responsible for their orbits and rotations. The spinning speed of the seven anticlockwise planets varies from 10 hours to 58 days, which defies scientific explanation. Furthermore, giant Jupiter spins one hundred and forty times faster than diminutive Mercury, even though it is thirteen times the distance from the sun, which is supposedly responsible for their spin. These anomalies suggest the planets gained their spin by some influence other than the sun.

	Orbit	Length of day	Speed	Tilt of axis	No. of moons	Distance from sun
	Length of year in earth time	In earth days— one rotation around own axis	Speed in orbit around sun	From North		Millions /kms
Mercury	88 days	58 days (11kph)	172,404 Kph	0.00°	0	58
Venus	225 days	243 days (6 Kph)	126,108 Kph	177.36°	0	108
Earth	365 days	1 day (1674 Kph)	107,244 Kph	23.45°	1	150
Mars	687 days	24.5 hours (866 Kph)	86,868 Kph	25.19°	2	228
Jupiter	12 years	10 hours (45583 Kph)	47,016 Kph	3.13°	16	778
Saturn	30 years	11 hours (36840 Kph)	34,705 Kph	25.22°	18	1,427
Uranus	84 years	17 hours (14794 Kph)	24,516 Kph	97.86°	15	2,870
Neptune	165 years	16 hours (9719 Kph)	19,548 Kph	28.31°	8	4,497
Pluto	248 years	6.5 days (123 Kph)	17,064 Kph	122.52°	1	5,900
Eris	557 years	26 hours (282 Kph)	12,370 Kph	unavailable	1	14,500

Figure 45: Current Planetary And Lunar Data

The Differences In The Planets

As mentioned in the previous chapter, it stands to reason that if the planets were created by a process of self-formation then all the planets should contain roughly the same levels of basic components. This, however, is not the case. Following is a short summary of each planet, beginning with the closest planet to the sun, Mercury, detailing their major differences.

Mercury

Figure 46: Mercury

The first planet, Mercury, is a rocky planet that travels at great orbiting speed. There are no clouds or moons and it has a limited atmosphere. The temperature varies from four hundred and twenty five degrees Centigrade during the day to minus one hundred and eighty five degrees Centigrade at night. Mercury orbits the sun in just 88 days but turns slowly, taking 58 days. Theoretically, it should be the fastest spinner in the solar system, given its proximity to the sun. It also has a large metal core that is out of proportion to its size. Most theories suggest it may have been in a collision that forced the outer core to break free into space, but there is no explanation or evidence as to the whereabouts of this material.

On its surface is a massive impact crater named Caloris Basin. It is so large that the assumed shock of this possible collision appears to have caused a mountain range to form on the opposite side of the planet. The expectation from any such collision would result in some tilt. Yet Mercury rotates with no tilt to its axis. Mercury's orbit is also odd. It has an eccentric orbit varying between forty six and seventy million kilometers. From an observer's point of view, there would be some strange effects to be seen in various locations on its surface arising from its slow rotation and relatively fast orbit—the sun might be seen to rise, shrink, stop, go in reverse, grow in size and shrink again in a relatively short time frame.

Venus

Figure 47: Venus

The second planet, Venus, is a rocky planet without moons. Venus is the night star, the first object in space visible from earth at sunset. It rotates on its axis slower than it orbits making a day on Venus longer than a year. It is the only planet to rotate clockwise, for which there is no valid scientific explanation. Some theorists believe it was struck by another object causing its orbit to reverse. A collision of such enormous force, causing an entire planet to spin the other way would have sent it out of orbit and tumbling off into space or into a collision with the sun. There is no evidence of such an impact on Venus or of any effect on its orbit. Neither is it spinning and rolling as one would expect, rendering this idea as unrealistic without any supporting facts.

Its surface consists of molten lava with a surface temperature of four hundred degrees Centigrade. There are more than fifty thousand volcanoes sites evident, some being over one hundred kilometers wide. None are currently active. There is no evidence of movement in the surface and it appears to be an inactive planet. It is covered by clouds about thirty kilometers thick and has a dense atmosphere ninety times that of earth at sea level, ninety five percent of which is carbon dioxide. Venus has only weak magnetic fields in some areas. There is no apparent planet-wide magnetic field.

Earth

Figure 48: Earth

The third planet, earth, is a rocky and active planet covered by vast quantities of water with many cloud formations. It has one moon at close proximity. Numerous life forms exist, ranging from microbes to fully developed species that have multiple variants. Life forms include those that are static (e.g. plants), mobile (e.g. animals/insects), airborne (e.g. birds/bats) and underwater (e.g. fish). There is low evidence of comet collisions. It has a multi-layered atmosphere that burns up most extra-terrestrial invading bodies and which holds in many gases. It is the only planet known to have tectonic plates on the surface. It is also the densest planet and has an abundance of elements not detected on other planets, such as gold, uranium and platinum. The image of earth from space reveals an abundance of water covering the planet. A more detailed analysis of the earth is presented in Chapter 7.

Mars

Figure 49: Mars

The fourth planet, Mars, is a rocky planet about half the size of earth. From Mars we name our third month of the year, March. It has two small moons, Phobos and Deimos, which have irregular shapes. Phobos has a diameter of about twenty two kilometers and Deimos has a diameter around half that of about thirteen kilometers. Both moons are covered with craters and are made of rock and iron. The planet itself is covered in desert and is constantly swept by dust storms. A Mars day is about the same length as an earth day. The surface in the southern hemisphere is littered with craters from impacting comets and shows virtually no sign of activity. However, the northern hemisphere is covered in lava and in some areas the surface has an appearance similar to the effects created by flooding water. Temperatures vary from around minus one hundred and forty three degrees Centigrade in the polar regions to thirty five degrees Centigrade at the equator.

There are four huge volcanoes all in the same general area in the southern hemisphere. The largest is Mount Olympus, which is about twenty five kilometers higher than the surrounding terrain. It is thought that these massive volcanoes occur from a regular build-up of pressure from underneath, which has been estimated to have continued for billions of years. As the surface of the planet is stable and doesn't move on tectonic plates like earth, these volcanoes always occur in the same place. Although the surface of Mars is very cold, it is thought to act as an insulating medium for the hot lava below. There are massive canyons up to seven kilometers deep and one hundred and eighty kilometers long that may have resulted when the volcanoes erupted.

The internal core of Mars is thought to be molten iron. Tests on the Mars soil by the *Viking* spacecraft indicate that it is highly corrosive and uninhabitable for any life form that we know about. It has a thin atmosphere, about one percent that of the earth, ninety percent of which is carbon dioxide. Measurements from the Mars Express Orbiter (which orbits the planet) indicated occasional minute traces of methane, in 2014 NASA's *Curiosity* rover revealed the atmosphere did not contain methane, dismissing the possibility that microbes have ever existed. Consequently, as far as science has determined, no life form that we know about can, or has ever been, on Mars.

The Asteroid Belt

Figure 50: The Asteroid Belt

The asteroid belt consists of rocks of varying sizes in the area of space between Mars and Jupiter. While thousands have been identified there are thought to be millions of asteroids within this belt loosely packed and spread over a huge width that exceeds one hundred and fifty million kilometers (the same distance as between earth and the sun). The speeds of the asteroids vary due to their vastly differing positions. The average speed is calculated somewhere around one hundred thousand kilometers per hour. At these speeds, there is insufficient attraction from the sun to hold all of them in orbit. The asteroid belt appears to be held together by some force other than gravity, as few are observed escaping the pack. The largest asteroid, Ceres, measures nine hundred and fifty kilometers across. One asteroid named 235 Mathilde has the most craters of any object ever observed in space.

There are many theories concerning the asteroid belt. Most theories assume that it is a planetary graveyard, whereby a large planet was broken apart by an undefined collision. However, a collision so powerful to smash an entire planet into millions of asteroid-sized pieces would be expected to leave evidence on neighboring planets and moons, yet there is no evidence of such a heavy bombardment. Prior to its discovery, the location of a planet at this point was mathematically predicted. The inference from these calculations is that the planets are not randomly located; they are set out in a mathematical plan that we can understand, interpret and predict. Scientists were relieved to find the asteroids but are at a loss to describe how they could have come to exist in such a format.

Jupiter

Figure 51: Jupiter

The fifth planet, Jupiter, consists mostly of hydrogen gas in similar proportions to the sun. Many propose it may have formed into a star if it had continued to grow larger. It has an atmosphere thousands of kilometers deep. It is the largest planet, being thirteen hundred times the size of earth by volume, with a mass two and a half times the total of all the other planets in our solar system combined. Computer calculations made by George Wetherill, Director Emeritus, Department of Terrestrial Magnetism, Carnegie Institution of Washington, also determined that the size, position and massive gravity of Jupiter limit the number of strikes on earth by comets and other space debris by one thousand times as would be expected if it did not exist.

Scientists have named the distance Jupiter sits from the sun as the 'snow line'. Here the temperature is so cold that water vapor and other gases freeze into solids (ices). It is theorized that when planets were forming from this distance, their composition included a great proportion of ice material and lesser amounts of rock and metal. Because the sun has little influence on its temperature, Jupiter appears to derive its heat internally, which explains its planet-wide even temperature. It has a huge magnetic field detectable up to twenty million kilometers away. A cloud layer comprising ice crystals, ammonia and water covers its surface. There is a constant, mammoth storm thirty thousand kilometers long (over twice the size of earth) that is called the red spot.

Jupiter also has an orbiting ring made up of debris that is difficult to detect. As of January 2009, Jupiter had forty nine named moons with a further fourteen pending consideration by the International Astronomical Union (IAU). The smallest moon is thought to be Leda, which is only sixteen kilometers in diameter. The outer moons orbit in the opposite direction to the larger inner ones, a finding that directly challenges the theory that they were made by gravity from local clouds: Jupiter's enormous gravity and spin should result in all moons orbiting in the same direction.

The evidence suggests that Jupiter's moons were sent into opposing orbits by another, unidentified force. As such:

The outer moons of Jupiter were not sent into orbit by the planet's spin; there was some other force operating.

The range of moons around this planet presents a series of contradictions for the Theory of Evolution. Some examples of the differing features of the moons are:

Io

Comprising mostly sulfur, iron and rock, its outer appearance has a metallic sheen and is covered with areas of yellow, red and white. There are so many active volcanoes on Io that the entire surface is covered in lava spills. Few meteor impact craters can be found. It is the approximate size as earth's moon, which is inert and too small to still have a molten interior, if self-created billions of years ago. A new theory called 'tidal heating' attempts to explain how a small body such as Io has a molten core. This theory claims opposing gravitational pulls from Jupiter and sister moon, Europa, alternately stretch and crush Io in 'tidal waves' and the friction created by these forces causes its surface to rise and fall by about one hundred meters. Molten matter which results from the friction and pressure breaks through the moon's crust along with gases causing volcanoes.

These gravitational forces are thought to be so enormously strong that they physically distort the shape of this moon, which is basically rock and iron, yet Io remains in orbit, neither pulled toward its sister moon nor its parent planet, Jupiter. This raises two alternatives:

1. If tidal stretching is the cause of the heat, another force is holding the moon in place to counter the natural effects of gravity from Jupiter and Europa as otherwise Io would be pulled toward, and collide with, one or the other. This force is not identified by science.

2. If stretching is not the cause, however, its hot interior means that Io is a relatively recent addition to our solar system as, due to its size, it would have cooled by now, as evidenced by our similarly sized moon. It must have been made in its current location at a far later time than most other planets and moons in our solar system.

With either scenario, there is another factor controlling the behavior of Io that is not explained by modern astrophysics.

Callisto

Callisto is mostly comprised of rock and ice. This moon is a classical representation of what science predicts all of Jupiter's moons should be. Covered in lots of craters and ice, it is cold and has not changed its surface in billions of years. However, science cannot explain how this moon can orbit Jupiter so close to Io, yet is so completely different, assuming it was made in the same way from the same materials at the same time. The observations of Callisto, in fact, also suggest that Io is a recent addition to the solar system.

Ganymede

This moon has a rocky core surrounded by ice. It is the largest moon in the solar system yet has few impact craters compared with its neighbor Callisto. This observation suggests that Ganymede's age is considerably younger than most of its sisters, yet the only explanation for this is that it was made at a more recent time. Along with Io, its existence suggests the influence of an unknown force other than gravity.

Europa

Europa has an iron core and silicate rock components. This moon is covered in ice and is very smooth, showing few signs of impact craters. There are many cracks in its surface. It is speculated that it has a volcano that emits water rather than lava.

Unfortunately, the current available evolutionary theories fail to explain the observable differences in Jupiter's moons. According to the theories, they should each have roughly the same surfaces, temperatures, orbits, contents, evidence of comet collisions, ages, and so forth, with only size being an accepted variable. Yet this is not so.

Saturn

Figure 52: Saturn

From Saturn we get our name for Saturday. The sixth planet is gaseous with a make-up similar to Jupiter, so light it could float on water. It has wild surface storms, one of which sits atop its North Pole and is twenty thousand kilometers across appearing as a hexagonal shape with a circular storm in the middle. The storm rotates every ten hours synchronizing with the planet's radio emissions and permanently remains at its longitudinal (east-west) position which is unique for any known storm on any planet.

Saturn is surrounded by thousands of small bright rings containing debris. These are not perfect circles, with some entwining each other and some radiating between concentric rings. A further, very narrow ring appears to be created by two small moons on either side of the planet forcing debris into a tight band. The theory is that the rings are from a moon that has either been pulled apart by Saturn's strong gravity or broken up by a collision with another unknown body. This is believed to have occurred around 50 million years ago.

However, if Saturn's gravity was strong enough to pull a moon to pieces it stands to reason it was strong enough to pull the moon or all of its pieces into itself. Yet this is not observed. Science cannot explain what force(s) held the moon in orbit in order for it to be pulled apart. The observations of Saturn's rings, as with Jupiter's moons, contradict the theories of solar system formation (whereby clouds of matter were gravitationally sculptured into the planets and the sun). The debris in these rings is remaining in orbit and is certainly not being pulled into either the sun or Saturn. Remember, the speed of orbiting bodies is claimed to balance out the pull of gravity so they remain in a constant orbit. The debris in Saturn's rings is virtually static, with no observable velocity. Without velocity to resist gravity, an unknown anti-gravitational force is therefore maintaining their position. The moons of Saturn also display some interesting features, of which there are approximately sixty in orbit, although not all are officially recognized.

Enceladus

Enceladus has craters on one side but is smooth on the other and is surrounded by a ring of ice debris. This presents a problem for any theories as bombardment from space comes from all sides, as evidenced on other planets and moons. Observations by the international Cassini spacecraft in July 2005 found water vapor over the moon's South Pole and zones of leaking heat. If correct, it will mean this moon is geologically active and may provide an explanation for the undisturbed surface.

However, this creates yet another complication in that, according to one Cassini spokesman, it is an astonishing find "equivalent to discovering that Antarctica is hotter than the Sahara desert". While there may be a reason for the smooth surface, the heat inside this moon is not scientifically possible for its assumed age. Like Jupiter's moon, Io, Enceladus is in the same category of recent additions to the solar system.

Iapetus

This is the third largest of Saturn's moons being about ten percent the size of earth, comprised about twenty percent rock and eighty percent ice with a thin crust. Its shape bulges at the equator and is compressed at its poles, showing only one face to its planet, as does our moon. It has many impact craters with the largest being five hundred and eighty kilometers across. There is a mountain range running around three quarters of its equator ranging from twelve to twenty kilometers high. The two hemispheres are different colors with the leading side being reddish brown and the trailing side bright. None of these features is expected from a self-formed body. Despite many ideas proposed to explain such anomalies, we really don't have a clear picture how it could form itself. The masses of impact craters indicate sister moon Enceladus may be a more recent addition to the solar system.

Titan

Titan is the only moon in the solar system to have an atmosphere. It has more nitrogen than earth and is denser.

Uranus

Figure 53: Uranus

The seventh planet, Uranus, is a gas planet that spins on its side like a barrel and is covered in cloud. Summer on Uranus lasts 21 earth years, as does its winter. Its magnetic pole was recorded by *Voyager 2* at sixty degrees to its axis. It has very narrow rings of debris that science considers virtually impossible to explain, as the material should have been pulled in by the planet's gravity. One ring is red, the other is blue. It is thought that particle size causes the color difference. The blue ring is only the second of that color identified so far, the other being around Saturn. These blue rings are similar in that they are both outer rings and house moons, whereas red rings do not.

There are thought to be twenty seven moons in orbit around Uranus. Oddly, of the five major well-known moons, only Umbriel shows no evidence of previous geological activity. The other four moons have disrupted surfaces from what is thought to be expanding ice causing enormous changes in surface structure. One of these moons, Miranda, is small, being just five hundred kilometers in width, and is covered in ridges, valleys and great soaring cliffs. Evolutionary Theory claims these five moons to have self-created from the same components, at the same time, in the same small area of space. Yet, Umbriel, with no geological movements, contradicts this theory. The conclusion is that Umbriel was made separately, either from different components and/or at a different time.

Neptune

Figure 54: Neptune

The eighth planet, Neptune, is besieged by extremely strong winds that constantly change the cloud formations. It has a surrounding ring that is more a series of clusters of dust and small rocks than a continuous ring. This scattered layout is more typical of what science expects from a random cloud formation, rather than the more organized rings around Jupiter, Saturn and Uranus.

Neptune has thirteen moons in orbit. The largest, Triton, is the coldest moon in the solar system at minus two hundred and thirty five degrees Centigrade. This moon has an ice cap made from nitrogen. When the *Voyager 2* spacecraft took photos of Triton it found large plumes of heated nitrogen gas were being expelled from the surface, an extraordinary observation for a body four billion kilometers from the sun in the freezing reaches of the outer solar system.

Triton's unusual orbit, displayed in Figure 55: Triton, is difficult to explain as:

- It rotates against the turn of the planet, whereas it should have slowed considerably or turned to a clockwise rotation, if it has been in place for billions of years.

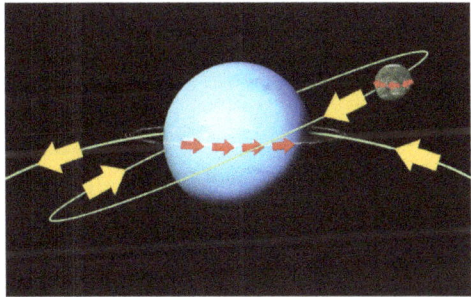

Figure 55: Triton

- Its axis is inclined to the planet's equator rather than in line with it. This results in differing pulling forces because it regularly crosses the equator then moves away, then back again. These pulling forces should destabilize its orbit.

- It revolves in time with its orbit although it is essentially going backwards, constantly showing the same face to its mother planet. Considering it is going in the opposite direction to Neptune, all of the gravitational and spinning forces should be acting against it maintaining this fixed position.

Evolution has no explanation for these anomalies with Triton.

Pluto

Figure 56: Pluto

The ninth planet, Pluto, is a dwarf planet likely to be made from a combination of rock and ice. Pluto is the smallest planet in the solar system, about two thousand three hundred kilometers in diameter, which is smaller than Neptune's largest moon, Triton. It also acts differently from other planets, having an elliptic orbit at a tilt to the other planets' orbits. This orbit at times carries it inside of Neptune for several years each cycle, seemingly more like an asteroid or comet than a planet. It has a moon, Charon, that is closer than any other moons' orbit. These two bodies are locked together and only ever show the same face to each other.

Pluto also has two other tiny moons—Nix and Hydra. In 2006, Pluto had its status as a planet withdrawn, albeit there is no unanimous agreement from the scientific community and debate continues.

Planet No.10—Eris

Figure 57: Eris

In 2003, astronomers announced the finding of another planet orbiting our sun. The discovery, made with the Samuel Oschin Telescope at the Palomar Observatory, places the new planet at nearly one hundred times earth's distance from the sun. This planet is claimed to be slightly larger than Pluto at two thousand four hundred kilometers in diameter but is more than twice as far from the sun. This planet was originally labelled as "2003 UB313" and was allocated the interim name of Xena, then officially Eris.

This is an important discovery for two main reasons:

1. It means that our solar system is much, much larger than we had previously been taught.

2. It is difficult to explain how this new planet could possibly have self-created under gravity because its distance from the sun means that it could not yet have built itself from nearby solar debris. It is currently another unexplained phenomenon.

Finally, it does raise an interesting thought: What if this planet is new and has remained hitherto unseen because it has only recently been made?

This would fit with the observations that the universe is expanding at an increased rate, allowing for the inclusion of a new planet. This is a problem for Evolutionary Theory, however, as it has no explanation for a new planet appearing from nowhere, just as it has no explanation for matter appearing from nowhere at the very beginning of the universe's creation.

Other Observations About Our Solar System

Neptune and Uranus are similar in size and composition. They are made mostly of ice, constructed from ammonia, water and methane. These elements indicate their formation occurred a long time after Jupiter and Saturn, which are also similar gas planets, meaning that they were made at different times. This suggests these two sets of planets were made by different processes from the one proposed by evolution.

If the theory of a self-created evolutionary solar system is correct, calculations reveal that the sun should have seven hundred times more angular momentum than all the planets combined. In reality, the planets have fifty times more angular momentum than the sun. This difference is huge: thirty five thousand times the opposite to what is expected.

To explain this point, we need to understand that when objects move around a point they are considered to have angular momentum. Through experiments, science tells us that angular momentum in the universe, like matter and energy, cannot be made or lost but can be transferred. The mathematical equation that defines angular momentum for the sun and a planet is $L = mvr$, where:

- L is the angular momentum

- m is the mass of the planet

- v is velocity or speed

- r is the distance between the two bodies' centers

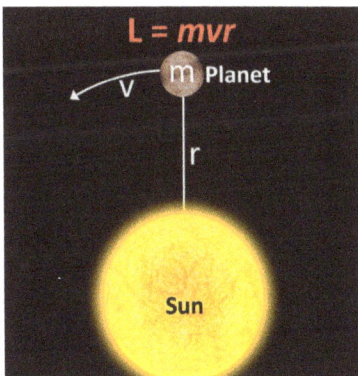

Figure 58: The Sun's Angular Momentum

In a practical way, we can display angular momentum by performing some simple experiments. Hammer a nail into a straight stick and attach a piece of thin cotton thread to it so it can rotate around the nail. Then tie the other end to a ball, say a tennis ball. Swing the stick above your head until the ball is whirling around with the cotton fully extended. That force pulling against your hand represents the gravity between the sun and a planet at that orbiting speed. So, that particular level of gravity will hold that particular planet in orbit at that particular speed. In this manner, the ratio of the speed of the planets to the pull of the gravity with the sun allows the planets to stay in orbit.

But, as detailed above, the angular momentum equations do not match the actual physical properties and speeds in our solar system, indicating that a strong external force is overriding the natural physics that are in place.

Further interest of note, as our solar system is such an infinitesimally small part of the universe, it is a reasonable assumption that all of the components making up the planets and moons should be virtually identical.

To visualize the scale and simplicity of this, imagine you are standing in front of a large, deep ocean. Now imagine you are inserting the head of a pin into the water. Assuming the ocean to be the universe, our entire galaxy would therefore be able to fit on the head of that pin. It is also reasonable to expect that the water in this small area would be the same general consistency. We certainly wouldn't expect the water to contain huge variations in constituents.

Likewise, the planets in our solar system should be made from similar materials with very little in way of variation and range. But they do, which suggests the universe was not created in the manner proposed by the big bang theory but by some other manner unexplained by Evolutionary Theory.

Chapter 6 Unraveled

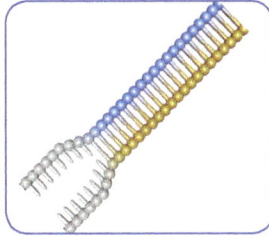

- The theory of self-building planets and moons is not substantiated by the observed operations of the planets in our solar system.

- Jupiter, Saturn, Uranus, and Neptune have moons and nearby debris that are not being pulled in by the gravity of either the sun or the planets themselves, yet they have more combined gravity now than during their assumed formative period.

- The asteroid belt between Mars and Jupiter shows minimal effect of gravitation pull from the sun or its neighboring planets.

- Many of the planets and moons show distinctly different features, components and ages; suggesting they were made at different times.

- The moons of Jupiter orbit their planet in different directions, which evolutionary science cannot explain.

- Natural physical forces should make the planets rotate at zero degrees to their axes, but only Mercury does this.

- The pull of the sun should make all planets rotate in an anticlockwise direction, but Venus rotates clockwise.

- The conditions on every planet other than earth will not support life as we know it.

- The recent discovery of a tenth planet in our solar system contradicts the current theory of how our solar system formed: It is so far from the sun it would not yet have formed itself from gravitational attraction of dust and gas particles. It also makes our solar system many times larger than previously thought.

7

A GENERALIZED VIEW OF THE EARTH AND MOON

According to Evolutionary Theory, earth is simply an accidental gathering of particles that formed themselves into a planet. Accordingly, the earth should be made of similar materials to the other planets in our solar system, all of which should also operate in roughly the same way, taking into account such variables as differing temperatures, size and proximity to the sun. However, the earth is a vastly different place, full of complex and wonderful things, and is the only planet where life is known to exist.

The Earth's Orbit

As discussed previously, a constant orbit is claimed to be created when the speed of a planet matches the pull between it and the sun, forcing it into a circular revolution. Most planets' orbits are not perfect circles. The earth's orbit, for instance, is an ellipse, a kind of 'stretched' circle. Its distance from the sun varies from between one hundred and forty six and one hundred and fifty two million kilometers. It also wobbles slightly off line with the pull of the moon's gravity causing an imbalance. The earth's axis is tilted at twenty three and a half degrees, whereas it should be zero degrees if normal, uninterrupted physical forces applied.

It is this tilt in the axis, however, that provides us with the four seasons. The most popular theory explaining this tilt is that earth was pushed off center following a collision with a very large comet or another planet a very long time ago. Likewise, similar explanations would be necessary to explain the tilt of every other planet in the solar system, except Mercury, whose rotation is vertical to its axis.

Yet this provides a quandary. Any substantial collision between two massive bodies in space would cause both to move off their existing paths. If all the planets (except Mercury) had been similarly struck, their trajectories would be heading in various directions, much different to the synchronous orbits in which they are observed to move today.

If this collision theory is correct and the earth was struck by another massive object, the impact would have been monumental. It is theorized that the impact was in fact so great the resultant debris was flung into space, which settled into orbit around the earth and finally accumulated itself to become a satellite, which we know as the moon.

However, the evidence for such a collision is non-existent. A collision of this nature would roughly follow the series of events below:

- The earth is spinning at zero degrees in an anticlockwise direction, because of the sun's gravity (see Figure 59a)

- It is hit by a huge object that causes the earth to start rolling from top to bottom, but which also maintains its anticlockwise rotation (see Figure 59b)

- Because of the change in orbital momentum it is knocked off its course (see Figure 59c)

- The earth either keeps rolling and rotating away from its orbit around the sun toward the outer reaches of the solar system, or it is pulled towards the sun on an impending collision course (see Figure 59d)

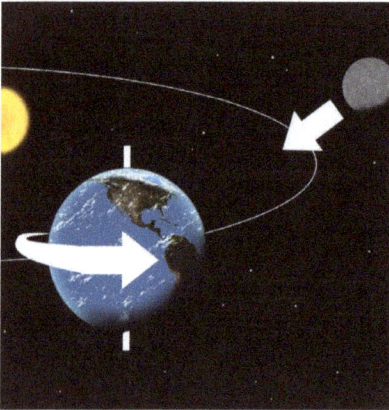

Figure 59a: Initial Earth Orbit

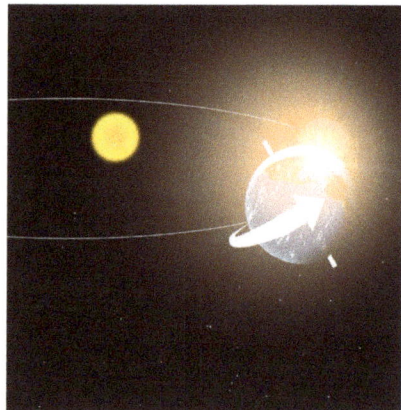

Figure 59b: Initial Planetary Collision

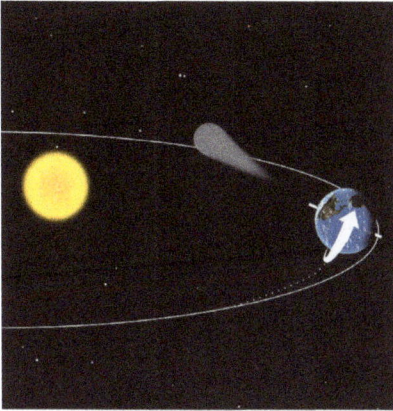

Figure 59c: Change In Orbit

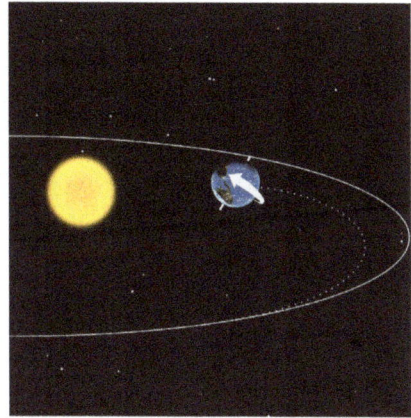

Figure 59d: Impending Solar Collision

The lack of evidence for the collision theory is such:

• There is no evidence of any impact crater large enough to suggest such a collision[9].

• The earth has no rolling motion.

• The earth is still in synchronous orbit around the sun, neither permanently moving away nor moving toward the sun.

Furthermore calculations by Sigeru Ida, Robin Canup and Glen Stewart (*Origin of the Earth and Moon*, Editors R. Canup and K. Righter, University of Arizona Press, 1997) indicate that about seventy percent of the material from this collision would have fallen back to earth. But such a great force would have created twice the current angular momentum that exists between the moon and earth today.

Earth's Atmosphere

The atmosphere is a very interesting and extremely complex structure. The earth is currently considered unique in having a multi-layered atmosphere (see Figure 60: The Earth's Atmosphere) that protects against the freezing temperatures of space whilst at the same time retaining the heat from the sun. It also provides protection from most of the damaging, incoming matter such as cosmic rays, the sun's rays and meteors, the majority of which burn up on entry through the process of friction.

---

9. Some geological scientists claim that most of the earth's early history was erased because of the hot, molten state of the planet during its formative years, thereby suggesting that such a huge collision with another planet-sized object would leave no geological fingerprint on the earth's surface. There would still be roll from such a hit but this is not evident.

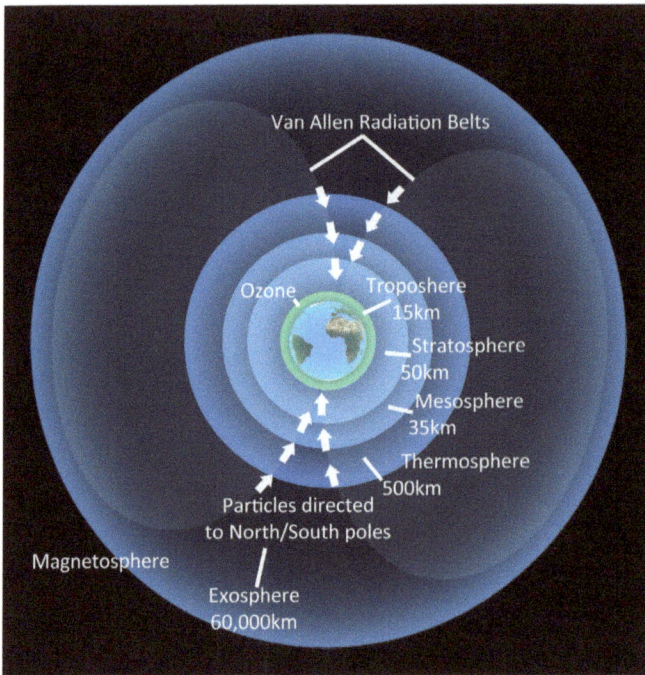

Figure 60: The Earth's Atmosphere

The atmosphere's main points of interest are:

- The troposphere, which contains most of the weather, starts at ground level and reaches up to fifteen kilometers.

- The stratosphere reaches from fifteen to fifty kilometers and has two levels. The lower band contains sulfates that are believed to be involved in the cause of rainfall. The upper band is mostly ozone that absorbs ultraviolet rays from the sun and is fundamental for life to exist.

- The mesosphere rises from fifty to around eighty to eighty five kilometers and is the part of the atmosphere responsible for burning up incoming meteors.

- The thermosphere is next rising to around five hundred kilometers. Within it is the ionosphere which reflects radio waves back to earth.

- The exosphere is the outer most layer that dissipates around sixty thousand kilometers from the surface of the earth. It contains a large radiation band called the magnetosphere that fans out into a thin tail of protons and electrons.

- The magnetosphere is a magnetic field that surrounds the earth and provides protection from solar radiation. It is also the realm in which most spacecraft orbit the earth, protecting astronauts from deadly radiation. It runs from one pole to the other through space and deflects the solar winds around the earth's perimeter as well as encapsulating the other atmospheric layers. There is also another layer of magnetic fields that directs any intrusive particles towards the poles (Van Allen belts) with the resultant auroras at the poles. These auroras can also be seen on Jupiter and Saturn.

There is a catastrophic level of radiation emanating from the sun and stars from which life on earth is protected by the operation of earth's atmospheric layers. The sun is not actually a fire as one sees when timber, gas or petrol burns. All stars are thermonuclear reactions that emit massive levels of radiation that can destroy life. Whilst the processes of an atomic bomb differ slightly from stellar nuclear reactions, atomic bombs do give a glimpse of this destructive solar force, not only the energy expended but also the impact of radiation.

Hundreds of millions of nuclear explosions happen in our sun simultaneously, every second, exposing us to a fearsome level of nearby radiation. Life survives because the earth has an electromagnetic field protecting its inhabitants from this harmful radiation. This electromagnetic field is generated by earth's liquid outer core moving around its solid iron inner core. The field is compressed on the sun's side by the incoming force of the sun's solar wind whereas it tails out longer behind the earth. The effect is shown as a red in Figure 61: The Earth's Shield Of Force.

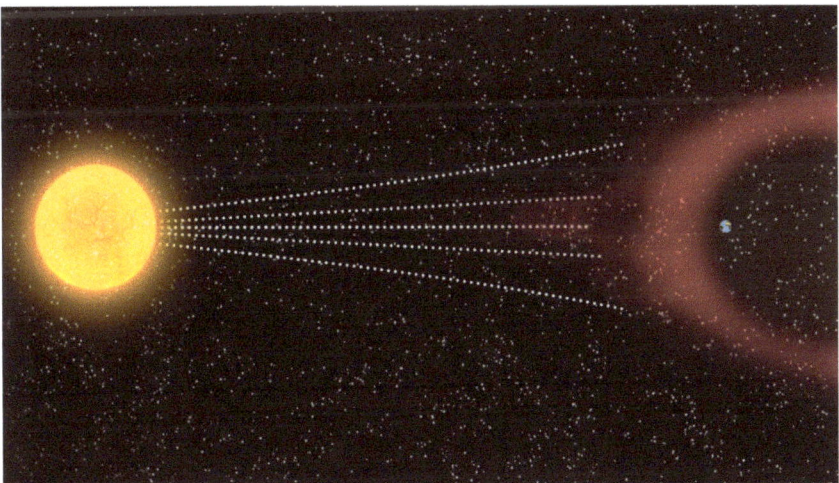

Figure 61: The Earth's Shield Of Force

The layers in earth's atmosphere operate in tandem with the electro-magnetic field as a series of barriers, deflectors and abrasive elements. These layers not only protect life from incoming destruction but also stop life-giving gases from escaping into space, such as oxygen and carbon dioxide.

The Age Of The Earth

Various branches of science use the fossil record to obtain information regarding past events. A fossil is a preserved remnant of a plant or animal, although often the original body has not remained intact. Commonly, the body of the fossil has decayed leaving skeletal bones or shells that may be petrified (changed to a stone-like state). Many fossil bodies will completely disintegrate leaving only an impression of the creature or plant. The fossil record indicates what was alive, where and when it lived, and what the environment was like at that time. Fossils are usually found in dirt or mud close to land surface and in rocks or geological deposits. The cumulative total of these individual finds is called the fossil record. Despite the common use of this general term, it is surprising to learn that there is no official record by any scientific body endorsing the dates of fossils, which means there is no official fossil record. The fossil record is only a database of the observations made about fossils scattered across a wide range of sources, of which any combination of these sources can be used to make a list of possible datings[10].

The fossil record reveals many things. One interesting fact it has revealed about the geology of earth is that for most of the planet's life it has maintained a constant even temperature of about twenty two degrees Centigrade. Yet this even temperature is in direct contrast to Evolutionary Theory because it is claimed:

- The sun has grown as a developing star increasing its temperature by around fifty percent since the earth was formed. However, the earth's temperature records do not reflect this increase in solar temperature.

- The earth was theorized to have shaped itself into a ball while in an initial hot molten state and then cooled down over time. It would therefore have had a much higher starting temperature, yet there are no observable signs of substantial cooling.

---

10. Section D at the rear of this book reviews the most commonly used tests to determine fossil ages. All of these appear to be reasonably challengeable. The conclusion is that datings, be they absolute or estimated, are neither provable nor reliable, and seemingly depend on unprovable assumptions. Consequently, any list of dates is an amalgam of estimates from various scientists using a range of methods, none of which are officially endorsed by science.

These observations indicate the timing of the formation of the sun and the earth is different from that proposed by Evolutionary Theory. Earth's constant temperature suggests that it formed after the sun was fully operational, as it is now.

So just how old is the earth?

Evolutionary Theory estimates the age of the earth to be around 4.5 billion years. Many variables are considered when calculating the earth's age. However, there are also many variables that indicate the earth could also be much younger than originally thought. Various sources have been used for the following information:

- The earth currently rotates at about sixteen hundred kilometers per hour. It is gradually slowing down because of gravitational drag from the sun and the moon. Based on its current speed and the level of drag, the earth of 4.5 billion years ago would have been turning so fast its shape would have been flattened and completely distorted. But if it had started with a relatively slower speed it would have already stopped spinning.

- The earth's core is molten but should have cooled significantly more by now given its age.

- The earth's magnetic field is gradually weakening with a decline of around ten to fifteen percent over the last 150 years and possibly up to thirty five percent over 2,000 years ago. Whilst these are viewed as within a normal range of variation, calculations on these rates show that just 7,000 years ago, the strength may have been as much as thirty times its current level. Going back 20,000 years, it may have generated so much heat that the planet would have been uninhabitable. Earth's magnetic field is weakest in an area called the South Atlantic Anomaly. Massive whirlpools of water have recently appeared there and in the Atlantic Ocean. These are considered to be indicators of a reduction in the strength of earth's magnetic field. The British Geological Survey, which was founded in 1835, analyses data about earth science. It has claimed that the magnetic field was double its current strength around the time of Christ's birth. Geological data also shows the magnetic field has reversed and changed from North to South poles and vice versa. There is conjecture that these previous magnetic swings caused molten magma to move from the earth's mantle to the ocean floor creating hot spots. Potential increased volcanic activity might occur in the near future at various locations, including Tristan da Cunha, Hawaii and Easter Island.

Our Moon

The earth has the largest moon in the solar system in proportion to its size (almost seventy percent the size of planet Mercury). It is also relatively the closest and has the most influence on its parent planet than other moons. For instance, the moon's influence on our tides and ocean movements allows a constant refreshing of water and nutrients. Its size and proximity also retains the earth's tilt, which itself maintains the seasons. A number of evolutionary theories have been raised to explain our moon:

1. *The moon developed as a nearby twin from the same material as the earth.*

If this were correct, calculations reveal that if both bodies were building themselves from nearby materials over time their combined gravity would increase sharply and they would eventually pull themselves together.

2. *The moon was flung from the earth in its early formation.*

This would indicate that the moon would have the same density as earth and an identical orbiting plane. Yet the moon has just two thirds that of the earth's density and it is inclined at a tilt of five point one four five (5.145) degrees.

3. *The moon formed somewhere else and was captured by earth's gravity as they passed each other.*

The physics of this theory require this body, when passing, to be relatively close and coupled with the perfect speed and placement to go into orbit, rather than just deviate from its path or crash into earth.

4. *A collision with a large celestial body knocked off part of the earth's outer layer which then formed the moon.*

While this might explain the density difference, it doesn't resolve why the earth's tilt is twenty three point four five (23.45) degrees but isn't still rolling or wobbling from the hit. The stabilizing influence of the moon's gravity, as suggested by some evolutionists, is not strong enough to stop the roll or wobble. Furthermore, the expulsed lunar material would require a precise amount of force to separate violently from earth's gravity and then stop, almost immediately, at a relatively close distance, commence a daily orbit around the now tilted earth, and reshape itself into a sphere.

This collision theory suggests the expulsed lunar material would have had some momentum after the impact, but the only force that could halt its momentum into space was gravity. Some theories suggest that the moon actually formed much closer to the earth than it is today, which had the capability of causing such mega tides that the cities of London and New

York would be submerged under a wall of seawater twice a day if it were still in the same position as it formed.

The changing distance from the earth, however, suggests that if gravity slowed and then ceased the moon's momentum away from earth, it would also have the power to reverse the momentum toward earth's surface. Either gravity was too weak to prevent the lunar body being flung further into space, or its strength pulled the moon back into a collision course with the earth. Either way, we would have no moon according the physics of this theory.

Another curiosity of the moon is that it always shows the same face to the earth (keeping its other face, its 'dark side', away from view), taking the same time to orbit and rotate. Under theories proposed by evolution, the rotation of the moon is the direct result of the rotation of the earth. Yet this would require the moon to have appeared after the earth was tilted with a stable rotation.

Lastly, the collision theory doesn't take into account the temperature variations between the earth and the moon. Samples taken from the first moon mission of *Apollo 11* show lunar rocks having roughly similar mineral composition to rocks found near volcanoes on earth. However, compared with earth rocks, they are devoid of water and contain few of the lighter elements. Even taking into account the lower atmospheric pressures on the moon, this finding indicates they have been subjected to much higher levels of heat, which has evaporated the water and lighter elements. This is a curiosity, as it means that the temperature of the moon must have been considerably hotter than earth if they self-formed from the same material at the same time; not a possibility as temperature is claimed to be the result of size.

These discrepancies in observable scientific evidence and evolutionary theories of solar system formation bring into question that the moon formed in the same manner, from the same localized material, and at the same time as earth.

Chapter 7 Unraveled

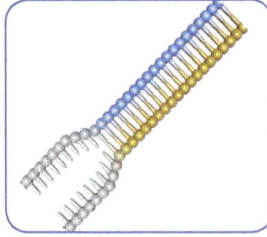

• The earth, its atmosphere and contents are constructed from materials that are different from the rest of the solar system.

• The evidence for a collision of the earth with another planet or large comet to give its tilt is non-existent.

• 20,000 years ago, based on the current rate of magnetic field decline, calculations reveal that the magnetic field may have generated so much heat that the planet would have been uninhabitable.

• The earth and the moon are different.

• Science cannot confirm that the earth and the moon formed in the manner claimed by the Theory of Evolution. Either:

    * They were not made from the same materials, and/or

    * They were not made in the same manner, and/or

    * They were made at different times.

Points To Ponder – Conclusion From Chapters 5, 6 and 7

What conclusion comes from these chapters regarding the theory that our solar system formed itself from a cloud of dust and gases?

- There appears to be only one theory about the formation of stars. Virtually all scientific references claim they formed from an incomprehensibly massive cloud of hydrogen atoms. However, there is no actual scientific proof that stars can make themselves from clouds of gas or planets from collections of rocks, just theories.

- If the planets formed from local dust and gases they wouldn't have been moving past the sun during this time; they would have been gathering size from particles in their nearby neighborhood while staying relatively still. Evolutionary Theory claims that planets get their spin from the combined gravity and spin of the sun. Because they had no initial speed, the only way for planets to move is to originally head toward the sun and then make a significant sideways turn to go into orbit, overcoming increasing gravitational pull as they grew closer. What made them turn? Science can provide no answer.

- The scientific evidence shows that the bodies in our solar system were made from different components at different times and operate in different ways. These observable differences are not compatible with the current theories of evolution.

- Therefore, the scientific conclusion regarding the formation of the solar system is:

The solar system did not use gravity to sculpture itself from localized clouds of matter, nor were the bodies within the solar system created at the same time as each other.

PART THREE

EARLY LIFE

THE START OF
LIFE ON EARTH

What is Life And How Does It Work?

To understand the magic of life, we need to analyze its components, how they fit together and how they allow living beings and plants to exist. The ability for living creatures to function requires complex interacting structures of chemicals, muscles, fluids, tissues, organs, sensors, electrical connections, and a myriad of processes, many of which are still not understood. There is an overwhelming complexity of both components and interactions necessary for life to exist.

There is no universal agreement on the definition of life. Studies in Australia have found minute nanobes (super small microbes) that appear to contradict the previous lower size limits at which life was known to exist. Whilst these findings have not yet received widespread acceptance as living material, they highlight the controversy in trying to define what constitutes a life form.

For the time being, it is commonly accepted a living organism can be identified by several basic characteristics:

1. It must be able to reproduce itself. This requires making a copy that has the same components as the original and assumes this copy can also reproduce itself.

2. It has to grow through metabolism. This is a process of increasing and breaking down of protoplasm (a substance like semi-transparent jelly). Protoplasm is considered as the physical base of life although its terminology has been superseded these days. The cells of all living things contain protoplasm but it is not found in anything that is inanimate (for example, nonliving things such as rocks). Although protoplasm appears similar in all cells, it is

unique to every type of living entity but it cannot grow into tissue for any other creature. As will be discussed further, this in itself is a major obstacle for any theory of biological evolution that claims one type of creature changed into another. It would mean that protoplasm has to change during cell development; something it doesn't do.

3. <u>Living organisms must be autopoietic</u>. That is, be able to adapt to different environments by changing. Observations suggest this only happens within the boundaries of that entity. A cell or organism will not change its structure to something else when exposed to an external disturbance. There seems to be an inherent limit or boundary within which any living organism must remain. It can adapt, but it can't change from its intended structure into some other type of life.

These three basic characteristics of life confirm that a living entity can only make copies of itself and cannot become anything else. Evolution on the other hand proposes all life has developed from some previous, different form. Which means Evolutionary Theory is proposing processes that are not known to be scientifically observable or even possible, and these will be examined in detail in the coming chapters.

All life is based on DNA (deoxyribonucleic acid). This structure is responsible for transmitting heredity. It can either change by mutating or replicating itself, thus allowing it to survive changes. Life is also composed of reduced carbon compounds, which are essentially carbon atoms surrounded by hydrogen atoms. Carbon is such an important element that, in combination with nitrogen, hydrogen, oxygen, sulfur, and phosphorous, these six elements make up almost one hundred percent of the dry weight of all living things. The percentages of these elements are similar in all life forms on earth, from very basic microbes to mammals. The levels of amino acids, genetic compounds, long proteins, and DNA macromolecules are also similar. Some scientists have drawn the conclusion that because of these chemical similarities all life has a common ancestry. For instance, a commonly quoted figure is that humans share ninety eight percent of DNA with chimpanzees so evolutionists claim, humans must therefore share a common ancestor.

Effectively, evolutionists claim that all life originated from the same source, whereby life emerged from one original lineage by accident and self-developed into a range of different variants with common components.

How Did Life Start On Earth?

Science cannot explain how life started by itself. New life is not observed to start by itself now and there is no documented scientific evidence for its occurrence in the past. Evolution therefore assumes that life commenced as a one-off event, an event that hasn't been repeated. There are a number of evolutionary theories, but they tend to fall into two main categories:

1. An extra-terrestrial method—for example, life was seeded from a meteorite or began as a colony from another planet.

2. It started here on earth suddenly.

If life came from another existing life form or from a meteorite as a bacterial cell, the same issue remains in determining how that life form first started. Pursuing such a theory, whilst fascinating, is not within the realms of this book. The issue is not where it started but how. As life is only known to exist on earth, this is where we will concentrate our investigation.

Understanding the evolutionary process of self-invention requires a step-by-step review of the processes proposed. The current theories will be examined using observations about how life works now, the fossil record and by looking at experimental work and scientific analyses.

The Chain Of Life

As an overview of the general evolutionary path followed by living organisms, we can say that the main groups thought to have been part of the line of life go roughly as follows:

- Non-living matter: chemicals, liquids, gases, rocks, and other solid materials.

- Protozoan: a single cell organism.

- Metazoan: a multicellular organism.

- Invertebrates: animals without backbones.

- Fish: cold-blooded aquatic vertebrates with gills and scales.

- Amphibians: vertebrates that lay eggs in water and go through a 'tadpole' stage, then metamorphose into lung breathing quadrupeds.

- Reptiles: cold-blooded vertebrates.

- Birds: warm-blooded vertebrates covered with feathers and wings as forelimbs.

- Quadrupeds: animals having four feet.

- Primates: monkeys and apes.

- Man: a primate animal at the highest level of development characterized by large brain capability.

The list below is one example of the timetables used to identify the arrival of various life forms on earth. These dates are gathered from various sources, predominantly the fossil record and the geologic column.

Life Form	Millions of Years Since Appearance
Microbial	3,500
Complex	2,000
First multicellular animals	670
Shell-bearing animals	540
Vertebrates (single fishes)	490
Amphibians	350
Reptiles	310
Mammals	200
Nonhuman primates	60
Earliest apes	25
Australopithecine (human ancestors)	5
Modern humans	0.15 (150,000 years)

The Geologic Column And Fossil Record

When working out the path that evolution may have followed, science has predominantly used the geologic column, which represents a build-up of sediments over time as its reference point. The hypothesis is that there is a sequence in fossils where lower, simpler forms appeared prior to more complicated higher forms. Therefore, simpler fossils mean older rocks. In the cross section of the column (see Figure 62: The Geologic Column), there is assumed to be a relationship between the ages of the rocks and the fossils found in them, where those species were alive in that period of time.

Period/Age	Start
Plaisocene	2Ma
Pliocene	5Ma
Miocene	23Ma
Oligocene	33Ma
Eocene	56Ma
Paleocene	66Ma
Cretaceous	145Ma
Jurassic	201Ma
Triassic	252Ma
Permian	298Ma
Pennsylvanian	323Ma
Mississippian	358Ma
Devonian	419Ma
Silurian	443Ma
Ordovician	485Ma
Cambrian	541Ma
Proterozoic	2500Ma
Archean	4000Ma

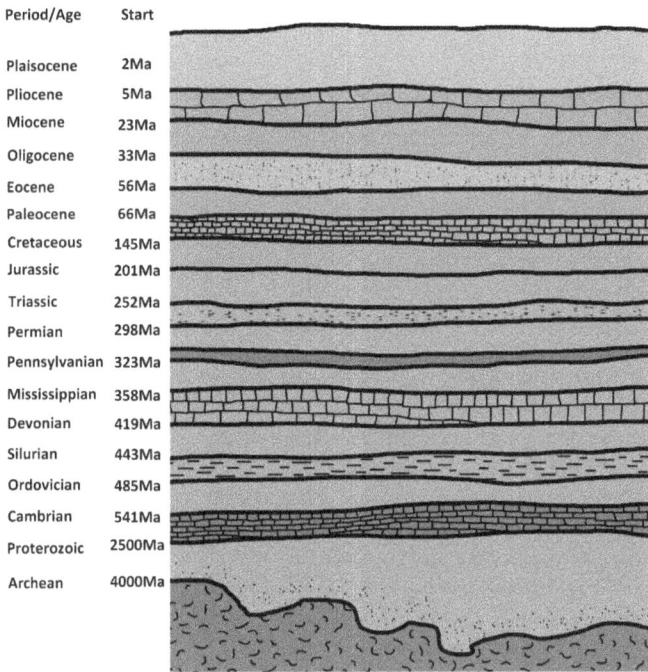

Figure 62: The Geologic Column

This assumption is on shaky ground because both simple and more complex forms are deposited in the lower levels of the geologic column, which means they were living together at the same time in the same environment. Because of the inconsistency in rock formations across the world, due to different environmental conditions–volcanoes, earthquakes, floods, and such like–the fossil record is relied upon as a standard, a consistent measuring stick with which to determine the age of rocks. For example, in places where violent earth movements have caused splits in terrain, where one side is completely upside down relative to the other side of the split, reference is made to the fossil record in those rocks to determine which side is the correct way up.

Unfortunately, this requires scientists to pre-determine the conclusion that fossils of a particular species were only to be found on earth at the same time in history, irrespective of the location. Which, by inference, refutes the possibility that certain species may have appeared at different times in different places, or even survived for differing periods of time, such as cockroaches, which have apparently remained unchanged for 300 million years. This assumptive law, by which much dating is calculated, could thereby result in incorrect dating.

What's more, the geologic column is not representative of successive layers of strata systems found across the world. It is essentially only a diagram made by geologists. There are only three or four locations that show these complete layers: central Poland, west Nepal and west Bolivia. Most of the earth's land surface and areas surveyed under the ocean display only approximately one third of the geologic column. Using this column to calculate dates is therefore not a reliable method.

The general impression of the fossil record is that it proves biological evolution by providing a vast array of examples of progressive 'improvements' in life forms over billions of years. Yet this is not the case. There are in fact no proven examples of anything changing into anything else. There does not exist one piece of scientific evidence to prove lower life forms changed into higher life forms. Even paleontologists (those studying fossil organisms) admit the lack of evidence is a problem. In fact, the fossil record displays only three types of scientific evidence:

1. Fossils of creatures and plants that once lived but are now extinct. These appeared fully developed without any ancestral trace and then disappeared without signs of changing into anything else.

2. Fossils of creatures and plants that are still living with us today. These appeared fully developed without any ancestral trace and are essentially the same today as when they first arrived.

3. Fossils of microbes appearing from nowhere and never changing into anything but microbes.

The scientific reality is that the fossil record has no examples of any intermediary life forms changing into anything else and this is known to the scientists who study it.

The following quote from Dr. Edmund J. Ambrose (Emeritus Professor of Cell Biology, University of London) gives us a clear understanding of what the fossil record really tells us:

"At the present stage of geological research, we have to admit that there is nothing in the geological records that runs contrary to the view of conservative creationists, that God created each species separately..." (Dr. Edmund J. Ambrose, *The Nature and Origin of the Biological World*, John Wiley & Sons, 1982, p. 164).

Effectively, Dr. Ambrose admits that there is nothing in the fossil record to justify evolution's proposals of continual changes. That doesn't mean there is some, a bit or lots of evidence. It means there is none.

When it comes to species development, the fossil record does not show any evidence of phyletic evolution; that is, there is no scientific evidence of one particular species or life form transforming into another species or life form. G K Chesterton (1874-1936) was a writer who dabbled in many genres and is responsible for many insightful quotes. One of his most famous remarks is that scientists "seem to know everything about the missing link except that it is missing". The point Chesterton is making is that the reason science can't find a missing link is because it's not there. Furthermore, in the case of biological evolution, there are, in fact, no links—they are all missing from the fossil record.

Summarizing the scientific findings:

• There is no evidence to support the theory of biological evolution in the fossil record.

• There are no physical records on earth of any creature changing into another creature.

• There is no scientific evidence to support the lines of biological evolution down the history of time.

Even Charles Darwin, the most famous of all evolutionists, agreed. He also said there is no evidence in the fossil record of biological changes from one creature to another. If the father of evolution reported no evidence in the fossil record to support such a theory, what conclusion should we draw?[11]

Cells

The main characteristics that we can observe about all living things are that they are made from cells.

• A cell is a basic unit of life.

• Cells are the common denominator in all living things.

This tenet was first proclaimed by German anatomist, Theodor Schwann, in 1839 and has been confirmed by numerous scientific tests ever since.

What Is A Cell And What Does It Do?

A cell is an extremely complex, but minute, entity. It is made of chemical compounds that follow normal chemical processes. Proteins make up the

---

11. Refer to Section B at the rear of this book under the heading, "Other Things You Should Know" which explains what Darwin actually said.

structure of the cell and are the key to its existence. Proteins themselves are made up of amino acids, which are molecules comprised of the elements carbon, hydrogen, nitrogen, and oxygen. Although these four elements could form a large number of amino acids, there are only twenty that are used to build proteins regardless of the type of organism—bacteria, plant or animal. These twenty amino acids plus nine other compounds (glucose, fats, sugars, and nitrogenous molecules) make up the totality of living cells. There are twenty nine compounds responsible for any life, as we know it.

There are two types of cells:

• Single cells are microbes but can also be called bacteria, protozoans and micro-organisms.

• Nucleated cells are those that make up living organisms such as creatures and plant life.

Cells were discovered by Robert Hooke in 1665 while using a microscope to look at organisms. He named cells after the Latin word *cella*, meaning room, because he thought cells looked like small rooms. The idea of cells was further explored by a Czech, J.E. Purkye in 1837. Finally, three German biologists, Schleiden, Schwann and Virchov concluded three rules about all cells:

1. All living things are made of cells.

2. The cell is the basic unit of structure and function in all organisms.

3. Every cell comes from another cell that lived before it.

The First Types Of Cells—Microbes

The first types of cells identified in the fossil record are microbes. There are many different types of microbes but there is no specific theory regarding how they came to appear. With no single explanation from evolution, the following discussion is a balanced amalgamation of many different theories.

The microbe is most commonly recognized as bacteria around the home. In wet areas such as bathrooms, the black 'mold' that appears is bacteria. It is on deteriorating foods and in areas with little ventilation.

This mold is made up of millions of individual cells. They are not joined together to make anything specific; rather, they are a group of separate cells growing in the same place through the process of replication. Microbes differ from other life forms in that their cells do not have a nucleus, and are collectively known as prokaryotes (from the Greek, *pro* meaning before and *karyon* meaning nucleus). Other life forms that do have a nucleus are called eukaryotes. Bacteria are essentially identical to the oldest form of living entity ever found. According to evolution, they are the first life form and subsequently responsible for all other later life forms including creatures and plants. It is theorized these microbe cells originated early after the earth's formation.

Parts and their functions of the single cell bacteria include:

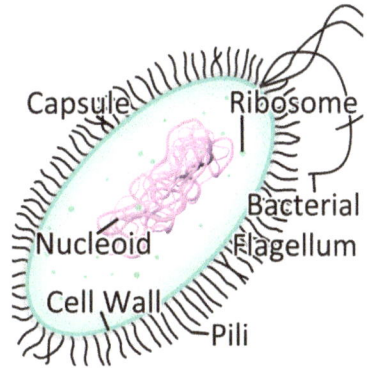

Figure 63: Bacteria

1. The cell wall: is strong, flexible, gives support and provides filtering. It surrounds the cell membrane and limits expansion when water enters the cell.

2. Ribosomes: synthesize proteins and link amino acids in a specific order set out by RNA which is a sister form of DNA[12].

3. The Nucleoid: contains the DNA, which sets out the cell's genetic coding. It is not surrounded by a membrane as in nucleated cells.

4. Bacterial Flagellum: The Latin meaning of flagellum is whip. These appendages are used for movement. They are made from differing protein components in nucleated and single cells.

5. Pili: attach to other cells and transfer DNA between cells.

As mentioned above, each cell has a database of information that determines its type, structure and operation in a component we know as DNA. A cell cannot operate without DNA; DNA knowledge must exist before the rest of the cell can form. Accordingly, this highly complex acid was the first component of life on earth.

By default, evolution also requires that:

a) DNA self-created by accident with an extraordinarily complex database of knowledge, which contains all of the information required for later life, at its first appearance.

b) Then every component in bacterial cells (shown in Figure 63 Bacteria) accidentally self-created from chemicals and compounds that were lying around in the surrounding environment.

---

12. DNA and RNA are explained in more detail in Chapter 11 when reviewing nucleated cells.

c) Later, an unknown event caused the combination of these parts to become life, which in theory meant they could reproduce, grow and adapt.

Interestingly, the research of French scientist, Louis Pasteur, contradicts every theory proposed that cells created themselves. In 1864, following the three rules of cells determined by Schleiden, Schwann and Virchov, detailed above, Pasteur conducted an experiment that revealed that all cell types can only be the product of other cells, including microbes. Pasteur's conclusions were, and still hold true today, that:

1. A living cell cannot be made by spontaneous generation from non-living matter.

2. Non-living materials such as chemicals and compounds cannot make themselves into life, that is, cells.

3. The only manner in which cells appear is from other cells.

Pasteur's experiment used two flasks, one of which had an S-shaped neck, the other straight. He poured identical nutrient broth into both and boiled the broths, killing all living matter. Some weeks later, new life was observed in the broth in the straight neck flask but not in the flask with the S-shaped neck. His conclusion was that germs (microbe cells) from the surrounding air had entered through the straight neck (and fallen into) and contaminated the broth but could not get past the S-shaped neck.

His findings were that spontaneous generation of life (cells) was not possible, otherwise both flasks would have been similarly contaminated. The conclusion was that microbe cells can only come from other microbe cells and has not been disproven since.

The steps he employed were to use a controlled experiment to test a hypothesis. This process parallels what is now known as the science method. If we then turn our attention back to the first cell in history, what Pasteur's experiments essentially tell us is that there is no known scientific process by which it could have self-created; and with no other cells existing, its appearance requires a non-evolutionary explanation.

Although this knowledge has been well documented by science for 150 years, it has not been widely broadcast to the general population. The curiosity is not that science hasn't been alerting the public to this understanding; rather, that countless studies and experiments have since been undertaken aimed at proving an unfeasible, if not impossible, outcome.

How The First Single Cells Formed Under Evolutionary Theory

Most theories on cellular evolution are founded on an assumed environment that existed on the early earth. Scientists theorize that lightning struck an area rich with chemicals, assumed to be the ocean, and fused several elements together to produce life. Scientists then assume a chain of accidental events that finally resulted in the first cell appearing. Based on these assumptions, considerable experimental work has been done to convert inert objects into living organisms in the laboratory. In 1953, Stanley L. Miller (Chemistry professor at the University of California, San Diego) applied an electrical discharge to a mixture of hydrogen, methane, water vapor, and ammonia. This was believed to replicate the primal environment of the early earth. Performed over a week, two amino acids and some organic substances were produced. Bacteria and fungi in normal air rapidly consume these laboratory-made chemical compounds making them quite fragile and requiring them to be retained in sterile conditions which weren't available in early earth. Further modified experiments based on Miller's have produced most, but not all, of the amino acids required for DNA.

Additional specific and planned laboratory experiments have produced most of the base components that make up the complex molecules of cells. Nonetheless, there is a significant gap between the most complex mixtures that scientists have been able to create in the laboratory and the simplest cell. This is not just a practical issue, however: there is no viable theory or practical example of how non-living components can possibly form into a living cell.

According to evolution, millions of years of accidental happenings with the right elements in just the right place, time and blend, resulted in living cells. Yet despite having at our fingertips the most sophisticated equipment and intimate knowledge of chemistry, biology and living cells, science cannot remotely come close to making a cell from these inputs. Which is not surprising, since Louis Pasteur has already shown that creating life from non-living ingredients is not scientifically possible.

Nonetheless, it is an interesting experiment to calculate the probability of the theoretical coincidental happenings occurring just once in the right area with the right mix of compounds in the right concentrations at the right electrical charge and temperature. Sir Fred Hoyle suggested that trying to grasp the unlikely chance of life self-creating could be imagined by throwing fifty thousand (50,000) consecutive sixes with a pair of dice; effectively impossible. The chance of one pair of sixes is one in thirty six throws. The chance of just ten consecutive sixes is one in three thousand, six hundred and fifty six trillion throws (1/3,656,000,000,000,000 - i.e. 36 multiplied by itself 10 times). The chance of fifty thousand consecutively is an impossibly large number from a practical perspective (36 multiplied by itself 50,000 times). The actual numerical chance of it happening is so small that it is irrelevant. If it were actually possible, life spontaneously self-creating by accident from inert chemicals, would be the second most extraordinary event in history, following matter making itself from nothing.

The Theory of Evolution proposes the concept of life self-creating because there are no other natural ideas that are even remotely tenable. There is no explanation of the actual steps that would be required and no supporting evidence as to how it could happen. It is just a concept to explain the scientifically inexplicable appearance of life.

Naturally, whilst experiments continue to fail to provide answers, there is considerable debate among scientists about how life originally formed. Because of the inconsistencies among scientific research, the majority of theories about life tend to focus at the point where the fossil record has identified the most basic life forms, such as microbes, and offer minimal if any detail on how the first cell was actually formed. The theories then endeavor to describe how these early cells developed into more complex life forms by mutation, cell division or combination.

Yet, even here, science encounters a major stumbling block, the Second Law of Thermodynamics, which is generally defined as:

There is a particular direction for all spontaneous processes, especially the flow of heat; the entropy of any isolated system can only remain constant or increase.

Entropy is a measure of the unavailable energy in a thermodynamic system, or alternatively described as the state of disorder in that system. Essentially, this law dictates anything organized will become disorganized over time. Any physical system that is left to itself will ultimately decay—this is true of the universe, the sun, the planets, and life. Any organized system will lose energy and structure, thus the entropy (amount of disorder) will increase. Worded another way: *Everything is structured to ultimately break down—not to develop into something more complex.*

For example, chemical processes will reach equilibrium then become inert. Therefore, for a life form (a combination of chemicals) to develop into something else it has to increase in information, which violates the law. The multiple billions of increases in the pathway of evolution proposed by Evolutionary Theory are strictly by accident, whereas science tells us that just one increase is against the natural order of things. The Second Law of Thermodynamics thereby discounts any potential for a cell to self-create.

The possibility of life spontaneously self-creating has been studied by Professor Steven Benner of the Westheimer Institute for Science and Technology in Gainesville, Florida. Speaking at the Goldschmidt 2013 conference in Florence, he claimed that only adding energy to organic molecules (such as lightning and/or sunshine) would not result in the formation of life as these molecules would only breakdown into a tar-like sludge (thereby following the Second Law of Thermodynamics). He believes there are two elements essential to first life being able to develop on earth—boron and the oxidized form of molybdenum—which inhibit organic molecules from decaying and provide catalytic reactions (that is, speed up the processes). He believes that both of these were absent from earth at the time first life has been recorded, dismissing the self-creation of life as impossible because not all of the necessary components were available.

Nonetheless, in support of the self-invention theory, Professor Benner suggests these elements might have been available on Mars and been blasted towards earth following a volcanic explosion. This Martian meteorite, assumedly and fortuitously containing both elements, would have to be rocketed in the direction of earth, allow for the movement of the planets, safely make it through our atmosphere and land in the same place where the original mix of chemicals were struck by lightning so that they could all combine together. This speculation about a Martian meteorite is not based on evidence; it is only conjecture to attempt to explain the improbable appearance of rare materials from another planet just when and where they were needed on earth so that biological Evolutionary Theory can continue to be debated rather than dismissed on current scientific evidence.

The message from this research is quite straightforward. The established facts are that the components required for life to spontaneously self-create did not exist on earth. Anything else is pure speculation.

Other Cell Characteristics

Evolution's main theory about cells includes the following developments:

- The early earth was assumed to be a harsh environment with an atmosphere of hydrogen, nitrogen and carbon, offering little protection from the sun's destructive ultraviolet rays. As these would have destroyed most organisms, it is assumed the early cells formed in water, which provided protection and a source of hydrogen. Later, the muddy areas at the edges of the oceans provided moisture and access to chemical nutrients whilst affording protection from the sun's rays, which would kill off the top layer of bacteria as further layers formed underneath. These mats of bacteria have been found in many parts of the world.

• Cells increase in number by copying themselves. Bacteria reproduce rapidly with ten divisions resulting in over one thousand new cells, which can happen in minutes. The process involves DNA replication and then cell division, resulting in two identical cells. The original DNA would also have been able to copy itself in full rather than new, different DNA molecules being made from the surrounding chemicals, but it is not known from where it got the energy or ability to split repeatedly. Science also cannot explain why it would actually do such a thing, or how the first DNA molecule was aware that it was fully formed to then begin the replication process. The best explanation offered by evolution is that cell replication, which is fundamental to life, was an unintended process.

• Different strains of bacteria are able to interact relatively quickly with each other's DNA, allowing them to do things that their own DNA isn't programmed to do. Bacteria are extremely flexible and, theoretically, can exchange and use each other's DNA. Eukaryotes (all other living forms), on the other hand, have to change form individually and pass changes to successive generations, often taking large periods of time to adapt to new environments.

• One form of fermenting bacteria is spirochetes. These move in a circular, twirling manner and are thought to be responsible for the first movement in life. One strain, *Desulfovibrios*, takes in sulfate and gives off sulfur gases while generating energy. During this process an astounding thing happens: porphyrin rings are manufactured. Porphyrins are light absorbing compounds that are colored and are vital to the use of sunlight as energy. Without them, the energy absorbed by a molecule will simply dissipate and the molecule will return to its normal energy state. The color that we see in an object is the visual representation of the amount of energy that remains after the object absorbs its pre-determined limit from sunlight. Without porphyrins, life would be colorless.

• Cells contain repair enzymes that remove the damaged portions of DNA molecules and, in the main, replace them with a copy of the original DNA. This extraordinary function is fundamental for the continuing life of most organisms, but there is no explanation about how they came to exist.

• Cells have small attachments called mitochondria that are encompassed by membranes and allow the conversion of oxygen into energy. These mitochondria are found in the cells of all living things that would otherwise suffocate for lack of oxygen. It has been proposed that a form of water-based bacteria was capable of 'breathing' oxygen from water as there was no (or very limited) oxygen in the early atmosphere. These cells then passed on this ability to successive generations of cells.

At this point in the evolutionary story, we have some theories about how life started on earth, although none of these is actually scientifically verifiable, and we are now at the stage where many types of cells can use oxygen. Life exists as microbes, which are replicating and mutating into various forms, performing a variety of different functions. In the next chapter, we will look at more of the activities of microbes and their importance in the continuation of life.

Chapter 8 Unraveled

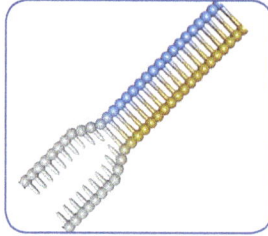

- *There is no scientifically viable explanation detailing how life could start itself in an evolutionary world.*

- *DNA is the basis of cells and would necessarily have formed first.*

- *Microbes were the first types of living cells.*

- *Life forms and other systems cannot build up their complexity from their original design; the Second Law of Thermodynamics states that it is the natural order of things to break down.*

- *Louis Pasteur discovered that cells have to be made from living matter; they cannot self-create.*

- *Science tells us that living organisms (life) cannot, do not and did not make themselves.*

9

WHY ARE MICROBES SO IMPORTANT?

A vast range of different microbes exist on earth. Although science cannot find any method that allows non-life to change itself into life (in the form of microbes), evolution now proposes that the multitude of different microbes resulted from changes to the original microbe by a process called mutation.

Mutation is an unintended, random, permanent change to the DNA structure of a living entity, which can then be passed on to following generations. Mutations affect the DNA sequence of a gene, which in turn changes the amino acid sequence of the protein encoded by that gene. As every living thing has a unique DNA structure, the result in this instance is a different microbe with a different, new DNA sequence.

Evolution claims that this process of random DNA mutation is the process by which microbes, being the first life form, have evolved into all the many different life forms on earth. This is the fundamental claim on which all biological evolution is based, yet, as we will continue to discover, there is no scientific evidence, chemical or biological, that supports this proposal.

Essential Microbial Functions

Microbes, or bacteria, live almost everywhere and require some form of water to survive, whether it is ice, liquid or vapor. They are under, on and above the ground and in almost everything you touch. They can tolerate varying temperatures, even above the boiling point of water, and they consume all manner of things, including wood, sulfur, iron, sugars, and oils. They can neutralize things deadly to other life forms. For instance, *Geobacter sulfurreducens* bacteria stabilize uranium particles and *Cupriavidus metallidurans* bacteria consume gold chloride, which is a natural chemical liquid toxic to humans, and as a bonus excrete ninety nine point nine percent (99.9%) pure gold.

Animals, plants and fungi depend on microbes, yet microbes help to protect us against fungal infections. They are in our digestive systems and on our skin, all over our body. They assist with food digestion, destroy harmful organisms and make some of the vitamins that the body needs. Although microbes can also result in illness or disease, many of the medicines we use contain microbes that cure diseases. They are critically important to our survival. They are marvelous entities without which the earth would not be habitable. Without their presence in the early development of the earth, life as we know it wouldn't be possible.

The main activities of microbes include:

• Turning food into energy and storing it.
• Maintaining the cycles of both inorganic and organic elements.
• Purifying water.
• Making soil fertile.
• Refreshing reactive gases.
• Putrefaction (decay, destroy or eat plant and animal waste).
• Fixing atmospheric nitrogen.
• Generating various fermentation processes.
• Photosynthesis.
• Making methane.
• Consuming nitrogen.

They are so flexible that they can exist:

• In freezing rocks.
• In boiling hot temperatures.
• Under enormous pressures and without sunlight.
• In space.

In terms of timing, there is still debate about when the first single cell microbes appeared on earth. Bill Schopf, an American paleontologist, has claimed discovery of the oldest fossils in Western Australia, where the remains of primitive microscopic algae have been measured at 3.5 billion years. Steven Mojzsis, an American scientist, has found grains of carbon dated at 3.9 billion years of age. It is thought that for the first 3 to 4 billion years following the first cell's formation, there were only single cell protozoa. Their existence led to nothing more than a thin covering of rocks and stagnant water. This was the only life as we know it. For billions of years the landscape was a barren and mostly colorless scene with no plants, trees, grasses or movement.

Around 800,000 years ago the oxygen level grew rapidly, although science cannot determine how this happened (see Chapter 10 for more discussion on this topic). The presence of oxygen would have been destructive to many microbes as they either did not have the enzymes for removing damaging oxygen radicals and died due to their build up, or their key enzymes stopped working resulting in cell death. Evolutionary Theory claims some bacteria accidentally adapted themselves to consume oxygen to remain alive. This process, called aerobic respiration, is a very efficient form of energy creation that uses the highly reactive nature of oxygen to maximum advantage. An example of this is the treatment of waste water, where bacteria convert complex substances into simpler ones. During this fermentation process, carbon dioxide and water are released and heat is produced. The heat is then utilized by heat exchangers. Oddly, the cyanobacteria that produce oxygen by photosynthesis are also the bacteria that consume this same oxygen by respiration.

All the newly available oxygen in the earth's atmosphere is then thought to have had another effect: some bacteria accidentally added a few molecules of oxygen to the fermentation process. This process coincidentally releases almost twenty times more energy than fermentation without oxygen. In effect, these bacteria retain the waste product of fermentation and combine it with oxygen to create more energy. This is called cellular respiration and is the basic process of how we, as humans, breathe. This development is claimed to be the catalyst that changed the process for all life on earth to be oxygen breathing.

Microbes And Oxygen

- Oxygen is critical to life development from this point forward. In the form of ozone, oxygen provides essential protection from the sun's ultraviolet rays, without which humans and most life forms could not survive. It is proposed that oxygen readapted to create this protective blanket of ozone. The process put forward by evolutionists goes something like this:

- Oxygen (O_2) accumulates in the atmosphere from bacterial waste. Oxygen, like most other things, is susceptible to the destructive effects of sunlight.

- When subject to intense solar radiation, the O_2 molecule is split into two single oxygen atoms. Some of these recombine to form the three molecular structure of ozone, O_3.

It just so happens that ozone has the capacity to absorb ultraviolet radiation that stops the splitting of O_2. If, however, radiation levels increase, or ozone escapes from the atmosphere, more oxygen molecules will be split into ozone, which builds up the barrier of resistance again. This process effectively maintains its own balance.

Most ozone is located in the stratosphere where it is called the ozone layer. About ten percent of the total ozone is in the troposphere (refer Figure 60: The Earth's Atmosphere). The major benefit of this new accidental ozone molecule was that it created a protective blanket from the sun's harsh rays, allowing higher life forms to exist on the planet's surface. Such a process, however, is purely speculative. Theories on microbial evolution and ozone creation are invented as an attempt to explain the sudden boost, and then stability, in oxygen levels that otherwise appears to be impossible to account for.

Microbe Summary

Thinking through the implications, early bacteria must have integrated as a highly developed group (of trillions of trillions) because they:

- Were the simplest original forms on earth, yet found a way to live on the ingredients available, being mostly carbon and hydrogen.

- Found a way to develop new strains that actually functioned in a way opposite to their original make up.

- Regularly altered the number of each strain across the planet to maintain the 'correct' ratio of oxygen, even though there is no correct ratio or level under Evolutionary Theory.

- Monitored the levels of oxygen in the atmosphere without having any sensory systems to measure it.

- Operated and communicated as a combined unit across land, water and altitude without any physical connection or means of communication.

- Developed themselves into many variant strains suitable for controlling the ability of later life forms to exist, even though there were no plans for other life forms.

Next we review two great mysteries for evolution to explain.

Chapter 9 Unraveled

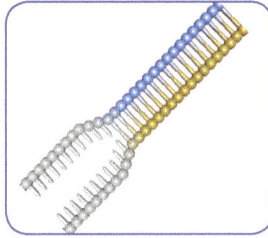

- *Microbes are extremely hardy and versatile.*

- *The regular appearance of new forms in the fossil record does not mean that one changed from the other. It only means that new types appeared.*

- *There are many strains of microbes, they are very adaptable (through mutation) and they have performed a staggering range of functions necessary for life billions of years before those later life forms existed.*

- *No scientific evidence exists now or in the past to show that bacterial microbes evolve into higher forms through mutation. The theory that they are responsible for all life has no basis in scientific fact.*

- *Rare, random, unexplained cell mutations are claimed by evolutionary scientists to be responsible for all of the different types of microbes on earth.*

10

WATER AND OXYGEN

Reviewing the evolutionary story to this point, it is easy to overlook two critically important issues: Where did water and oxygen come from?

Here is the paradox: Both these components are needed to exist on earth before the other. It's a chicken and egg conundrum as, under any natural process, water was needed to make oxygen and oxygen was needed to make water.

Let's examine how water arrived on earth from an evolutionary perspective.

Water

A challenging issue for any scientist is how water is actually made. On a molecular level, two hydrogen atoms combine with one oxygen atom to make a water molecule, giving the chemical formula H_2O. Water does not make itself in the natural course of events. This is because oxygen and hydrogen atoms require enormous amounts of heat and energy to combine into water, which raises a number of questions:

- What was the original heat source for our water?

- Where and when did it happen?

- How long did it take?

- Has the process stopped?

About seventy percent of the earth's surface is covered in water. There is no scientific record of gradually increasing water levels around the planet, which indicates all of it arrived at once. Through geological records, we know this massive quantity of water existed as far back as can be determined, with no corresponding evidence of oxygen levels in the atmosphere. Therefore, water appeared on earth before oxygen, an essential molecule for its creation. The issue is: Where did earth's water come from if there was no oxygen on the planet to make it?

The most common evolutionary proposal vaguely refers to water "appearing in forming stars" but fails to elaborate on the actual process required. This concept is suggested because stars are the only known natural source of intense heat and pressure, as well as a natural reservoir of plentiful hydrogen and assumedly oxygen. Water is thought to be

Figure 64: The Earth's Water

synthesized in a nuclear fusion process, much akin to the creative process of higher elements (as previously discussed).

There are immediately obvious problems with this theory. Firstly, there is no explanation how water could escape the enormous gravitational pull of a star. Secondly, the immense heat (ten million degrees Centigrade) would vaporise any created water molecule, possibly back to its original elements. Furthermore, even if stars were the source of water, the water would then need to be transported to the developing early earth, for which there are two main options tendered:

1. One concept has water arriving in the form of icy asteroids or comets which were assumed to be part of the original components that were pulled together to make the earth. The theory is that once the sun had stabilized and the earth had cooled and formed a crust, it released gases containing water vapor or possibly water seeped out through rocks.

However, there is a contradiction of process with this theory. Firstly, any water released by the newly formed crust would have evaporated quickly into space because there was no oxygen in the atmosphere to protect the earth's surface from the high solar temperature. Secondly, Evolutionary Theory claims that oxygen molecules in the atmosphere built up only after being separated from water by bacteria. Yet there would have been no stable water accumulation on earth without oxygen in the atmosphere, which itself requires water to exist beforehand.

2. The second proposal is a similar theory in that it involves a process of water accumulation after the earth had cooled and formed a crust. Water is claimed to have arrived when comets/meteors/asteroids, laden with ice, crashed into earth, the ice melted and stayed on the surface. The number of comets required to cover the earth's surface with water as it does today would be in the order of trillions of trillions of trillions. As yet, however, there is no evidence of such a bombardment of comets. Even if the history of the early earth was erased through massive geological temperatures to forever hide evidence of such a bombardment, analysis of water from most comets indicates that it differs from common earth water, containing more deuterium (known as heavy water). Measurements by the Rosetta spacecraft in December 2014 show that the water on comet 67P/Churyumov-Gerasimenko has three times more deuterium than water on earth. If more than fifty percent of the water in animals' bodies was replaced with heavy water, cells would be impaired and result in death. Having the exact type of water on earth is critical to life but there is no evolutionary explanation where this came from.

The geologic record gives some degree of certainty to determine the levels of hydrogen and oxygen gases in the atmosphere at various times in history. In order to accumulate the amount of water currently on earth, there should be evidence of incredibly vast amounts of these two gases, evidence that would stand out like a beacon in the earth's geological history. But there isn't any. There is very little hydrogen in the current atmosphere and there is no scientific record of any significant levels in the earth's history.

To summarize the scientific evidence for the accumulation of water on earth:

• There is no scientific explanation for the lack of oxygen or hydrogen to make the vast amount of water we see today.

• Notwithstanding the lack of hydrogen and oxygen, even if they were available there is no available scientific method by which water could have made itself due to the requirement of enormous heat and force.

• On earth, there is no record of the molecular components, no record of any gradually increasing levels of water, no record of masses of ice-laden asteroid collisions, nor any record to verify the evolutionary theories.

Where Did Oxygen Come From?

At this stage in our story of evolution around 2.5–3 billion years ago, historical records reveal certain things:

1. Earth had a harsh climate.

2. There was barely any oxygen in the atmosphere.

3. Microbes were spreading across the globe.

4. Water existed in plentiful amounts and was available from almost the very beginning when the planet was made, remaining on earth by defying the sun's heat, which should have evaporated it into space.

The environment at this point in earth's history was unsuitable for future living creatures due to extreme temperatures and little oxygen. In order to create the conditions to enable higher life forms and plants to survive, something had to happen. What was needed was oxygen.

It is calculated that there was very little oxygen in the early atmosphere, which was predominantly carbon dioxide and hydrogen. As detailed above, all of the water already existed despite requiring oxygen in the atmosphere to stop it being evaporated. Around 2.5 billion years ago, oxygen started appearing in relatively small quantities mostly binding with existing elements and compounds into oxides. The mineral record shows that free (atmospheric) oxygen levels exploded in huge volumes around 900 to 300 million years ago settling at twenty one percent (at sea level), the concentration that is ideal for air breathing creatures to exist, which is where it is today.

This happened even though it wasn't an evolutionary requirement. Evolution happens by accident, with no plan. It is a theoretical process of opportunity and adaptability. Yet, fortuitously, the appearance of oxygen at just the right levels for higher order survival was a critical event for all life on this planet.

The relatively sudden appearance of oxygen in vast quantities raises an aspect of life's early development that is not satisfactorily explained by science. The issue stems back to the nature of oxygen, in particular its reactivity or proclivity to cause oxidization.

- It reacts with organic matter creating chemicals that hinder the compounds that support life, such as hydrogen, nitrogen and carbon.

- It combines with the elements critical to cell reproduction.

- It oxidizes minerals creating new oxygen-bound compounds.

- It can cause spontaneous combustion of living organisms if concentrations are higher than current levels[13].

Some estimates claim the maximum amount of free oxygen actually rose to about thirty five percent. This high level appears to be a calculation based on the massively increased size of insect fossils in the Carboniferous period. During this period, flying insects were as large as birds, centipedes as big as crocodiles. The assumption is that because insects do not have lungs and can only absorb oxygen through their external shells, the greater concentration of oxygen in the atmosphere led to greater absorption of oxygen, which in turn caused insects to grow to hideous proportions. The estimated level in the atmosphere to enable such an increase in insect size is calculated to be around this figure of thirty five percent.

However, this calculation is just an assumed level as there is no actual evidence supported by geological measurements that oxygen ever got to that level. Even though this higher oxygen level is assumed to be a short term spike in the history of the earth, it fails to adhere to the natural controls in place under the carbon – oxygen closed loop system (explained below), making this assumption mere speculation without scientific evidence.

The Carbon – Oxygen Cycles

The ability of oxygen to remain around the stable level of twenty one percent is due to the carbon – oxygen cycles, which create a closed loop system in the environment (detailed below under the subheading: *The Gaseous Cycles*). As around seventy percent of oxygen comes from the oceans, these gaseous cycles are dominated by ocean life forms. The process is that, if atmospheric oxygen levels increase, marine zooplankton will respond by increasing their consumption of organic matter[14], thus reducing the level

---

13. Organic material can self-ignite without external ignition (such as flames, sparks, lightning) through heat generated internally in the presence of water. When localised levels of oxygen are too high (even 24%) they can cause chemical reactions to occur quickly, generating the heat necessary to ignite combustion. Metals, like copper, can lower the combustion temperature of surrounding material and can attract oxygen from air, releasing it into the material and allowing combustion to occur.

14. Produced by algae.

of this matter that would eventually be buried at the bottom of the oceans. Zooplankton will also increase their total use of oxygen through respiration (breathing), negating the increase in oxygen levels.

Conversely, a reduction in oxygen levels would limit zooplankton food consumption, resulting in an increase in algae-produced organic matter in oceanic sediments as well as a reduction in oxygen consumption, leading to increased oxygen levels and the negation of the original oxygen decrease.

This closed loop system creates a problem for theories proposing a naturally caused increase in oxygen levels (detailed below under the subheading: *The Timeline of Oxygen Creation*). Some scientists believe that plant photosynthesis would not have provided enough oxygen of its own accord because organisms that can't produce their own food (*heterotrophs*) breathe in oxygen, using it to metabolize organic food. Today, animals consume new plant growth while dead plants are broken down by decomposition, effectively balancing both carbon and oxygen levels in the environment.

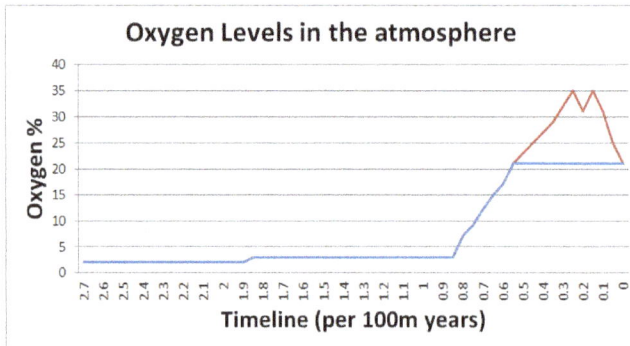

Figure 65: Oxygen Levels In The Atmosphere

Over time, various events have affected the earth's atmosphere, such as:

• Massive volcanoes spewing destructive gases and ash across the landscape.

• Human clearing of oxygen-creating forests and plants.

• A massive increase in the number of living creatures consuming oxygen.

• Forest fires of enormous size lasting for months, which consume vast amounts of oxygen and also destroy both the plants and trees that generate oxygen.

Yet despite these events, stand-alone or in combination, the oxygen levels in the atmosphere have not depleted and remain in an equilibrium state of twenty one percent.

Of their own accord, there is no proven method for substantial increases in oxygen levels in the environment.

The Timeline Of Oxygen Creation

There are several evolutionary concepts on the creation of oxygen. Each is based around naturally occurring processes, but all follow a similar line of thinking that goes roughly as follows.

About 3 billion years ago, an organism (cyanobacteria) appeared on the scene, an appearance that completely changed the way that bacteria had acted up until that point. This organism did something that no other bacteria had done before: it spliced hydrogen from water molecules, effectively splitting water into its components hydrogen and oxygen. The hydrogen was then used in photosynthesis, combining with carbon dioxide to make organic food chemicals, while the oxygen was released to the atmosphere as a waste product.

Under this proposal, oxygen is purported to have derived from water that already existed on the earth's surface. This explains why there is oxygen in the atmosphere and where the hydrogen went. However, there is little or no explanation regarding the amount of water required to generate the oxygen levels we see today. Furthermore, there is no explanation of the lack of evidence of reduced water levels caused by the separation of its two components.

There is also the issue of time. The process of making oxygen would have needed to occur in a relative hurry. The rapid accumulation of oxygen raised the level in the atmosphere by a staggering two hundred thousand times. Considering the size of the atmosphere, massive amounts of oxygen were needed in a short space of time to get to the levels of twenty one percent and higher.

Science estimates about seventy to eighty percent of current atmospheric oxygen is made by algae, which are single cell marine plants. As the oceans cover about seventy percent of earth's surface, land production of oxygen is needed to cover the remaining twenty to thirty percent. Considering about forty five to fifty five percent of the earth's land mass has been covered in plant life for only the last 300 – 450 million years, the original burst of oxygen could only occur if marine production increased dramatically and/or much of the earth's surface was covered in masses of cyanobacteria.

However, the effect of the sun's radiation would have been lethal to any bacteria that were too far from the safety of water, so it is unlikely huge colonies of bacteria could have survived on barren land.

For mathematical simplicity, we can assume the algae and water splicing bacterial organisms increased the oxygen level in the atmosphere by one percent every 25 million years. It would therefore account for the 500 million years or so it took to reach twenty one percent (that is, twenty lots of 25 million years). Ignoring the disputed readings over twenty one percent, once at that level most of those organisms on land and in water that had been producing oxygen for 500 million years suddenly stopped doing so and reduced output to a rate that then only maintained the level at twenty one percent.

This is a point that's worth repeating: Once the oxygen level in the atmosphere had reached twenty one percent, most of the organisms that had been producing oxygen for an incredibly long period of 500 million years suddenly stopped doing so.

Evolutionary Theory, however, fails to provide any valid reason for this sudden stoppage or why a large percentage of oxygen producing organisms suddenly died off or disappeared. Although cyanobacteria were claimed to have appeared 3 billion years ago, the level of oxygen levels didn't commence its rapid rise until about 800 million years ago. The reasons for the delay in the appearance of cyanobacteria and the rise in oxygen levels are not known.

Further complications arise because fish breathe through gills, which separate dissolved oxygen from the water in which they live. Consequently, oxygen has to exist in enough volume as a gas in the atmosphere to allow absorption across the vast quantities of global water sources. This requires additional oxygen production from that required solely for the use of land life. There is no explanation as to how these levels are either known, monitored or controlled by plants, algae and bacteria that have no connections to each other.

Further, and just as importantly, there is no matching record of hydrogen levels increasing. When water molecules split they release twice as much hydrogen as they do oxygen, but their existence in unaccounted for and should be particularly evident over a period of 350 million years (800 – 450 million years ago) until land based plants appeared, elevating the consumption rate of hydrogen in photosynthesis.

Following this, plants and trees appeared. These produce significantly greater levels of oxygen than microbes[15], yet the record shows no impact upon oxygen levels in the atmosphere following their appearance. On top of all these issues, there are some critically important points regarding the blend of gases in our atmosphere and the structure and design of earth.

There Are Three Main Gases:

Our atmosphere has about seventy eight percent nitrogen. This gas combines with oxygen during thunderstorms and the compounds created are carried to ground by rain and used by plants which absorb the water and use the nutrients.

As previously discussed, oxygen is extremely volatile. Should the level of oxygen increase by just three percent to twenty four percent, some objects (e.g. hay stacks) would be more prone to spontaneously self-combust across the planet, highlighting the extremely fine balance supporting life.

The rest of the atmosphere (less than one percent) includes carbon dioxide (about zero point zero four percent - 0.04%). Plants absorb this gas and release oxygen as a waste product. In turn, animal life forms consume oxygen and exhale carbon dioxide as a waste product. In effect, each life form supports the other and balances the gases.

The Earth's Crust:

The thickness of the earth's crust is ideal for the level of oxygen at twenty one percent. Should the crust be thicker, it would hold oxygen in the form of oxides below the surface. If it was thinner, there would be more earthquakes and volcanoes with the resultant ash thrown into the air, possibly suffocating many life forms on a continual and regular basis.

The Earth's Size:

The size of the earth is important in the balance of gases. Hydrogen, which is a lighter gas, escapes from our atmosphere. If the earth itself was slightly larger, hydrogen gas would not be able to escape because of the increased gravity and the build-up would eventually suffocate life. If earth were smaller, oxygen would escape with the same suffocating result.

The Renewing Cycles Of Earth

The earth is virtually a self-contained recycling unit full of creatures and systems that ensure it is regularly refreshed. Each cycle relies on the energy of the sun as the primary generator. In brief, these are:

---

15. Plants, for instance, produce oxygen at an average of 5 mls/hr.

The Hydrologic Cycle

Water is recycled and moved around the earth by a system called the hydrologic cycle (refer Figure 66: The Hydrologic Cycle). The heat from the sun evaporates open water from lakes, reservoirs, dams, and oceans as well as plants and creatures. Water is drawn into the atmosphere as tiny droplets leaving any impurities behind, effectively cleaning and purifying the water. Condensation in the atmosphere causes precipitation (such as rain, snow) which is pulled to earth by gravity with water being stored as snow or ice on higher mountains or at the poles. Much of the water eventually makes its way back to rivers and streams, with some seeping through the land surface into underground streams, then into the oceans and lakes for recycling again.

In this cycle, water can go through various phases of being liquid, solid (ice), and gas (vapor). These phases of water also transfer energy: evaporation draws energy from the local environment, thereby cooling it. The energy is released during condensation, warming the surroundings. Minerals are moved around by liquid water and ice, and can travel across the world.

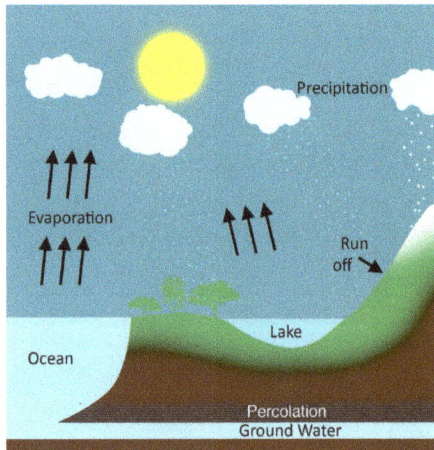

Figure 66: The Hydrologic Cycle

The Gaseous Cycles

These systems use the atmosphere and the earth (soil and rocks) to store various gases which include oxygen, carbon and nitrogen.

Oxygen in the air is consumed by creatures who release carbon dioxide as a waste product. When living forms die they decompose and release carbon dioxide into the soil and atmosphere. Carbon dioxide is absorbed by plants, which animals can eat as food, and release oxygen back into the atmosphere. This cycle is shown in Figure 67: The Oxygen Cycle and Figure 68: The Carbon Dioxide Cycle.

Figure 67: The Oxygen Cycle

Figure 68: The Carbon Dioxide Cycle

The Nitrogen Cycle

Nitrogen constitutes seventy eight percent of the earth's atmosphere and is essential in many life giving components, such as proteins, acids (RNA and DNA), and is important in photosynthesis. Nitrogen gas is required to be fixed (chemically processed) into other forms, such as ammonia, so that it can be used by living organisms for food production. Bacteria (and some legumes) are involved in most nitrogen fixing, whereby the gas form is converted by enzymes into ammonia then organic compounds. These processes are displayed in Figure 69: The Nitrogen Cycle.

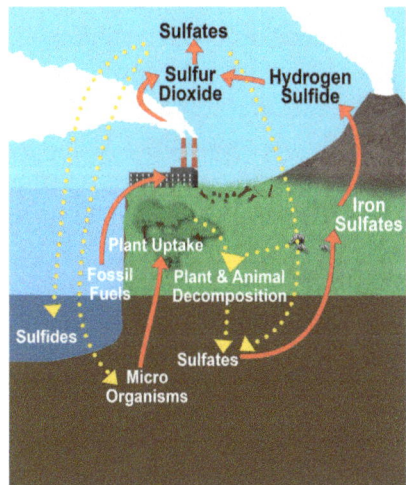

Figure 69: The Nitrogen Cycle

One further issue that is raised when examining the above systems is the original appearance of nitrogen that dominates our atmosphere at almost four times the level of oxygen. It is thought that during the earth's self-creation process, the building materials included a large proportion of interstellar clouds containing ammonia, methane and formaldehyde, which abound with nitrogen compounds. However, when the sun exploded into life, these molecules would have been swept away into space prior to the atmosphere developing to its current form. To counter this, it is assumed that a second wave of nitrogen compounds arrived by meteorites, presumably formed from the same interstellar clouds. The volume of nitrogen is so immense that the number of meteorites now expected to have landed on earth (e.g. delivering water) is incomprehensible, rendering the self-existence of nitrogen as a mystery without explanation.

The Sulfur Cycle

Sulfur has an important role in the operation of the amino acids (proteins) in soil where it is converted into sulfates by microbial interactions. Plants absorb sulfates through water in the soil. When animals eat plants they consume sulfur which assists in maintaining the health of the organism.

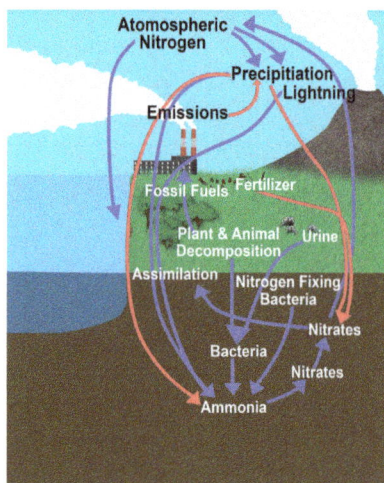

Figure 70: The Sulfur Cycle

The sulfur cycle involves sulfur gases (dioxide and trioxide) mixing with chemicals and water in the atmosphere to produce sulfur salts and acids. These return to the ground by gravity or as acid rain and begin the cycle again.

The Pyramid Of Life

The term 'Pyramid of Life' has been coined for the food chain. It is the process by which energy cycles through our ecosystem. If we start with plants, we can follow the process.

Plants trap the sun's energy by photosynthesis. They use approximately ninety percent of this energy in their own building processes. The remaining ten percent is available for the creatures that eat plant life, such as herbivores. Carnivores (animals that eat other animals) and omnivores (animals that eat both plants and animals) eat herbivores and similarly only have access to about ten percent of the herbivores' energy, which is left over for their own use. Effectively, this limits the number of carnivores and herbivores relative to the quantity of plant life, without which they would starve and die. When these creatures die, the decomposing bacteria return the basic elements and minerals to the ground, which is in turn used by the next generation of plants for food. Plants then start the cycle again with the aid of the sun's energy (Figure 71: The Pyramid of Life).

This eco-system feeds on itself and controls its own size naturally. Oil based products predominantly result from decaying plants, effectively allowing later use of their latent energy fuels. Similarly, sunlight causes water evaporation and purification and many other important functions such as ripening foods.

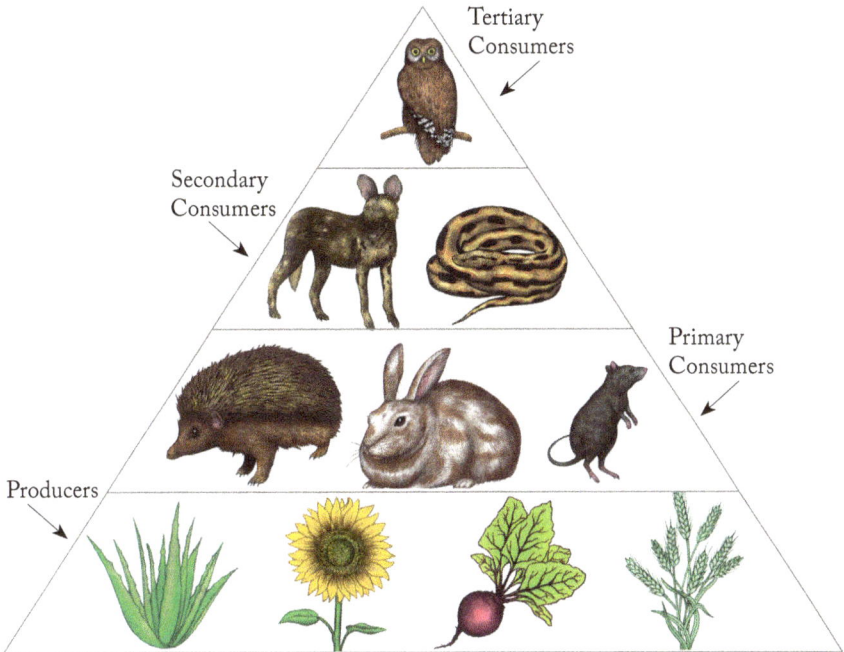

Figure 71: The Pyramid Of Life

There are other systems that operate on earth. For instance, the sedimentary system releases particles from rocks through weathering and erosion (water, wind, abrasion, chemicals, sunlight), which are transported across the globe by natural processes (rivers, storms, insects and animals).

The systems mentioned above are essential to life on earth. They convert, diffuse, control, and utilize the sun's thermonuclear radiation emitted millions of miles away into food and energy sources here upon earth, enabling constant renewal processes that allow and ensure life to be maintained.

Chapter 10 Unraveled

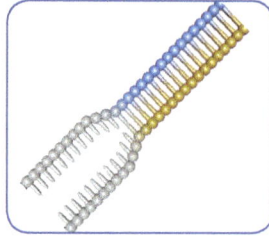

- *There is no record of the earth's atmosphere ever having enough oxygen or hydrogen to make all the water now on earth.*

- *There is no scientific evidence for the self-creation of water in stars or elsewhere in the universe.*

- *There is no scientific explanation how water arrived all at once on earth.*

- *There is no feasible scientific or biological explanation for the massive expansion of oxygen in the earth's atmosphere.*

- *Science cannot explain how or why, if microbes were the source of oxygen production, they suddenly stopped making oxygen at the perfect level for life to flourish on earth.*

- *Earth has multiple complementary renewal systems that allow purifying and cleansing of base elements that are crucial to the maintenance of life.*

11

THE APPEARANCE OF NUCLEATED CELL LIFE

So far, the evolutionary journey has seen many twists and turns, with many of the proposed events scientifically invalid and unprovable. The three previous chapters detailed propositions that:

- Water and oxygen appeared on earth by themselves at different times by methods that science cannot explain.

- A random bolt of lightning or electrical charge struck a soup of chemicals in primal waters that accidentally resulted in complex, life-forming molecules that eventually assembled as DNA molecules, which would then become the basis of all later life forms—a process that is scientifically unverified, lacked the basic components to achieve a life outcome, and remains unproven by numerous experiments to replicate the self-creation of life.

- Early DNA molecules would not have been able to survive the deadly radiation of the sun and the harsh elements of early earth, yet science cannot validate or explain the claims of Evolutionary Theory that this happened.

- Science cannot explain the evolutionary theories about the self-formation of single-cell organisms.

- Science cannot explain why single-cell life forms began to divide into two cells and replicate without any intended purpose.

- The multiplication of single-cell organisms continued, with occasional mutants, until a vast variety of microbes existed. Many of these performed different activities that accidentally resulted in the perfect level of a wide range of gases and complex chemical processes necessary to support later life forms, which appeared accidentally hundreds of millions of years later.

- Under the principles of evolution, only a few types of bacteria should predominate across the globe. This is because evolution, being a principle of random events, favors adaptability, especially to the ever changing environment. Yet bacteria across the world appear to collaborate to maintain the balance of the ecosystem, including breathable gases in the atmosphere. However, this ability to collaborate without any communication or sensory systems contradicts the underlying principle of evolution: that everything is random and unconnected.

The story of evolution now brings us to around 1.45 billion years ago. At this time, following billions of generations of single cells, their evolution came to an abrupt halt and there have been no new forms since then. A further 700 million years passed with no new life forms until 760 million years ago when sponges appeared, followed at 570 million years by arthropods. These completely new and different organisms were comprised of nucleated, eukaryote cells. Evolutionary Theory proposes that single cells developed by transforming themselves into highly complex eukaryote cells. This is despite the lack of scientific evidence for such a process. Science is unable to prove how such a massive transformation from single cell to nucleated cell could occur. For this theory to be workable, trillions of single cells mutating to more complicated cells needed to have occurred, but there is not even one piece of evidence to show how one of these single-cell organisms mutated to a nucleated cell, let alone trillions.

Nevertheless, to continue this theory forward, it is a biological necessity that at least one nucleated cell appeared after building itself from microbes. Then, two amazing things had to happen:

1. The process of microbes self-building into nucleated cells stopped. All single cells around the world somehow received a signal to stop increasing in complexity. How do we know this? Firstly, the fossil record has no evidence of any changes from microbes to nucleated cells despite the theory requiring there to have been trillions of trillions of trillions of single cells mutating part-way to other more complex forms. Secondly, there is no scientific evidence of single cells self-changing into more complex forms, even today; all they do is produce and replicate single cells.

2. A nucleated cell was somehow aware that it had all the parts it needed to later build a living being—even though there was no such thing nor was one planned—and so it reverted to the system

of dividing into two, just as the single cells did originally and still do now. We know this because the fossil record shows us that living creatures containing nucleated cells existed from this point on followed by plants at 450 million years.

Evolutionary Theory proposes this series of events because single cells are recorded as existing before nucleated cells. A theory therefore has to explain how one type of cell could change completely into the other, even though science is unable to provide any explanation how this might be possible. Yet consider this: if the order was reversed, it would make sense and fit with the scientific laws of entropy (as discussed in Chapter 4), in that nucleated cells gradually broke down into single cells. Nonetheless, evolutionary science still adheres to the scientifically unverified, and knowingly impossible, process of single-celled organisms developing into nucleated cells.

The Nucleated Cell

Before we examine some ideas proposed by Evolutionary Theory about how nucleated cells self-created, we need to understand the innate complexity of these types of cells.

At the center of a nucleated plant or animal cell is a nucleus that includes all genetic information unique to the body in which it exists. There is a vast array of organelles such as filaments, tubes, ribosomes, cilia, mitochondria, and lysosomes that perform a wide range of functions. These include controlling metabolism and protein synthesis, neutralizing toxic compounds, production of adenosine triphosphate (ATP), and the ability to consume oxygen (breathe). Each cell maintains the dynamic balance of conditions—such as the regulation of temperature and pH—within the cell's own environment. This is called homeostasis.

Without getting bogged down in their function or structure, it is important to have at least some understanding of these complex and intricate things that govern, direct and control life. As such, in order to comprehend how important cells are and how they are structured, Figure 72: Animal Cell sets out the make-up of a typical, nucleated animal cell.

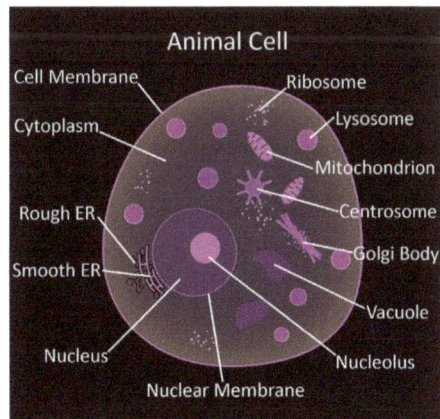

Figure 72: Animal Cell

Nucleus

The nucleus manages how a cell operates. A single nucleus contains all the information needed to control the synthesis (combination of simpler structures into a more complex one) of proteins in that body. For example, humans have approximately 100,000 proteins. The nucleus determines how the cell is structured and what it can do through controlling which proteins are synthesized, in what proportions they are synthesized, and the timing of protein synthesis.

DNA

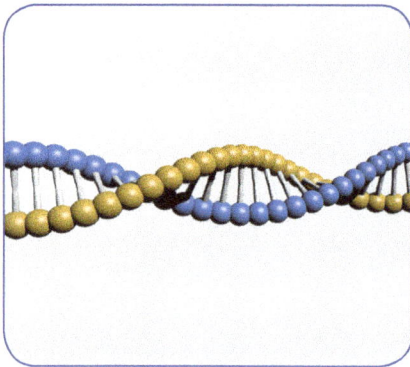

Figure 73: DNA Structure

Deoxyribonucleic acid (DNA) is a database of vital information within the nucleus. DNA is structured into complex patterns called chromosomes. These contain DNA strands to which are attached special proteins called histones. These are coiled together so that they take up only a small space. The arrangement of its four bases—cytosine, guanine, thymine, and adenine—serves as a coded set of building instructions for the whole organism. There is an almost infinitesimal number of variations possible allowing, in theory, the same infinitesimal number of any particular life form. In the human body, every cell with a nucleus has forty six chromosomes[16] in a spiral structure.

RNA

Ribonucleic acid (RNA) is a similar coded nucleic acid to DNA. RNA acts in a complementary way, assembling proteins using the information communicated to it by DNA. It contains long molecules that all cells need to reproduce and function. It is a simpler version of DNA. These two acids must communicate so that RNA can assist DNA. These act in unison giving a cell its own uniqueness.

ATP

Adenosine triphosphate, ATP, carries energy for cells in chemical form.

---

16. There are rare exceptions, such as trisomy 21 (Down's syndrome), and other trisomy variants that have 47 chromosomes.

Proteins

Proteins speed up chemical reactions and give directions for biological reactions while protecting cells against outside chemical influences. Proteins are essential to the operation of cells and there can be thousands of different types in a single cell.

Nucleotides

At the molecular level, DNA and RNA nucleic acids are constructions of monomers or subunits of organic molecules called nucleotides: adenosine monophosphate (AMP), guanosine monophosphate (GMP), thymidine monophosphate (TMP), and cytosine monophosphate (CMP). Nucleotides carry ATP energy packets within the cell. They also give cellular signals on protein levels.

The assumption is that nucleotides will give the same signals or readings for all cells in that body, but they actually give different signals or readings for different types of cells. This is considered an anomaly because it creates a complexity that suggests a nucleotide can decide its own impact, depending on the circumstances and environment in which it finds itself, acting as if it has a unique preset program with a broad range of variables it uses for assessments. In other words, it is not simply registering a level, it is making decisions about what level is appropriate depending on circumstances that can vary and it could not do so unless it was pre-programmed to make those assessments.

Eukaryotes

Within nucleated cells, both nucleic acids and proteins have highly complex structures. DNA and RNA nucleic acids need the assistance of proteins to be synthesized. Similarly, proteins are only synthesized when the corresponding nucleotide sequence is available. Both nucleic acids and proteins require the other to exist. Consequently, in order for Evolutionary Theory to be correct, these complicated, interdependent components would through necessity have had to appear from nowhere at exactly the same time, by accident, for no reason, and be perfectly coordinated from the beginning of their existence. Mathematically, their complexity is such that the likelihood of just one of these components appearing as a result of some chance combination of microbes is virtually zero. Evolutionary Theory therefore needs to explain:

- How many different nucleic acids and proteins appeared at the same time and place with the correct interlinking structures required for life.

- How random chemicals created complementary components necessary for later complex life without an awareness of what they were ultimately intended for.

In our body and many other animal species, cells are surrounded by an interstitial fluid (which simply means 'between the spaces') containing thousands of ingredients required for cellular and bodily function, such as acids, sugars, vitamins, hormones, and salts. Each of the trillions of cells in our bodies constantly takes from the interstitial fluid the specific amount of these substances they need to stay healthy. They have the remarkable ability to identify and absorb the correct compounds, measure the correct levels, then stop absorption when appropriate levels have been reached. Cells also have a very thin plasma membrane, within which the cell's contents are wrapped, like a cocoon. The plasma membrane has specialized protein molecules that provide channels to enable the movement and transfer of substances across the membrane into and out of the cell. Each cell has a recorded, pre-set level of its components as well as inbuilt sensory and corrective systems that maintain cellular equilibrium—specialized functionalities that allow them to absorb what they need.

Nucleated cell organisms, eukaryotes, differ from single cell organisms, prokaryotes, in that they:

- Are larger, being up to twenty times bigger.
- Have a nucleus (with a specific program of operation).
- Have many oxygen consuming microbes attached (mitochondria).
- Have complex structures that hold them together.
- Are surrounded by cytoplasm.
- Display a high level of active movement within the cells, with components such as mitochondria and ribosomes appearing to have predetermined paths to follow.
- Are only found as part of living creatures and plants; they do not exist on their own.

The last point is very critical for any Evolutionary Theory concerning the self-development of living bodies. Nucleated cells are generated from a developing or fully formed body and do not exist on their own (Protozoa are classified as single celled animals but differ from normal animal/plant cells having tails that give them mobility – refer Chapter 17:2); they are part of a living creature or plant and die quickly when separated. Any theory proposing that nucleated cells appeared separately and independently from a living body therefore contradicts the known science regarding these cells.

This is a difficult issue for evolution to overcome. Nucleated cells only exist as part of a body; they are not independent cells lying around and looking to form themselves into a body. They only exist as part of a body. Without the body, there are no nucleated cells. Naturally, this creates an enormous stumbling block for any theory regarding their development from single cells. Why? Because a body made of nucleated cells has to exist first, then they can replicate. A body can't self-create from individual nucleated cells because these types of cells can't survive independently. A full body with a variety of different and interdependent cells therefore has to be made first by some other method—science tells us so, yet it can't, or won't, agree on what this other method could be.

In summary, Evolutionary Theory states that:

• Single cells, as the first form of living entity, have changed or developed into nucleated cells, a process that is not scientifically verified or proved.

• Nucleated cells increased their complexity and functionality against the scientific principles of entropy.

Eukaryotes And Evolution

As we have discussed, eukaryote cells are extremely complex life forms that operate a vast range of different and specific functions. They have much higher levels of proteins and huge increases in DNA—up to one thousand times higher—than prokaryote cells. To get an understanding of what's required to change from a single cell to a nucleated cell there has to be a progressive development of complex features, such as cytosol, microtubules, cilia, centrosomes, ribosomes, peroxisomes, and so forth. Yet there is no scientific evidence today or in the past of such a stepped progress towards these complex components.

Nonetheless, some estimates calculate that the self-developmental process of evolving the first nucleated cell would have taken 0.5-1 billion years to occur. Other calculations estimate 4-5 billion years, which means it couldn't have happened yet. Taking the first calculation, assuming trillions of trillions of replications every year over a period of 1 billion years, the number of partly developed cells down the generations should be enormous, yet there is no evidence of a stepped change from single to nucleated cells in the fossil record or any observed evolutionary changes today.

Other ideas put forward on the emergence of nucleated cells also contradict the scientific evidence. One such theory is that one prokaryote cell entered another and they grew together as one. This is only suggested as a way to explain why eukaryote cells are much larger and more complex. However, single celled organisms don't join together; they do the opposite, dividing themselves in two.

About 1.45 billion years ago, science explains that, at a specific point in time, the frantic evolution of all single cells stopped in its tracks because these organisms had reached their full potential. But this is a counter-evolutionary argument. Under evolution, there is no set limit to the capabilities of a single cell. Evolution claims these cells were really extraordinary, that they couldn't stop replicating themselves by the trillions. Other than mass extinction, there is no evolutionary reason for the process to stop.

Even the evolutionary proposal of niche fulfillment isn't scientifically valid. The proposal goes something along these lines: if a niche exists in which no other life form has exploited, the most adaptable life form will fill the void. For instance, in the early oceans there were no sea creatures. A niche therefore existed, and the first creature to adapt itself to the ocean environment would dominate, such as fish. Because fish have now filled the niche, there is no longer any requirement for them to evolve further, which is why we don't see new creatures evolving from fish today.

Some evolutionists use the same argument as to why single cell organisms stopped evolving 1.45 billion years ago. Just as the waters of the oceans were filled with fish, the earth's environment had reached its capacity of bacterial life forms. Therefore, because the niche had been successfully occupied, there was no longer any need or driving force for bacteria to evolve further.

But this argument is not scientific. Firstly, the evolutionary drive is claimed to be based on continual, blind, accidental DNA mutations, not on any internal need to evolve or on the requirements of an outside environment. Secondly, the niche proposal infers that there is an underlying awareness that, in the case of bacteria and fish, these organisms knew when their species had filled the environmental niche to its capacity and then somehow stopped their DNA from mutating and evolving further – thereby bringing control to processes that are random and uncontrolled according to evolution.

Another inconsistency is the age of the oldest confirmed nucleated cell fossil—sponges—at 760 million years. This was the first known living entity other than bacteria. There was a period between 1.45 billion years and 750 million years ago when no new life forms appeared on earth. No new types of single cells appeared and nucleated cells did not exist. One theory says that after lying dormant for 700 million years, some single cells burst into mutating activity and evolved into nucleated cells, then stopped again once these were made. On that timeframe, one could expect them to start up again soon.

But here is another example of the inconsistency of current Evolutionary Theory: one group of scientists claim that single celled organisms had reached their potential and stopped evolving, whereas another group claim these organisms continued to develop into much more complex nucleated cells. Either single celled organisms hit the wall of maximum development or they didn't. The only valid conclusion we can make at the moment is that the current theories are just guesswork at best, because they are certainly not based on scientific evidence. Until the science is confirmed, the question is: should they still be taught as scientific fact in our schools and educational institutions?

Chapter 11 Unraveled

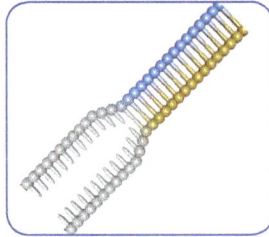

- *There is no scientific evidence that single cells self-developed into more complex forms.*

- *There is no acceptable explanation detailing how single cells could become nucleated cells. There is no evidence or supporting test data that even points in the general direction of how it might be possible.*

- *The fossil record has no evidence of anything other than either single cells or living things made from nucleated cells. There is no record of any link between the two forms or of any part development.*

- *Nucleated cells can only exist as part of a body.*

- *Scientific evidence suggests that nucleated cells did not come from single cells; they were made independently as part of living bodies.*

Genes are a set of instructions that tell cells how to make specific proteins that are essential to a body's functioning and structure. Each gene is comprised of sufficient DNA to code one protein. Proteins affect the layout of the body, its components and their operations, such as digestion, hair, eye and skin colour, shape and size of ears, and so forth.

The now familiar layout of a DNA molecule (displayed in Figure 73: DNA Structure) is made up of nucleotides, each of which has three parts: a molecule each of sugar and phosphate and a nitrogenous base that holds the genetic information. The long sides of DNA molecules are made up of alternating sugar and phosphate molecules linked to each other. The cross members joining the sides comprise two linked nitrogenous bases.

Scientists have studied human genes in detail and formulated a summary of their structure, called a genome, which is the sum total of DNA contained within an organism. This enables identification of specific markers that can identify susceptibility to different types of diseases, likelihood of physical ailments, personality traits and such like. These markers can be traced back through ancestral lines in history. The position of markers can determine which genes an individual has or had and the relative importance of those genes.

Genome modelling of species other than humans has revealed that the variations from human DNA are relatively small. Mathematical modelling of assumed 'evolutionary cell mutations' (that is, changes to DNA) forecasts major variations between species. Effectively to change from one type of creature to another requires masses of DNA changes so that different body structures, sizes, components, systems, and functionalities can appear. But these large variations do not appear in the actual DNA genetic codes. For example, there should be enormous differences between humans and chickens (perhaps only a few percent the same) but anywhere up to seventy five percent of coding regions are conserved or shared between the two species[17].

17. As discussed in the journal article by David Burt, Chicken genome: Current status and future opportunities, Genome Research, 2005. 15: 1692-1698.

To try to explain the lack of variation between species that should occur due to randomness of mutations, evolutionists have deferred to the Complexity Theory. This theory postulates that the manner in which the components of a system behave collectively allows an interaction with its environment that should otherwise become chaotic. That is, a system is so complex it doesn't become organized (thereby breaking the second law of thermodynamics) or chaotic, it sits between the two, on the edge.

As Evolutionary Theory claims that the highly complex genome of any creature is the result of mutations, it should break down rather than increase its level of order. Complexity Theory forecasts that it does neither because its history of accidents is locked into its structure, but that the changes to DNA (in this example about 30%) will be allowed for by other components recognizing those changes.

The latest idea is that, in the process of life forms changing via mutation, if there aren't enough differences in DNA between species then some other factor has to be involved. In this case, it is theorized that it isn't in fact the DNA coding that is of importance (since we've recently discovered the same DNA is shared so widely between species), rather how the RNA *reads* or decodes the DNA sequencing. It is this difference in RNA decoding that determines, in theory, the differing species.

However, this places the responsibility for the evolution of any new creature now onto RNA. But here's the dilemma: The RNA must know what the DNA sequences are for all creatures *before* they have mutated from something else, otherwise it wouldn't know what it is reading and wouldn't know what changes to make. Prior knowledge or awareness, however, is not permissible under Evolutionary Theory because everything is the result of accidental chaos.

INTERIM SUMMARY
PREVIEW TO CHAPTER 12

At this point, there are still many unanswered questions of evolution. As we have seen, every step proposed by Evolutionary Theory so far remains scientifically unproven. There are even issues that evolutionary science ignores, for which no explanation is provided or is scientifically possible.

What we know of the evolutionary path so far is:

1. It is a scientific truism that matter cannot self-create from nothing, so the existence of matter cannot be explained through evolution.

2. There is no scientific evidence that matter congregated together to ignite a massive big bang explosion. The universe is cleverly structured into a specific but unnatural layout with many functions contradicting the laws of physics.

3. The timing proposed by Evolutionary Theory for the self-formation of our solar system suggests that ninety percent of the material required to build the planets was expulsed before the planets could have formed. Our solar system is also made up of objects of differing ages and components. The scientific evidence does not confirm or validate the theory that our solar system was made from localized clouds of matter in a relatively short time.

4. The earth is the only place that can sustain life as we know it and it is in the exact, specific and perfect place to do so. The earth's structure, components and placement are exactly such to allow life to flourish.

5. The scientific evidence is highly suggestive, if not proof, that life cannot self-create. The elements that make physical life possible are highly complex. They earliest life forms included complicated sets of genetic instructions and cellular processes that laid the foundation for highly developed life forms to exist billions of years in the future.

Even though science shows us that no evolutionary processes or theories are able to work in practice, now or in the past, it is what happens from this point onwards along the evolutionary path on earth that most people associate with the term 'evolution', this being a process whereby living things change from one form to another entirely different form. This is the reason why we began our examination of evolution right at the very beginning of the universe, to establish how evolution and scientific theories had arrived at this point.

To proceed with the evolutionary path, we now have to accept that nucleated cells self-created by some scientifically unknown and inconceivable process and then mutated into the life forms we see today, even though science tells us they cannot do this. This brings us to what is actually meant by biological evolution, which we can summarize as:

Biological evolution proposes that life forms can increase in complexity from one level to a higher level by accident.

This of course assumes an increase in complexity as an improvement. This point needs clarification before we move on. We are familiar with the concept of 'trial and error', which is a method that has an intended end point whereby various trials are made in the hope or expectation that one will work to reach that end point. When one trial fails, further controlled changes are made to get closer to the end point. There is a specific, clear, process that happens; it is controlled and there is an expected outcome.

This is not what happens in evolution. There is no end point or target in evolution as there is no underlying awareness or consciousness guiding the process. The process it undergoes is, theoretically, just random unconnected variables, some of which may be advancements. Technically, there are no errors under evolution. Everything that happens is just part of a massive range of variables with some working better than others, but none are right or wrong.

The process of biological evolution involves billions of small positive increases that are in a distinct linear order, one after the other, accompanied by billions of billions of billions of failures that are going in other directions. In just one stepped change, cells theoretically mutate into thousands or millions of different forms, of which only one, or a few, may adapt better to a particular environment. Of the remaining mutations, some would continue on for extended periods of time in following generations. These later generations would also experience a range of other, different, mutating cells and so on. Eventually, some of these would survive and these lines of mutants would continue on in different but acceptable forms, for endless generations. There is no evolutionary perfect shape, design or layout for anything. Any different step of any entity that lives and reproduces has every expectation that the following generations will also reproduce indefinitely.

Evolution is, therefore, a concept where new organisms are constantly being created with some adapting better to changes in the environment. These will form a new structure from which to build the next level whilst the existing levels can maintain life for many generations (possibly forever) without necessarily changing. Evolution is not a one-off occurrence with a specific end. It goes on randomly forever and, by definition, is out of control, natural and unstoppable.

According to evolution, over billions of years, an original single cell changed through billions of steps to become you. The bacterial mold in a bathroom or on a piece of cheese are our distant relatives as we both share the same ancestors. In fact, it is the evolutionary premise that bacterial cells evolved into every type of living thing on earth including plants, insects, birds, animals, and humans.

Unfortunately for evolutionary theorists, there is not one piece of scientific evidence to support such a claim. It isn't even remotely feasible from known biological processes.

Furthermore, the existence of bacteria today is absolute, indisputable proof that living things can and do survive in the same form for billions of generations regardless of the environment. They are also proof that life forms do not automatically disappear from the fossil record after they (theoretically) mutate to a higher form, which is an important point to understand. We have records of microbes as the first life forms and they exist in the fossil record down the ages. They are still with us today, unchanged—for over roughly 3 billion years. Yet there are no fossil records of any of the countless billions of billions of different mutations that should have occurred along this claimed evolutionary path from bacteria to humans. There is also no evidence of unusual, living mutations that biological evolution would predict to be alive today similar to bacteria.

The question Evolutionary Theory must tackle is this: what scientific examples or proof exist of any particular living organism mutating into anything else? If the evidence of bacteria can survive over 3 billion years, scientific evidence must also exist in the fossil record for these organisms, or there must be proof these organisms are alive today. But there are none, and as there are no mutational changes either in the fossil record, or any evidently alive today, biological evolution must therefore explain the reason for these gaps in the proposed evolutionary pathway.

Nonetheless, we will now attempt to do it for the evolutionists, to explain in step-by-step detail what has never been done before: how the first living body in history made itself from one nucleated cell, the self-existence of which is an impossibility.

12

THE FIRST LIVING CREATURE ON EARTH: NUCLEATED CELLS SELF-FORM INTO A LIVING ENTITY

The world is overflowing with a vast variety of over one million different life forms. Evolution suggests these came from one original ocean-based creature that self-created from one original nucleated cell. Its development is critical to evolution as virtually every creature on earth shares some basic components that must have originated from this creature, such as heart, blood, neurological systems, digestive/excretion systems, muscles, and so forth.

The challenge for evolution now is to explain how a living body could make itself starting from just one cell. There should be a clear, straightforward explanation describing the processes because, as we will see, the concept conflicts with known biology. Of all matters relating to the self-formation of the universe, it is how living creatures appeared that is the most important to us. Unfortunately, it is here that evolution faces its greatest failure through its total inability to provide any detail.

Living bodies, comprised of nucleated cells, appeared in the oceans 700 million years after new types of single cells stopped appearing. The evidence from the fossil record is of previously living creatures, not individual nucleated cells. There isn't any evidence that single cells formed into nucleated cells, which formed into a living body. The scientific evidence is such that *the appearance of nucleated cell life has no known link to single cell life.*

Despite the lack of evidence, evolution still proposes that nucleated cells evolved from single celled life forms. This is probably because there is no room in Evolutionary Theory to acknowledge that this new life form appeared in full as a complete living body.

To understand the scientific reasons why a living creature cannot self-create, a logical process is presented below, starting with one cell. Essentially, the process we will follow is that nucleated cells multiplied, grouped together, then gradually developed more complex forms and body parts through random mutation, resulting in the various characteristics of life that we see today.

The oldest animal life on earth at 760 million years ago is generally considered to be the sponge, which is multi-cellular. Its place in evolutionary history relies on the definition one applies to animal life, as most definitions require animals to have sensory organs and a nervous system, but sponges have neither tissues nor nerves.

Recent genetic studies indicate that the comb jelly(fish) was a genetic offshoot of a previous but unspecified life form claimed to exist before sponges. However, no fossils of this life form have been found making this proposal speculative.

The first positively identified fossils with the characteristics of modern animals including developed eyes are the trilobites (about 520 million years ago), becoming extinct about 250 million years ago. Such is the sophistication of these creatures that most trilobite eyes had double lenses (human eyes have single lenses) with an incredible fifteen thousand separate lens surfaces in each eye[18]. The trilobites however, were arthropods having external skeletons and are not claimed to be related to fish.

These various early life forms make it difficult to find a starting point in the self-development of ocean life as none of them appear to be related to normal fish that populate the oceans in abundance. However, for our purposes, fish have been chosen as an illustration of how cells could form into the first living being. Fish appeared fully developed and are brilliantly suited to their individual environments, with evidence dating back hundreds of millions of years. While it could be argued fish were adapted from a more simple line of creatures, the internal and external components of fish still have to be accounted for by accidental mutation.

---

18. There are three recognized types of trilobite eyes: holochroal, schizochroal, and abathochroal. The majority have two compound eyes that comprise of a number of separate units with rigid, crystalline lenses having two internal layers each with differing refractive indices. These rigid eyes were not flexible enough for focussing on different lengths but allowed for good depth of field in reasonable focus.

The exact sequence and manner in which they formed is not at issue and will always be a source of debate because it had to be a random series of events. The following processes are no more or less valid than any others. It is the principle of how the process could work that is examined, not the correct sequence.

The issue is: how could a fish form itself from an initial, random collection of cells and over what period of time? The evolutionary process requires a cluster of cells aimlessly floating in the ocean to assimilate and become a fully functioning fish capable of sight, smell, digestion, excretion, movement, reproduction, as well as being able to judge distances and make decisions about direction of travel and food collection.

The assumptions used are:

• Cells mutated to a better adaptation than their parents.

• The first mutation was successful.

Scientists have estimated the probability of one in a million for each and every cell mutation. In reality, the process described requires billions of failures for every small step forward, but these are ignored for the sake of simplicity.

A Detailed View Of How The First Living Creature May Have Formed

The course of self-invention is presumed to have happened in a process along the following general lines:

Floating in water, the first nucleated cell began replicating itself and forming a cluster of cells, which would become the first 'proto-fish'. These cells have no blueprint from which to develop, nor do they have a plan toward which they are working. There is no particular structure that is right or wrong. Their components can change at any time as they are not deliberately doing anything. They are replicating and mutating randomly.

Figure 74: The First Living Creature displays the start of the self-creation of a living body, the first in history. It is a blob of nucleated cells floating in water.

Figure 74: The First Living Creature

These free-floating cells are assumed to have bound together as a group. At this point, the most important thing for their survival was food for energy. As there were no other available food sources (it was the first of its kind), there needed to be accessible chemical nutrients they could absorb. Rather than growing in size individually, these cells used their energy to replicate.

However, to develop from a primitive functioning cell cluster to a fully functioning organism requires billions of positive mutations over an enormous time frame. This means the original cells would have had time to form only some of the parts of a fish before full reproduction was necessary. The developing part-fish needs billions of offspring over many generations until one finishes up as a complete fish in which all the interlinked, interacting and interdependent body parts exist in one body at the same time.

Early Internal Organs

As bodies are made from many tubes and organs, the focus of our starting point is the first tube. We will assume the first proto-fish gathered around a nutrient source and linked its cells to form a tube. These first cells are branded type 'A'. Nutrients floating in water could then enter this tube of A-cells.

Continued replication made the tube longer and thicker. Some cells became surrounded by others, thus preventing direct access to food. These starved cells therefore relied on other cells to allow them access to the nutrients within those cells, which transferred the necessary nutrients through the process of absorption. With less work to do, these new types of cells began to store any excess nutrients they received. These are B-cells.

The outer layer of cells would have been subjected to the erosive forces of water movement, presumably shearing them from the tube of A- and B-cells. One mutant cell with a stronger surface bond, however, would have resisted the erosive forces better than others and reproduced to develop a strong covering around the tube. These are C-cells.

The tube now has a range of cells:

- A-cells directly absorb nutrients.

- B-cells store nutrients.

- C-cells protect the inner A- and B-cells.

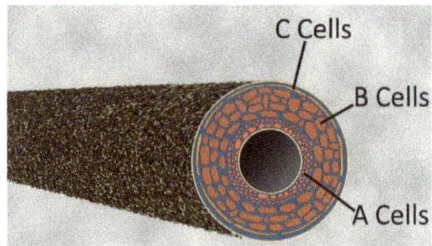

Figure 75: The First Living Tube

Over time, the tube grew in many dimensions. As nutrients drifted down the tube, concentration levels decreased as the lining of A-cells progressively absorbed what was available flowing down the tube. The cells taking in food at the front collected more nutrients while those in the middle became more reliant on storage. Near the tail end, there were relatively little nutrients available causing some B-cells to shrivel up, forming a small internal hole. Other

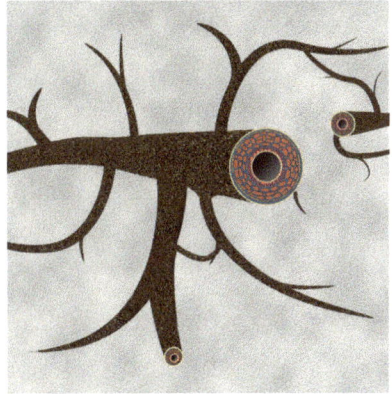

Figure 76: Extensions Develop

cells formed around the outside of the hole making a new tube forking from the original. Over time, more forks developed, as shown in Figure 76: Extensions Develop.

Some cells would have been robbed of nutrients and become waste. Eventually a cavity formed that could pass waste cells to the outside. This process was effectively a filtering mechanism, possibly the first primitive kidney.

The tube now has a food source and waste system. If the rear end closed up, internal pressure would have forced more forks and cavities to develop. Over time these cavities would each develop special characteristics that fulfil a range of internal functions that we know about today, such as:

Spleen:

This organ is intimately associated with immune cells and assists in the body's ability to fight infections.

- It filters and removes contaminated and dead blood cells and bacteria.

- It holds reserves of iron, necessary for aerobic metabolism.

- It produces cells that eliminate contaminants in the blood stream.

Liver:

This organ is known to perform hundreds of functions.

- It stores vitamins, sugars, minerals and blood and manufactures bile.

- It takes in nutrients and oxygen.

- It produces proteins and fats.

The Intestine And Stomach:

- Store single foods.

- Absorb vitamins.

- Enact chemical digestion through enzyme activity.

- Develop compounds such as acids that breakdown more complex foods.

Bladder:

- An expanding pouch that stores waste fluids until released.

At this point, there is a now a biological system with basic functions that can take in food, utilize nutrients and separate and store waste, which it can expel in a primitive way. It is almost a sealed system in that it won't let in any more than its capacity can handle.

Early Vascular System

There now exist free floating cells within the tubes and forks that transport nutrients to other cells and carry waste to the excreting holes. Effectively these are primitive blood cells or 'proto-blood' moving through primitive veins.

However, now that the constant flow of water has ceased inside the primitive tube, the free floating cells become stagnant. For the organism to do anything other than consume food and grow, the free floating cells are required to flow continuously through the veins. The solution is a self-driven muscle with chambers, valves and veins that can pass and receive proto-blood at a regular rate. For this, muscle tissue is required. Muscles are cells and fibers that contract together. One explanation may be found within cells we see today, an inclusion called a spirali, which is worm-like in shape and wriggles. If a strain of spirali cells formed in one of the cavities of the new organism and pulsed together in a rhythmical peristaltic (wave) motion, proto-blood would have moved through the veins making the consumption of food and expulsion of waste more efficient. A lack of peristaltic co-ordination would have achieved no progress.

This first developing creature would at some point require a heart as its pump. For survival, the heart muscle would need to have formed very early in development. However, there is limited possibility of any cells accidentally forming themselves into this structure and then pulsing as a group of their own accord because heart muscle cells twitch in unison under neurological control. To function properly over an extended period of time, even primitive hearts need neurological monitoring to maintain heart rate and pressure, but at this stage of development the proto-reptile has yet to develop any neurological cells or systems. Without neurological input, the heart will beat out of rhythm and fail to pump adequately. The appearance and operating processes of the 'proto-heart' are critical for the future development of the cells into a fish. However its presence and functioning are unexplained without neurological monitoring.

Today, a modern fish has a heart with two chambers as shown in Figure 77: Early Two-Chamber Heart. The heart pushes blood through the gills where it is oxygenated and sent to various organs where needed.

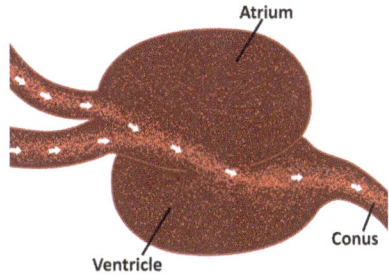

Figure 77: Early Two-Chamber Heart

The processes of separating oxygen from water, passing it into blood cells and targeting specific organs involves extremely complex biological mechanisms and would have needed to work from the very start of life.

From this point, the cells have to eventually form all the body parts of a complete, fully functioning, mobile fish. The major components are shown in Figure 78: Fish Skeletal Anatomy and Figure 79: Fish Internal Organs. There is a comprehensive array of differing parts, functions, shapes, sizes, connections, and locations. All of these have to be set out in a stream-lined, waterproof body, capable of floating, separa-ting oxygen from water as well as processing nutrients and waste from an ever increasing range of inges-

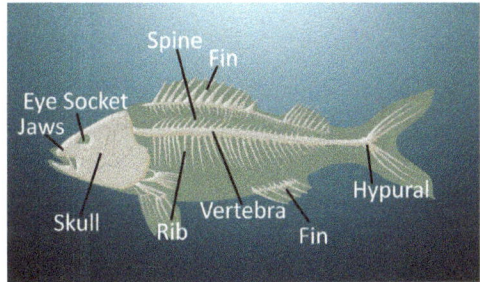

Figure 78: Fish Skeletal Anatomy

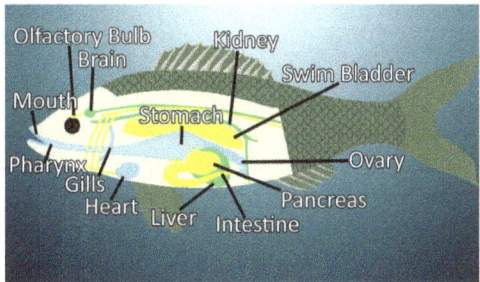

Figure 79: Fish Internal Organs

tions. Each part has to be in the most appropriate location; eyes, mouth and teeth at the front, internal organs, and external parts.

Rather than attempt to map out a process for each of these, our review will focus on the operations of the brain, fins, eyes, sense of smell, and the slimy coating that covers the scales of a fish. As well, we will review the most amazing ability that modern fish possess—the mysterious 6th sense.

Brain And Nervous System

The brain is the command center for living bodies, controlling all body functions. Without it, the body will die. Hearts, however, have their own neurological networks that allow them to continue beating for a short period if the brain dies. With external medical intervention this period can be extended, but at this point in history there is no external intervention, there are merely cells gathering randomly together. The most likely option appears to be that firstly a heart self-created and then started its own pumping action. Then the brain, the circulatory system, which carries blood around the body, and the blood self-created. For a body that invents itself without a mother producing an embryo or egg, the brain and the heart need to co-exist so the creature can live while developing new parts. This is because the brain controls the rate of heartbeat which would have to change as the creature invents more parts. Otherwise the body will just be a conglomeration of cells making body parts that have no controlling center (brain) or engine (heart) to give the body life.

All mobile creatures appear to have some form of electromagnetic charge. It is assumed each cell holds a small proportion of this charge. We know that the brain acts on receipt of electrical messages from all parts of the body. In most skeletal bodies, these messages get to the brain through the nervous system via the vertebral column. These messages are processed, interpreted and a deliberate, specific response is sent back through the vertebral column to the appropriate areas. The vertebral column in a proto-fish would need to develop from calcium excesses into a perfect symmetrical layout that assists in holding muscles in place and helps maintain the overall shape of the organism. Vertebrae have to form with many identical joints, evenly spaced, moving in unison with muscles. A strain of cells that responds better to electrical charges has to grow inside the vertebral column, which protects these new proto-nerves from damage. A series of 'connectors' begin to branch off from the proto-nerves to every part of the body taking electrical impulses to and from the proto-brain. All of these things had to occur of their own accord without a plan or reason.

The proto-brain has accidentally given itself the purpose of receiving electrical charges, recognizing them, identifying their location, sorting, prioritizing and producing a positive response, aimed at prolonging the life of the body. Accordingly, brain cells are structurally and operationally different from all other cells. Chemical ions across the end points (synapses) of brain cells (neurons) transfer messages from brain cell to brain cell. These chemical ions have positive and negative charges. When they move between cells, they create polarization of charges or what's known as an electrical potential, which is why scientists say the nervous system transfers electrical messages. But it's not really an electrical flow as we know it in and around our house, boiling our water or turning on our lights. The electricity of the brain and nervous system is more like the chemistry of a biological battery.

The proto-brain cells of our proto-fish needed to develop the specialized ability to use chemical ions to transfer messages between themselves and the rest of the body, via proto-nerves. The cells receiving the messages— heart cells, stomach cells, muscle cells, and so forth—would need to develop the means to translate and understand the message being sent, otherwise the message would go unheeded.

The newly forming brain cells would also need to interpret, sort and understand specific electrical messages returning from the body as it gradually formed. The other thing to consider is that the first messages were initially coming from each body part before they were fully developed and would have been different from the messages when fully developed, requiring the brain cells to recognize the differences. The proto-brain would also need to categorize them as sight, sounds, tastes, touches, pain, and so forth. It would need to understand three dimensions, size, shape, color, movements, and textures. Then, it would need to respond appropriately and understand when a successful correction had been made.

Suddenly, at this point, the proto-brain attains awareness. It might not be awareness on the level of humans, but it is awareness nonetheless. This is a huge leap in evolution. No longer are life forms on the planet just reacting to chemicals and sunlight, but now they are reacting to perception. Perception requires a sensory system to detect external and internal environments and to have some form of knowledge of what constitutes a problem and the correct course of action to take to ensure survival. Most importantly, it involves memory. Chemicals don't have memory, but fish and other life forms do. Somehow (and scientists still don't know how it is achieved today) chemicals inside a brain are used to store memory. Memory in living organisms is not stored like a computer stores memory in data bits; it is a different process involving the area of the brain known as the 'memory center', which somehow uses chemicals in the brain to remember past events.

The issue that confronts scientists is that the chemicals used to store a particular memory are fluctuant. That is, they come and go. They are replaced by other chemicals, but despite this coming and going the memory remains the same. This means one particular chemical molecule either passes on its memory to the next molecule that takes its place, or memory isn't established at the chemical level but by some other means. Nevertheless, the brain cells of the proto-fish had to develop a memory center and utilize this new remarkable process of memory in order to function and survive.

This, therefore, is the evolutionary conundrum of the brain: with all its remarkable workings, science cannot explain how the first proto-brain self-created, cell-by-cell and function-by-function, without a pre-existing program from which to operate. (We will look at the operation of a brain in much more detail in Chapter 25.)

Fins

Figure 78: Fish Skeletal Anatomy displays a typical fish with six different types of fins. On virtually all fish these are different yet perfectly shaped for their location, size and function. At some point, the proto-fish has become mobile. To survive, it needs to be able to push forward to allow greater access to food. It needs muscles, flesh, skin, and scales to proceed. It needs fins to stabilize movement and prevent rolling and crashing into rocks and other objects. However, at this stage it does not know that it is moving or that it is colliding with objects. It probably has no feelings, sense of balance, sight, or co-ordination of its systems. It does not know which way is up or down as it is the first living creature in the history of the world.

In the evolutionary process of random mutations, not just one but six fins now emerge. According to evolution, fins are an unintended mutation. Perhaps a group of cells stuck out from the main body of the fish causing it to move only in one direction, possibly around and around in circles. Perhaps many outcrops formed on the fish's surface. Over time, some of these may have fallen off, others grown bigger and, purely by chance, finished up in the most appropriate positions to allow the fish the flexibility to maneuver.

Modern fish have fins with different shapes and sizes. The fins are specifically located to allow that particular type of fish the ability to move and change direction while maintaining stability. Evolution proposes that fins self-created and positioned themselves by chance mutations. Although fish need specific fins in specific locations, evolutionary mutations are blind and random. The chance that fish will have fins is no more or less likely than any, or every, other type of living thing having fins, even humans. The obvious dilemma for evolution is that fish are the only creatures with fins *because they need them*.

Slimy Coating

Fish have a slippery surface over their scales that is a film of mucous slime. Mucous is made by tiny glands on the skin lubricating the body surface. As fish also breathe through their skin, too much or too little mucous causes gas efficiency problems and can result in relatively quick death. This slime reduces water resistance when swimming and is also antiseptic, keeping fungal and bacterial infections in check. This resistance is assumed to have occurred early in the process of self-development, meaning the nucleated cells made a substance that fought against the microbe cells from which they were originally formed.

Eyes

Fish eyes are different from animal eyes in that they are not interconnected. Positioned on each side of the skull, each eye provides a separate image to the brain and operates independently of the other. The eye is probably among the most important body organs after the heart and brain. Without the heart, there would be no functioning life. The brain allows beings to do things independently. The eyes allow freedom of choice and safe mobility. They also serve to identify possible dangers and, for the first proto-fish, some idea of its surroundings.

Sight is generally accepted as normal because most creatures can see. However, the makeup of the eye is extremely complicated and intricate, involving the exact number and placement of many millions of different parts, all operating in unison to allow the brain to see (refer Figure 80: Fish Eye).

The eye can move and focus of its own accord and transfer a massive amount

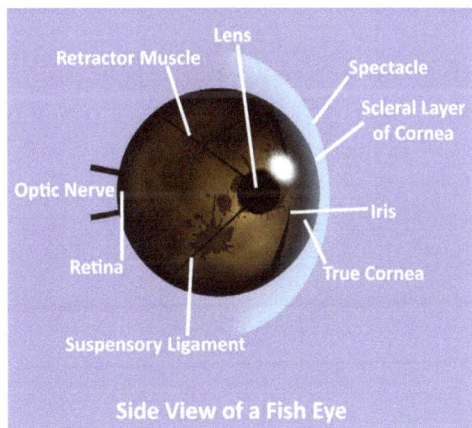

Figure 80: Fish Eye

of information to the brain. It has a number of layers including muscle, retina, sclera, and choroid. At the front it has a cornea, iris, pupil, lens, and chambers. There are numerous blood vessels. It sits in a socket that protects it from damage and aids movements through the use of ocular muscles. Within the retina wall there are rods, cones, a pigment layer, and a range of different cells—horizontal, amacrine, bipolar, ganglion—all used to detect colors and movement. There is a central artery and vein, optical discs and nerves.

The eye is a fantastic and extremely sophisticated organ. It is, however, somewhat fragile. Minor damage to any one of a number of parts can severely limit sight, if not cause visual loss altogether. If the eye is not almost perfectly constructed it will not function and there will be no sight.

The developing proto-fish at this point has no sight. It lives in absolute darkness because the construction of its sight organ has not started. Nevertheless, although it doesn't need sight to survive, it begins to develop the organs of sight, which are connected to its proto-brain by a new type of special nerve. The proto-brain also begins to develop the difficult task of sorting, translating, recognizing and remembering the images it receives, even though the proto-brain has developed from a tangled mass of uncoordinated cells by accident without reason or explanation. The eyes themselves must now form millions of cells in the correct circular shape with a multi-layered covering of different cells and a lens that will focus incoming light on specialized light cells, photoreceptors, on the back of the eyeball. More so, the proto-eye will need to be located exactly where it is needed, at the front of the creature in a protective socket, even though, under evolution, it could have easily finished up on any body part, or even internally.

Even if a miracle occurred causing the basic shape of an eye to form, it would not work. The biology of the eye is such that it has to be complete in every part with a fully functioning lens and nerves connected at both ends for it to function at all. As cells replicate randomly, mutants would have to form over dozens of highly complex but different parts of the eye without any effect happening. They would have to remain in that form without further change until the next million accidental replications caused the next layer to build up. Otherwise, any one of the new components could be lost to eternity.

This of course is just one eye. But our proto-fish has two. Being the first creature with sight, it would need to simultaneously develop both eyes at the same time, symmetrically lined up at the front of the skull, complete and functioning with intact optical nerves attached to its proto-brain. Note, however, that only two eyes developed, not three, four, or five, or even more.

The optic nerve is essentially an extension of the brain. Sight occurs in the brain, not the eye, in the area of the brain called the optic center. It is similar to the operation of a remote camera that sends the image via a cable connected to a television screen. The camera is like the eye, receiving the image. The television screen is like the brain, displaying it. Fiddling with the camera settings distorts the image on the screen. So it is for eyesight. The eye only needs a minor distortion to throw out the brain's picture. If the cable is cut, the image would disappear from the screen even though the camera was still receiving it. Scientists have attempted to copy the process of sight by developing complex computers but are far from close to matching this brilliant function, despite knowing how it worked beforehand. Yet evolution claims the first eye in history self-created identically twice without plan or purpose. As eyes are such complex organs, we will have a closer examination of the make-up of the human eye in Chapter 24.

Sixth Sense

Ironically, fish have a characteristic that is probably of more advantage than eyesight. They have a system of minute canals in their skin—on their head, face and sides—called the lateral line. These are super sensitive organs that allow the fish to detect movement and currents in water (see Figure 81: Fish Canal System and Figure 82: Fish Lateral Line).

The behavior of fish shows how brilliantly this sixth sense operates:

• In muddy water, fish can turn away from danger instantly even though they cannot see anything.

• Around reefs they can quickly find a safe spot in coral with no chance of seeing it beforehand.

• In shoals, thousands of fish will mysteriously hold their formation in unison as the group constantly moves around.

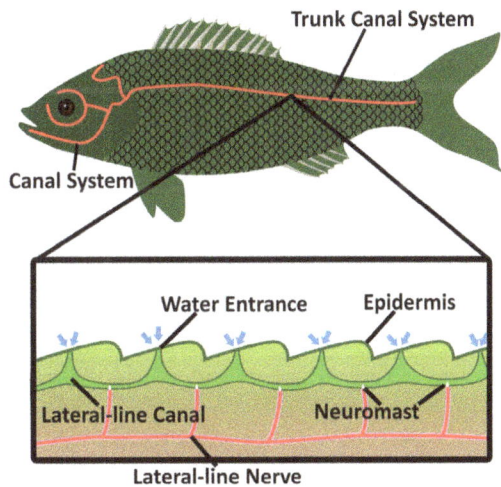

Figure 81: Fish Canal System

The lateral line consists of hairs in a jelly like cap (Figure 82: Fish Lateral Line) that sense variances in water pressure such as depth or waves coming from objects or other sea creatures. Called neuromasts, they act like the hairs in vertebrate ears. For instance, neuromasts change mechanical bending energy on the hair into differing electrical impulses, depending on the direction and force of bending. As a fish swims, it creates a bow wave ahead of itself. This wave becomes distorted by approaching objects and the lateral line senses the changes, alerting the fish to change direction. The lateral lines in some fish can also detect electrical charges and magnetic fields.

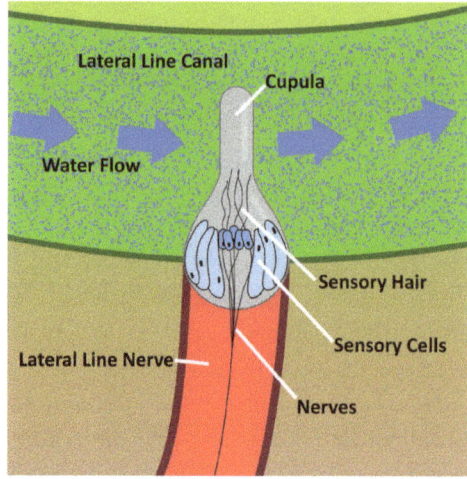

Figure 82: Fish Lateral Line

Nutrition

Another aspect to consider is that during its development, this proto-fish will need a proper nutritional food source. This requires some form of food to be available, yet there is no scientific explanation about what this was, how it appeared, why it appeared at the corresponding time, nor why it contained the appropriate nutrients for a fish. It had to be available and accessible. The fish needed systems to recognize it, consume it, separate waste from energy, and to use the energy.

Full Circle

The cells have now come full circle. Interestingly, at some point after the first living creature had formed itself from independent nucleated cells, these cells became so dependent on one another they could no longer exist without being attached to the body. They then re-coded themselves to die if separated, even though, according to Evolutionary Theory, they had actually existed as independent, single cell organisms in the first place.

Chapter 12 Unraveled

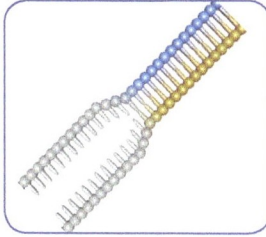

- In any scenario about how the first creature on earth might self-create from a blob of nucleated cells, there is no scientific evidence to justify how any parts developed from accidental mutations.

- Science cannot prove that nucleated cells formed into any type of living creature.

- The brain is a complex organ operating via positive and negative charges from chemical ions and can only function effectively if it contains complex, predetermined programs of operation. These programs had to self-invent virtually without fault from the beginning otherwise life would not have continued

- The eye is a highly complex and intricate organ with many differing and functioning parts, almost all of which must be present in order for basic sight to occur.

- The scientific evidence shows that nucleated cells cannot exist independently from an entire body.

13

THE FIRST FULL BODY REPRODUCTION

For life to continue, the first proto-fish would need to start reproducing itself in full, as a complete body regardless of its level of development, resulting in a new generation. Up to this point, nucleated cells have been reproducing themselves as a group and changing by mutation to form various body parts. No matter how many different groups of these cells exist, they can't live forever, necessitating at least one proto-fish to make a full copy of itself. There is no scientific evidence to indicate how many body parts may have formed by the time this happened.

Although Evolutionary Theory is quiet on this matter, there are three main possibilities that explain how this full body self-reproduction might have occurred:

a) Each of the cells making up the proto-fish replicated, separated from the body and grouped together in the same layout to form a new, identical body.

Whilst this appears to be a logical progression, as duplication is the limit to what cells have been able to achieve so far, there are two major challenges to this process: the challenge of how cells inside the body would exit the parent body; and the challenge of how these new cells would know where and how to align themselves in a new body.

b) The proto-fish developed all the body parts and operating functions required to become a complete fish. It then made a full copy of itself by a process involving ova (eggs).

If one fish developed all the life-supporting functions such as breathing, ingestion, excretion, sight, brain, nervous and skeletal systems, scales and muscles, the period of time involved has been estimated by various scientists as many hundreds of millions, if not a billion, years. One fish could not live this long, let alone survive without fatal injury from natural events. If it did, such long life would be passed onto successive generations, which fails to occur in everyday life.

c) A limited number of developments occurred in the original proto-fish which were passed on to the next generation by some form of primal ovum, or egg. Subsequent generations repeated this process of gradual development until a complete fish was formed.

Effectively, changes came about by small increments over successive generations. This is mainstream Evolutionary Theory, which claims this type of biological progress has occurred in all life forms. However, herein lay some huge logistical problems.

The partly developed proto-fish was made of many different types of cells performing different roles. When the cells replicated they were only making the same type of cell, plus some mutants that were shaping its future. Perhaps it was 10, 100 or even 1,000 years old with a few parts formed when the first-ever reproductive process emerged. Modern fish reproduce by sperm and ovum. The assumption is that an ovum, at least, was somehow involved in the reproductive process as there were no male or female genders at this point in time. This would be the first ovum in history—the single most important development for the continuation of life—an unplanned event of extraordinarily good fortune. And with this new process come second, third, fourth, and fifth events that are perhaps even more extraordinary.

Reproductive Event 1: The First Ovum In History

As Evolutionary Theory assumes that all events are unplanned, the appearance of an ovum, or egg cell, can only occur by accidental mutation. To develop into a full copy of the existing proto-fish, this new egg cell had to collect every single bit of cellular information—their structures, positioning, functioning, and interactions—in order to create another fully replicated body.

This new ovum needed to contain information such as:

• The exact numbers of every type of cell.

• The specific locations of each cell.

• The arrangements by which the cells were connected together as organs, tubes, bones, coverings, muscles, and fluids.

• Each cell's exact roles and functions necessary to keep the body alive.

- Some type of security check or failsafe measure to ensure that each replicated cell was exactly the same as its parent cell, rather than something different.

The figure 83: Proto-Fish Replication shows that even with a few basic body parts consisting of perhaps ten million cells, the process required to transmit and replicate such voluminous and complex levels of information is extraordinary.

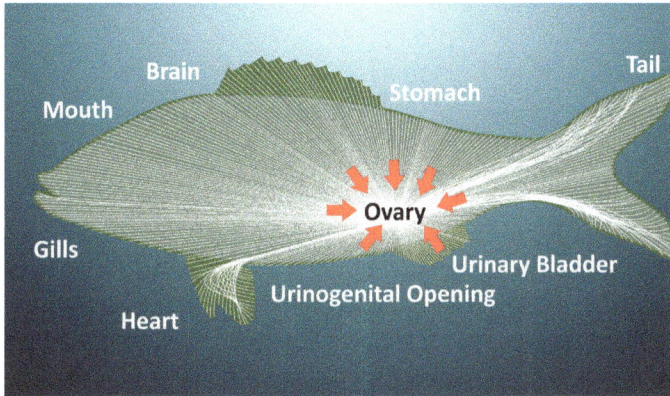

Figure 83: Proto-Fish Replication

For replication of the proto-fish to be possible, each of the ten million cells is required to transmit information about itself to one specific egg cell. This would need to be in a coordinated manner so it was clear, concise and uncontaminated. Assuming the primal egg cell developed as a result of gradual evolutionary change within the developing proto-fish, the egg cell can only receive information about other cells in the body in this manner.

However, this also raises several issues:

- What triggered the body cells to transmit their information, particularly as the proto-fish was only partially developed?

- By what process did the cells become aware of what information to transmit about themselves?

- With what method did the cells communicate their information, considering no physical connections existed through which to transmit this knowledge to the egg cell?

- Why was this data sent to only one specific cell?

- By what means did the egg cell organize and store the information it was now receiving, and how did it know what it was for?

This process of transmitting cellular information from one particular cell to another does not occur in any known biological system today. In fact, it is not thought to be biologically possible. If this process had actually happened it would be a one-off occurrence, something that never happened again, as future reproductions of complex biological organisms are created from egg cells that already contain the genetic program for making a new creature, which they receive at the moment of fertilization. They do not get that information from their own body, as described in the process above, because it has not yet formed.

Reproductive Event 2: The First Reproductive Program In History

At this point, the primal egg cell held all the new cellular information in some unknown form. It would also have to be sure that all the information had been received in full from every other cell before it started replicating, otherwise it could not make a complete copy of the parent proto-fish.

Then, by accident, it had to do something even more extraordinary. It had to invent both sequence and timing programs from the information provided by the ten million cells in order for every new cell to replicate in identical form and in the identical location as in the original proto-fish without putting at risk the survival or location of any other cells. This would become the first cell replication instruction program in history and it would need to have:

1. Ingenious creativity—to invent its own instruction codes.

2. Impeccable timing—to specifically time the appearance of cells so that they appeared in their correct locations.

3. Flawless perfection—to make not one mistake and be without error.

This new cell would need to sort millions of instructions in faultless order without knowing why or what it was forming.

Previously, cells had only duplicated themselves and changed by chance mutation. From this time forward, when each new duplication occurred, it would be a generic cell that would change into a specific cell as instructed by the program—as happens in complex biological organisms today. There also had to be a process for communicating these program instructions to each newly replicated cell so each one knew what to change into once it appeared. The source of knowledge to write such a complex program with this mass of data from millions of cells is a scientific mystery. The method of transmitting instructions to newly formed cells is still unidentified.

For biological evolution to continue forward, the proto-fish would have to know that its new egg cell contained all the necessary information it required for the next generation before being cast out of its body. Yet, according to evolution's principles, the egg expulsion was itself an unplanned chance event, thereby relying on good fortune for its timing. This new egg cell, surrounded by water, had to then start replicating again, ultimately creating the exact number of replications, in this example ten million, in the exact location as in the mother proto-fish, in order to get one copy of the parent. Once at that point, some trigger had to advise every cell to recommence randomly replicating and mutating in order for newer body parts and functions to develop.

The process described above required a sequential program for each new cell to invent itself at the correct time and in the correct place. Whilst in itself this is unable to be explained by evolution, there is another very, very challenging complication that had to occur.

Reproductive Event 3: Making A Miniature Version

When today's fish reproduce they make a fertilized egg that develops the same body parts as its parents. However, the new fish is not a full-size version; it is a miniature version—a baby. When living entities reproduce they make baby versions that grow into adult size. This is a standard biological process for all reproductive life forms. One of the early versions of our proto-fish must have produced a baby fish as shown in Figure 84: The First Baby Fish. Exactly when this process happened

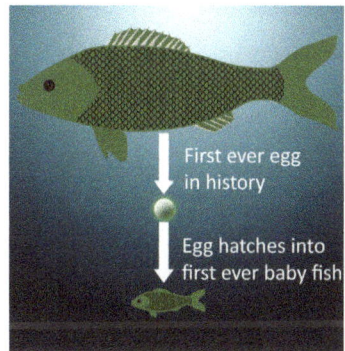

Figure 84: The First Baby Fish

for the very first time is not known, but it had to be very early in the history of life's formation because it is the template from which all life forms operate, including our own.

Reproductive Event 4: Two Matching Genders Appear

The proto-fish is genderless yet almost all fish reproduce by sexual input from a male and a female. Most fish lay eggs but some fertilize internally and give birth to live offspring. There are a number of fish with the capacity for hermaphroditic reproduction, whereby its own sperm can fertilize its own eggs. Possibly this may explain how the early proto-fish could have replicated itself, but hermaphroditic replication still requires the presence of both sperm and egg and there is no explanation from evolution about how or when these two differing components appeared, nor how they were perfectly matched to make replication possible. Males and females have differing chromosomes as do their sperm and egg cells.

Chromosomes are made of protein and one DNA molecule and exist as matching pairs in cells. Males and females each supply identical numbers of chromosomes from their sex cells, known as gamete cells—sperm in males and ovum in females—which need to match and fuse at fertilization in order for the embryo to replicate each new cell with chromosome pairs. The total number of chromosomes varies

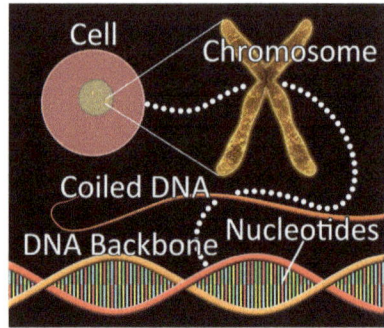

Figure 85: Chromosome

depending on the type of creature. Not all fish have the same number of chromosomes. Chromosome have tightly bound structures that restrict the longer DNA strings into tiny areas, as if wrapped in threads (refer to Figure 85: Chromosomes).

Taking humans as an example, there are normally forty six chromosomes made up as twenty three pairs in each cell. However, when the reproductive cells (the testes in males that produce sperm, and the ovaries in females that produce ovum) are made during embryo development, they only receive one copy of each chromosome, totaling twenty three, not forty six chromosomes. When united in fertilization at a later time, these twenty three chromosomes fuse with the partner's twenty three chromosomes into a single cell that contains twenty three matching pairs. This process is repeated each generation. When a cell divides it makes twenty two pairs of autosomes (non-sex chromosomes) and one pair of sex chromosomes, being two X-chromosomes for females and one X- and one Y-chromosome for males. Hence different genders are formed with twenty three chromosomes from each parent.

According to evolution, having two genders is an accidental, chance development that is highly unlikely to have happened in the same manner more than once. As the number of chromosomes had to split evenly from a genderless entity, the two genders most likely would have needed to come from one embryo that divided into two exact halves. These new embryos then self-created as matching male and female. In short, this means:

1. Chromosomes had to appear from nowhere without the intended purpose of forming genders.

2. The new male has to self-create including the first appearance of a penis, testes and sperm, which contain the coding for its offspring (in particular, the Y sex chromosome).

3. The female has to halve the chromosomes she had previously produced to match the male's chromosomes.

Without an awareness of the other's existence or functionality, these new developments had to happen the first time around, at the exact same time, otherwise reproduction would have been incomplete and the continuation of life would have ended. The chance that there would be a fifty/fifty split of chromosomes is only one in fifty. The male and female chromosomes have to be different from each other yet fully complementary so when combining later they can produce offspring that can be either male or female with all chromosome pairs.

This self-invention of genders appears to be an enormous backward step for evolution as it requires separating half of the genetic information between two fish only to join together again later to reproduce. They would then both have to independently and accidentally release sperm and eggs into water in the same location at the same time to ensure fertilization occurred. However, the process of natural selection, proposed by Charles Darwin, should have eliminated two-gender development as too complicated and too risky to continue.

To explain, natural selection at a cellular level involves genetic adaptability. This is the flexibility of the gene structure (alleles) to adapt to a different environment. If a member of any group has alleles better suited to a new or changed environment they are more likely to survive and reproduce offspring with the same genetic structures, naturally selecting themselves as more adapted. This is why genetic diversity (essentially more allelic variants existing in a species' gene pool) is considered an evolutionary advantage—the more genetic variation, the greater adaptability the species has and the higher chance it has to survive.

However, the evolutionary emergence of two genders, which had self-created both physically and genetically separate creatures, is not naturally selective because the new sexual fertilization process of combining sex cells in moving water is of higher risk and more complicated than internal self-reproduction. This process, which relies on a chance meeting and is subject to many unknown and uncontrollable factors, fails the low complication and low risk criteria of natural selection. Despite its potential for greater genetic diversity, two-gender sexual reproduction at this stage of evolutionary history would not have been as adaptive as other reproductive techniques and thus failed to continue as an evolutionary process. Furthermore, if the two genders had originally formed independently of one another in separate embryos they could not have known about the other's existence further limiting the chance of their sex cells randomly meeting.

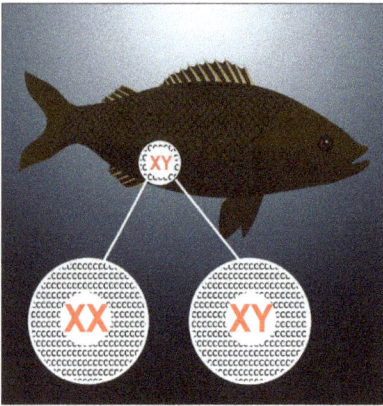

Frame A

Figure 86A: Genderless fish invents chromosomes and splits into two separate embryos

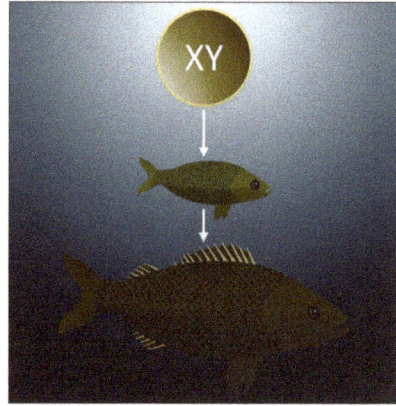

Frame B

Figure 86B: An embryo invents male sex organs and sperm

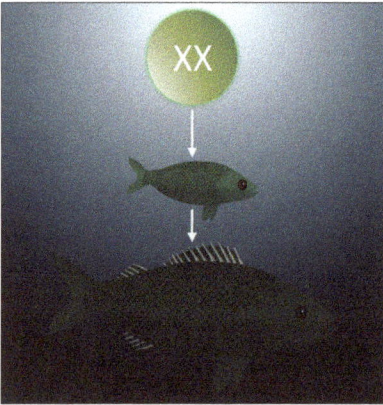

Frame C

Figure 86C: An embryo invents female sex organs and ova

Frame D

Figure 86D: First ever male and female release their sex cells to combine in water

There is no scientific explanation about how or why the first male fish (Figure 86 Frame B: The First Male Fish) initially made or released sperm nor how such a process could self-create. If the timing was wrong on the first occasion, then reproduction would not occur and there would be no more fish or other life forms. Other fish such as sharks, guppies and redfish have internal fertilization, which is a more efficient form of reproduction and one would expect, as discussed, this process to dominate in an evolutionary world. In all probability, even the fish of today are not actually aware that their sexual releases result in new fish. It is simply a pre-programmed natural act like breathing. This concept of a natural act is driven by some internal desire, which as far as we know causes a pleasurable feeling in the fish's body. This feeling is meant to be an accident, yet the continuation of any fish species relies on the perfect timing of these feelings in both parents who must be in the same preferred location for that species at one specific time.

The conditions required for modern fish to reproduce are quite limited. Most fish will only spawn in a particular season when the temperature suits them. They also have specific areas—spawning grounds—where they congregate every season, such as migrating salmon. They appear restricted by circumstances and show little adaptation to changing environments. Most modern female fish cast millions of eggs into the water, whereby the chance of connecting with a male's sperm is subject to many variables, such as tides or predators. This random, ineffective and seemingly wasteful process does not appear to have changed (evolved) in millions of years.

Reproductive Event 5: Passing On Mutant Cell Information

To summarize, the processes necessary to make a second, new living body so far are:

- All the cells in the proto-fish body send information about themselves in an unknown manner to the first ova cell, and only to this new cell.

- The ova cell then creates its own program so that when it replicated it would produce each cell in the correct place and correct time with the correct functions to become a full copy of the parent.

- This program self-created in two phases. In the first phase, a baby version was made that could survive and function in miniature. In the second phase, the baby had another different program that made all parts grow progressively in unison until a full-size copy of the parent was made.

At this point, evolution has conjured the first full body reproductive process. Next,

- Male and female fish appeared by an unknown method. These were perfectly matched both in genetics and physical characteristics from the very first moment.

- Male and female released sperm and eggs, modifying the reproductive process as a combination of the two sets of chromosomes, which grew by another new program making either a new male or female fish through the phases of newborn baby to mature adult.

- To form a new creature, somatic (body) cells now appear and change under a pre-set sequential program of instruction rather than the previous method of random mutations.

Evolution now includes two types of full body reproductive processes, the first being an unfertilized egg from a genderless parent, the second requiring the existence of two separate and distinct genders with a combined fertilized egg.

The second process, as outlined above, has the effect of producing the same type of incomplete proto-fish every time reproduction occurs, which are simply products of the same template from the same replication program. However, to eventually produce the types of fish we encounter today, Evolutionary Theory claims that different and improved body parts and functions continued to occur as a result of mutations which resulted in minor changes across many generations.

As observed today, the differences caused by mutations do not occur after birth, they happen in embryos prior to birth or hatching. Biologically, mutations passed down to following generations actually occur in the development of the gamete cells–sperm or ova–in the embryo. This is explained in more detail in Chapter 14. Suffice to say, Evolutionary Theory is suggesting that mutations shifted from occurring in individual cells (which then sent this newly changed information to the ova cell) to now happening in the developing embryo. Although difficult to theorize how such a change could occur, we will also revisit this process in more detail in Chapter 24.

Suffice to say, in these developments above, there is no explanation from evolutionists about the appearance of elaborate and structured chromosomes without which there would be no genders.

The Theory of Evolution has now reached the point where the first complex life forms, proto-fish, have started to reproduce in full through embryonic development. In the following chapter, we will examine in detail the scientific validity of inherited evolutionary mutations.

Chapter 13 Unraveled

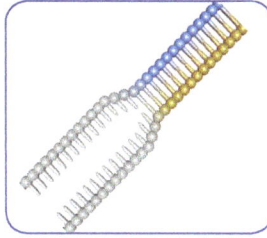

For the first living body in history to replicate itself and then make two genders, a variety of unknown scientific processes had to occur. These unexplained processes include:

- *How the first proto-fish coded its entire cell information into one specific cell, the ova.*

- *How that ova self-created a program to replicate cells in a lesser number, and in a sequence that produced a miniature version of the parent—the first baby in history.*

- *How a second program self-created allowing the infant to grow proportionally into a copy of the larger parent.*

- *The chance appearance of chromosomes that were necessary for two genders that hadn't yet appeared nor were planned.*

- *How a non-sexual proto-fish split itself into perfectly matched and complementary male and female genders, each with exactly half of its chromosomes.*

- *How male and female genital organs self-created without an awareness of their intended function, what opposite sexual genitalia they had to match and be compatible with, or how they actually functioned to successfully achieve a new form of reproduction.*

- *How the male and female coordinated the expulsion of their gamete cells in the same place at the same time without an awareness of the reproductive consequences.*

- *How another new program blended the male and female chromosomes into matching pairs and then self-created yet another new program to make either a male or female proto-fish with matching chromosome pairs in every cell except its gamete cells.*

14

MUTATIONS

The mechanics of cell mutation are vital to the Theory of Evolution as it is based on the premise that mutations are natural, constant and unstoppable processes responsible for every new development in living things, processes which ultimately resulted in every different life form that exists and has existed on our planet. We will now begin to discuss these processes.

What Is A Mutation?

In DNA there is a specific order of the base components (nucleotides) of the nucleic acids. A mutation is a permanent change in that order, which is then passed onto subsequent generations. The DNA sequence of each gene determines the amino acid sequence for the protein it encodes. If the DNA sequence is altered by mutation, the protein it codes for can be affected because its amino acid sequence has been altered. The result is that mutations can occur in any type of cell and can be caused by external factors (such as exposure to carcinogens or gamma radiation) or internal faults (such as errors in nucleotide duplication).

Female creatures are born with the total number of eggs (ova) they will ever have in their lifetime already in place; this includes human females. A female does not manufacture ova during her life. Rather, her body simply releases them from an existing store within her ovaries that she had at birth. In humans, the store of ova usually depletes by mid-life, around late-forties to mid-fifties, the age known as the menopause[19]. When a female embryo has completed its development into a baby girl, her ova contain the hard coded design for her own babies, the next generation.

---

19. Recent research with mice has indicated it might be possible that mature age females can generate new eggs, but the evidence is not conclusive at this time.

Males, on the other hand, make and release sperm regularly throughout most of their lifespan via their testes, which are also pre-coded at the time of birth. Once an ovum is fertilized by sperm, it undergoes a process of cell multiplication allowing it to grow into an embryo. This is when physical mutations can occur. Mutational changes happen during cell multiplication whilst in the embryonic state, after the ovum has been fertilized by sperm. Science suggests a mutation rate of about one in a million embryos. However, this is just a convenient number; it is not a measured statistic.

In the newly formed embryo, there are two basic types of cells:

1. Gonadal cells (testes or ovaries).

2. Somatic cells (all other cells in the body).

Mutations in gonadal cells are passed onto following generations because the gametes they produce—sperm and ovum—code for the offspring of the next generation. Spontaneous mutations[20] in humans usually involve either deletions or duplications of genetic material on parts of DNA gene sequences. An example of gene deletion is Kearns-Sayre syndrome, which affects the central nervous system and muscles. Examples of genetic duplication or repetition are fragile X syndrome, one of the commonest forms of inherited mental retardation, and Kennedy syndrome, which results in a spinal and muscular atrophy. More commonly, a form of adult muscular dystrophy known as myotonic dystrophy is also the result of gene sequence duplication, which can be passed down to successive generations.

On the other hand, mutations in somatic cells are not passed on to the next generation because they still have the normal genetic code in their gonadal cells, and thus sperm/ovum; only their physical appearance and/or functionality has been altered (for example, six fingers in a human baby instead of five).

To highlight the difference in gonadal and somatic cell mutation transmission, in Figure 87: Female Fish Forming From Embryo, the somatic cells are those that develop into her body, brain, organs, blood, and so forth. But it is her ova cells that will be fertilized by male sperm when she has matured at some future point and develop into her own offspring. These gonadal cells are the means by which mutations will be transferred to her descendants. Because the ova cells are already present in

---

20. Spontaneous mutations are so called because they occur within the living organism and have not been passed down or inherited from a parent.

her body when she is born, any spontaneous mutations they carry will be passed onto her offspring, even though she will not have the mutations manifesting in her own body.

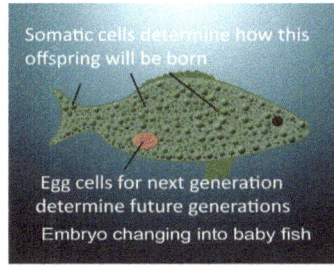

Figure 87: Female Fish Forming From Embryo

We know offspring differ from their parents due to the combination of the father's sperm with the mother's ovum. The genetic code in both the mother and father will blend in the embryo to form a unique combination of its parents' genetic codes. Genes affect the characteristics of offspring, such as eye and skin color, size and shape of body parts, and so forth. Basic body components are not changed by genetics despite occasional faults.

Although there are significant changes to living bodies during an average lifecycle, mostly as a result of puberty or ageing, evolutionary mutation of living bodies after birth is not evident. When a developing embryo is subject to contamination from an external source, the appearance of new arms, legs or other body parts can occur, as can stunted development of any body part. These mutants are not evolutionary progressions, however. Rather, they are limited faults that rarely last for more than a few generations. Metamorphosis is seen in the changing of caterpillars into butterflies and tadpoles into frogs, but these are normal developmental processes of these particular species, not mutations. A living body does not technically mutate after birth and, even if it did, the changes would not be passed onto successive generations as the mutation is not present in the genetic code of its sex cells.

What process occurs in order for mutations to pass from one generation to the next? Following is an example of how evolution claims mutations are passed down the generations.

Generational Transfer Of Mutations

Consider for a moment a male and female proto-fish, which we shall call Generation 1. Each comprises ten identical 'components' that we shall number 1-10 (for example, two fins, two eyes, two gills, one digestive system, one skin, one set of muscles, one set of teeth). These components have been passed down at birth from their own parents. They are also coded into the genes of their gamete cells, their ova and testes, which will also be transferred to their offspring. Each of these ten components consists of millions of individual cells. We shall now explain the genetic transfer process by using these body parts as a whole, rather than tracking the millions of individual cells.

The starting point for these two proto-fish is Generation 1.

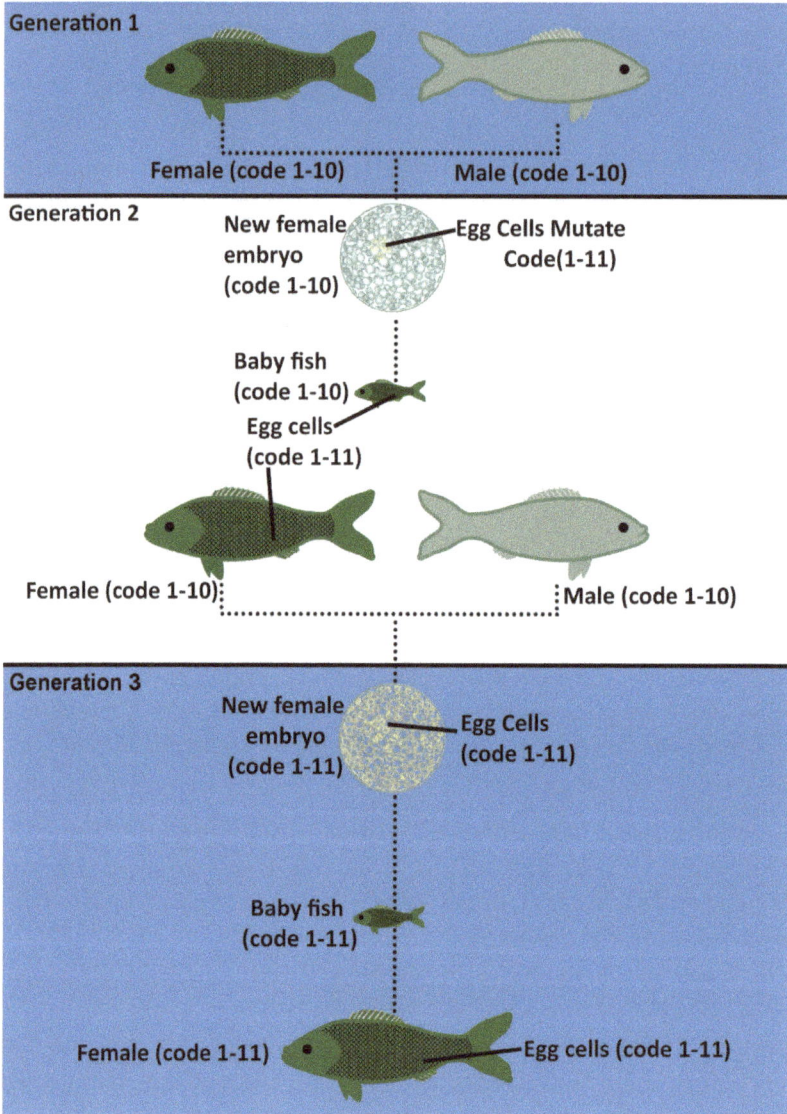

Figure 88: Three Generations

Generation 1

The Generation 1 female releases a batch of eggs (coded 1-10) that are fertilized by sperm from the Generation 1 male (coded 1-10) creating a

Generation 2 embryo (coded 1-10). We shall assume this union creates a female embryo. As the new Generation 2 female embryo is forming inside the Generation 1 mother, we will also assume some of her gamete cells mutate, starting the new development of scales (component 11). When born, however, the Generation 2 female will not display these scales on her body as her somatic cells have not mutated, only her sex cells, so she will only form the same 10 components as her parents.

Generation 2

Although physically the same as her parents, her ova are different as they carry the genetic program for the eleventh (11th) component of scales. This new development will only be passed to her off-spring (Generation 3) when her own eggs are fertilized by sperm (as shown in Figure 88: Three Generations) and to successive generations.

Generation 3

It is only in Generation 3 that the fish and its ova cells are both coded with components 1-11 and exhibits the new component of scales.

Of note in Figure 88: Three Generations is that in Generation 2, the male's sperm has a code of 1-10 whereas the female's egg code is 1-11, as she is the only one to have the mutation. Evolutionary Theory assumes there must be a way the two cell programs can adjust to these newly created differences in genetic coding.

If a particular mutation is not an advancement for a living form, however, Evolutionary Theory assumes it will be lost to extinction forever as it will not survive. Most spontaneous mutations happen to be silent and do not affect the survival of a living creature, and genetic variability means these mutations eventually disappear from the genetic pool, yet there doesn't appear to be any scientific reason why a fully functioning line of fish could not have a mutation that is just a nuisance or is useless while still continuing to function and replicate as they had previously done. For example, suppose a genetic mutation meant that a new generation of fish developed a new useless fin on the end of its tail and a new useful development of scales. This would mean that the fish now had both a regressive and a progressive mutation at the same time. If the new scales allowed it to adapt significantly better than its relatives, the useless fin would continue to be part of its descendants' coding. According to science, there is no known method for a genetic coding to disappear simply because one of several mutations was not advancement.

Sperm And Ova Contact

Fortuitously for evolution, sperm swim, enabling them to move towards the egg and they do so in a predetermined direction – not just in any random heading. When joining with an ovum at conception, there is a chemical reaction at the surface and the ovum opens up its protective casing allowing the sperm to enter. Suffice to say, this chemical liaison must have occurred the first time a sperm and ovum united, otherwise no embryo would have formed.

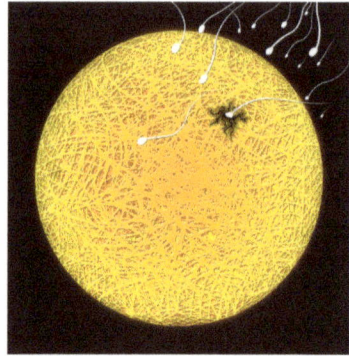

Figure 89: The Moment Of Conception

Once the sperm head enters the ovum, it drops its undulipodium (tail) and the surface of the ovum closes behind it, thus limiting access to only one sperm. This creates a fusion of two cells into one. Scientifically speaking, this process of single sperm fertilization would of necessity have occurred at the beginning of embryo development. If it didn't, fish would have become extinct because of one simple reason: fertilized embryos would have excess male chromosomes as a result of multiple sperm penetration, which would have resulted in massive deformities and dysfunctional life forms.

This new process where gamete cells each have half the chromosomes and must locate and unite with a matching partner (female and male, not male-male, or female-female) to complete fertilization is significantly more complex than single cell reproduction by division, as discussed previously with bacterial replication. Evolutionary Theory suggests that sexual reproduction diversifies the genetic pool, as briefly mentioned in Chapter 13, increasing the amount of genetic material available for reproduction and at the same time also limiting the vulnerability of a species to genetic annihilation through catastrophic mutation. Yet the biological leap from single-cell division to multi-cell sexual reproduction is counter intuitive to scientific laws, which crave simplicity and favor processes that lean toward equilibrium[21], because it is highly complex and consumes an enormous amount of energy. There is no scientific explanation as to why nucleated cell organisms would change the reproduction method they inherited from single cells, particularly as it is a far more inefficient and risky process.

---

21. See Chapters 2 and 8 on the discussion of the Laws of Thermodynamics.

Disappearing Mutations

One of the mainstays of Evolutionary Theory is that life changes into other forms by mutations (for example, all species of fish came from one original but unidentified sea creature). Progressive developments are those that allow superior adaptability, such as the development of lungs in fish enabling survival out of the water. Progressive developments continue in a species, whereas regressive developments disappear due to lack of adaptability.

Since the proposed beginning of evolutionary development on earth, countless billions of mutations would have occurred over many generations of life forms that were not progressive, yet there is a distinct lack of evidence of these failed mutations in the fossil record. Furthermore, science would expect today's fish to exhibit countless billions of variations and all kinds of characteristics and developments in the process of improving their ability to see, smell, swim, hear, live and survive a variety of different temperatures, environments and predators. The seas should be teeming with these variants and the fossil record should be overflowing with examples of evolutionary mutations. But neither is the case.

The issue therefore stands: if there aren't any examples or scientific evidence of fish evolving through mutation, how did they form?

Chapter 14 Unraveled

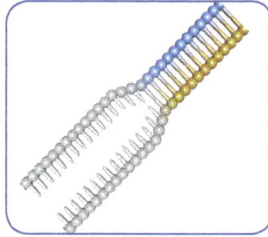

- *Mutations in embryonic gonadal cells are responsible for passing mutations through the generations.*

- *Mutations in all other (somatic) cells in the embryo only affect that particular offspring—they are not passed on.*

- *There is no scientific evidence or reasoning why single-cell replication accidentally became overly complex and risky by dividing themselves into two sexes and finding an alternate process for later intercourse (that is, separate sexes originally came from the same sexless origin).*

- *Science reveals that the theory of a self-created creature, such as the proto-fish we have been considering, is incomplete and unexplained. As it stands, Evolutionary Theory does not offer any explanation about how life changed from one step to another, nor does it comply with known biological processes.*

Using the evolutionary scenario just reviewed, following is a theoretical calculation to determine how many different original and mutant fish would exist if we tracked the mutations from the start of a cycle.

Today, we observe fish can produce millions of eggs per generation. Assume one pair of fish produces one million embryos of which just one thousand survive and that only one of these mutates (that is, one in a million). In a group of ten pairs, producing ten million embryos and ten thousand survivors, the first generation of offspring would have nine thousand, nine hundred and ninety normal progeny and ten mutants out of the total of ten million embryos for these ten pairs. If this insignificant ratio of one in a million continued for just five generations, how many mutant fish (not normal ones) would exist in the fifth generation? A staggering ten trillion mutant fish would have been born.

Figure 90: Ten Trillion Mutant Fish shows these numbers but only includes the normal fish reproducing one-in-a-million mutants. It does not include the mutants reproducing more mutants, bringing the total to over ten trillion.

Generation	Normal Fish	@ 1 Million Eggs	Survivors 0.1%	Mutant Survivors 0.1%
1	10	10,000,000	10,000	10
2	9,990	9,990,000,000	9,990,000	9,990
3	9,980,010	9,980,010,000,000	9,980,010,000	9,980,010
4	9,970,029,990	9,970,029,990,000,000	9,970,029,990,000	9,970,029,990
5	9,960,059,960,010	9,960,059,960,010,000,000	9,960,059,960,010,000	9,960,059,960,010

Figure 90: Ten Trillion Mutant Fish

Taking these figures back hundreds of millions of years ago, as the generations continued up to today there would be countless trillions of trillions of trillions of mutations, some adapting better than others but most still more than capable of living in their general environment like their ancestors had done. However, the reality is that no evidence of mutating fish exists in the fossil record nor is there any evidence of it occurring today—raising the question as to how it can be proposed by evolution if there is no evidence?

INTERIM REVIEW OF EVOLUTION'S PROPOSALS ABOUT LIFE SELF-CREATION

In the previous seven chapters, evolution has presented a conglomeration of ideas about the origins of life on earth that can be summarized as follows:

- The first living thing (a single cell) self-created from non-living matter.

- It reproduced in full making two cells. This process continued to form trillions of cells.

- Single cells mutated into many different varieties, which performed a range of activities necessary for later life forms to survive.

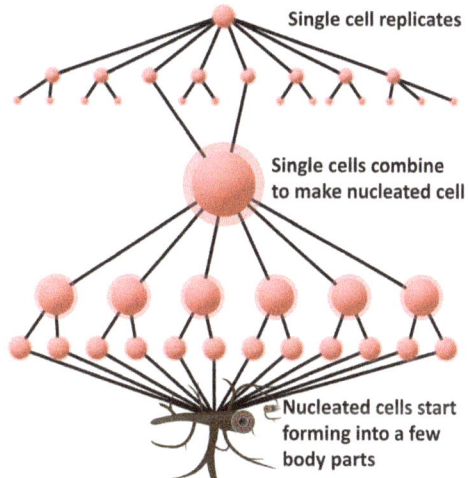

Figure 91: Early Life

- Single cells made themselves into vastly more complex nucleated cells by combining together.

- Nucleated cells replicated billions of times with many mutations, all of which combined to make the beginnings of the first living creature in history.

These steps are set out in Figure 91: Early Life.

- Cells continued to develop body parts until a partial creature was formed.

- This 'proto-creature' produced an ovum containing all the information about its own cells, which replicated itself until a new copy of the creature appeared as a miniature version, itself becoming the first baby in history.

- The baby grew into a copy of its parent.

- At some stage, male and female genders self-created and were perfectly matched from the start.

These steps are set out in Figure 92: New Proto-Life:

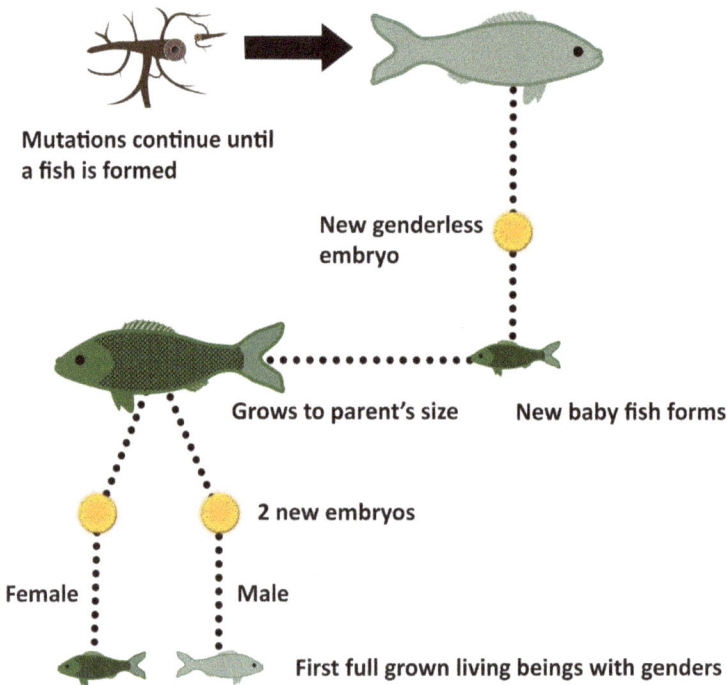

Mutations continue until a fish is formed

New genderless embryo

Grows to parent's size   **New baby fish forms**

2 new embryos

Female   **Male**

First full grown living beings with genders

Figure 92: New Proto-Life

- The two genders released their sperm and ova in water.

- These met and, by chance, matched to produce a baby.

- Babies grew into male and female adults and continued gender based reproduction for every other life form that was to follow.

- After a period of time the fish died.

These steps are set out in Figure 93: First Sexual Creatures.

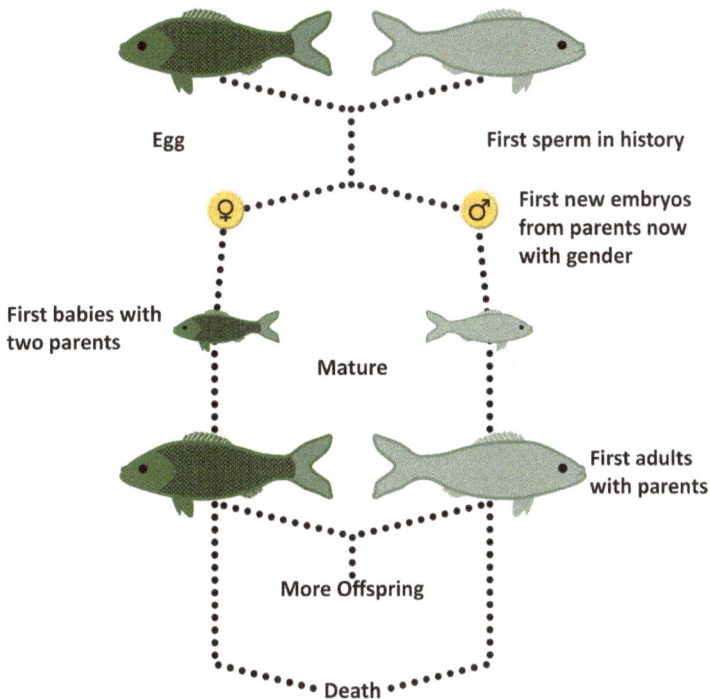

Figure 93: First Sexual Creatures

At the point where two parents first combine sperm and egg, only one cell (i.e. the newly fertilized egg cell, or zygote) is required to contain all the necessary genetic information transferred from the two separate parent creatures. This tiny zygote is required to reorganize two previously independent sets of billions of instructions into a totally new, third set of instructions, then duplicate billions of times in a very specific and immaculately timed sequence to make another new creature. This new creature is a paradox; it is not an exact match for either of the two parent creatures who provided the initial genetic inputs, yet it is also a miniature version of both parents. In effect, the evolutionary leap from full self-replication of cells to a new sexual process of genetic transfer requires:

1. Significantly more time.

2. A specific sequence.

3. The life form to begin from a billionth the size of the original.

4. The creation of a blended miniature creature—a baby.

For any cell development program to be successful, a predetermined chronological sequence is required in which these new cells will duplicate before the embryo starts forming. Eye cells won't develop until a skull is present to house them. Feet and toes don't appear before the legs. In fact, every cell has to appear in a predetermined coordinated order. This is a scientific, biological fact. It stands to reason then, that a sequence program would have been by necessity coded into the very first embryonic cell in history, or reproduction by this method would fail at the very beginning when complex life forms were establishing themselves along the evolutionary line. Furthermore, when the first parents combined, another new process was required, a process making one of two genders. This information would also need to be coded into this first fertilized embryo cell, yet there is no scientific evidence as to where or when the exact sequence and timing came from. This information is new. It didn't exist previously and, in accordance with Evolutionary Theory, there was no need for it to be invented.

These new processes are not the only ones that required inventing and pre-coding:

1. The zygote cells have to duplicate until a smaller, infant size creature is formed rather than the adult size from which they originated. It has to be proportional in every body component, all of which have to be in the correct and matching locations.

2. Another phase of the program is then required to proportionately grow the infant into adult size.

3. Once grown into adult size, the cells peak at some arbitrary point of 'maturity', then deteriorate towards old age and death.

To summarize, Evolutionary Theory assumes:

• A primordial living cell self-created from non-living matter,

• Then changed into a nucleated cell,

• That multiplied into a living entity,

• Then copied itself in full size,

• Then split into two sexes which matched each other perfectly from the outset.

• Then their sex cells combined and made a miniature version one of the two new genders that then grew into in adult size.

Yet there is a lack of detailed scientific explanation how these things could possibly happen as science has already determined they are not biologically possible on their own accord.

Science knows:

- Living matter cannot self-create from non-living matter.

- A primordial cell cannot change itself into a nucleated cell.

- Nucleated cells cannot live outside a body other than for brief periods and cannot make body parts without a mother's uterus or ovum.

- A body cannot copy itself in full.

PART FOUR

COMPLEX LIFE

15
THE FIRST NEW SPECIES IN HISTORY

Now that the first living creature has formed itself into a complete entity, for evolution to proceed there needs to be a process for it to change into a different type of creature. To identify these, scientists have named these classifications 'species'. When a creature fits a certain species description, we know what it is, we know what it can do and we generally know what its limits are. Some species are familiar, such as humans (Homo sapiens sapiens), and others not so familiar, such as the Carob Moth (Apomyelois ceratoniae).

There is no single definition of a biological species universally accepted by science. Generally, a species can be described as: *A group of similar living organisms capable of reproducing fertile offspring from interbreeding*. In this particular area of science there is a lot of confusion and debate because definitions fail to reflect the exact reality of how life works, as not all things have the same boundaries or operate in the same ways. The confusion can be observed in the following example.

There are denoted thirty six species of non-domestic cats, including the big cats which are generally categorized into five groups:

1. Lions

2. Tigers

3. Cougars

4. Cheetahs

5. Jaguars and Leopards[22]

---

22. Panthers are normally grouped into category 5. They are not a species as they are black versions of any of these cats, most commonly jaguars and leopards, that have alleles causing dark skin pigment, known as melanism.

Each cat reproduces with its own kind to make its own fertile kind, qualifying them as individual species. There are DNA similarities between lions, tigers, some leopards and jaguars, indicating they have come from one original source. Despite the DNA connection and although not 'natural', only lions and tigers can interbreed, making a **liger** (male lion and female tiger) or **tigon** (male tiger and female lion). These offspring are called hybrids and are often, but not always, sterile. Second generation hybrids can be produced, for example when a fertile female tigon is mated with a male lion (a **litigon**) or male tiger (a **titigon**). Whilst their combined offspring are neither a lion nor a tiger, common definitions do not exclude lions and tigers from being one species because they are similar via structure and DNA, and can *reproduce fertile offspring by interbreeding*, even though they produce a variety of types.

For many, this situation is confusing because lions and tigers are clearly separate creatures, that is, species. It is not the animals themselves that create the dilemmas, it is our need to categorize things with definitions. These definitions do not necessarily match the complexities of life and can cause confusion about things that science and evolution propose. The disparity between the real world and our definitions needs to be considered when looking at how evolution proposes different species evolved.

21,000 Fish Species

At this point in the history of life there is only a single form of living thing—an undetermined type of fish. There are about twenty one thousand species of fish living today. Naturally, all fish have some common characteristics. To begin with, they all live in water. They have gills to aid the transfer of oxygen from water to their cells. They have fins, eyes, mouths, and the ability to smell (discussed below). Evolutionary Theory assumes that all fish species were mutations of an original proto-fish, much like the example we have discussed in the previous chapters. This theory arises because in the beginning of early life development on the planet there were not twenty one thousand different clumps of proto-cells which formed into the same basic structure of a fish at the same time, in the same way, by accident. Therefore, extrapolating the theory, there was some point in the evolutionary development at which the basic features were hard-coded into the genetic structure of just one original type of fish, genetic codes which then transferred to later species.

The major similarities in fish are:

• Gills: these are symmetrically placed on both sides and are common. There are a few exceptions, including lung-fish that can breathe air as well. These exceptions may be caused by our definitions of what constitutes a fish. Whales, dolphins and porpoises are not fish; they are mammals that live in water.

- Fins: on their backs and generally a greater number on their stomachs. There are side fins that have matching fins symmetrically placed on the opposite side of the body. These are in different positions in each species. There are supports for fins that are embedded in the body.

- Eyes: two at the front—with interesting examples such as the hammerhead shark.

- Mouth: at the front although not all have jaws.

- Smell: all fish have an incredible sense of smell; anatomically they have a relatively large olfactory bulb and nerve for detecting smells.

Most fish also have a similar type of flesh, vertebral columns with spinal nerves, brains, kidneys, bladders, stomachs, livers, hearts, tongues, muscles, skulls, blood and blood vessels.

A fish could not live in water unless it had most, if not all, these components at a minimum. The first 'complete fish' would therefore have been expected to develop these components before it evolved/mutated into other fish species. The evolutionary assumption is this: a family of waterborne creatures with these critical features was able to survive and replicate as the first species of fish; they could swim, see, smell, eat, swallow, extract oxygen from water, convert food into nutrients and rid themselves of waste; they had a protective coating, heart and blood vessels, vertebral structures and an operating brain; they had two genders and could reproduce by combining their sex cells and making a baby fish; all these features developed by accident over hundreds of millions of years. Evolutionary Theory also assumes that later fish species appeared with the same functions but fundamentals such as size, shape, fins, eyes, scales, tails, teeth, external covering, coloring and so forth mutated from their original structures and placements. Most components appeared in different locations on their bodies while still maintaining symmetry on both sides.

In the fossil record, there are examples of fish as far back as 500 million years ago. These are classified into the five following groups shown below: (refer Figures 94, 95, 96, 97, 98). Fish then appeared in large volumes 350—400 million years ago.

Figure 94: Jawless Fish   *Figure 95*: Armored Fish   *Figure 96*: Spiny Fish

Figure 97: Higher Bony Fish   *Figure 98*: Cartilaginous Fish

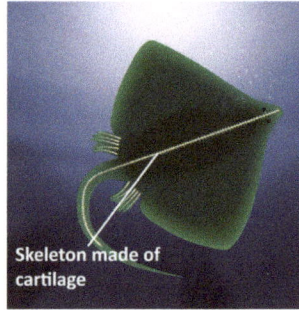

These five groups not only appeared 500 million years ago but also at various later times in the fossil record. Each has similar characteristics to modern fish and, even after reproducing millions of generations, appear to be essentially unchanged today. Almost endless generations have managed to exist unchanged and there is evidence of them. Following are some images of today's sea life that are identical, or closely similar, to the oldest fossils finds. Age is expressed as millions of years ago (m.y.a.):

Figure 99: Sponges—760 m.y.a.   *Figure 100:* Jellyfish—505 m.y.a.

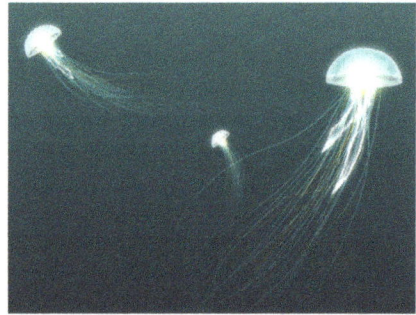

Figure 101: Nautilis—500 m.y.a.

Figure 102: Horsheshoe Crab—445 m.y.a.

Figure 103: Coelcanth—360 m.y.a.

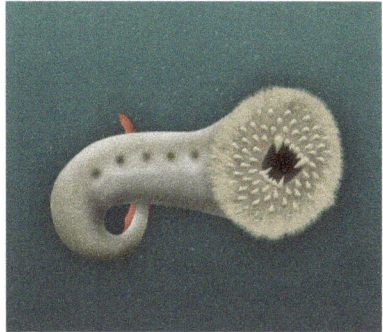

Figure 104: Lamprey—360 m.y.a.

Figure 105: Hagfish—300 m.y.a.

Figure 106: Tadpole Shrimp—220 m.y.a.

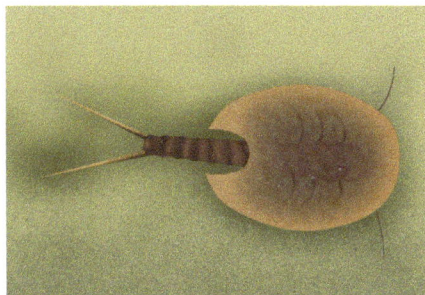

Figure 107: Horseshoe Shrimp—
200 m.y.a.

Figure 108: Sturgeon: (Several
Species)—200 m.y.a.

As discussed, Evolutionary Theory suggests different species of fish appeared through the process of improved mutations. It is also assumed that if the original fish is able to survive unchanged for so long, then so would some of the mutant descendants, as they would theoretically have the same ability to survive. Consequently, versions of these mutant descendants will be with us today, although none are specifically identified by evolution. There would also be many other mutants that survived for varying periods and should be evident in the fossil record. Evolutionary Theory further suggests that the original fish should have died off because they were inferior to the new improved mutated versions, and thus less likely to compete for food and territory or adapt to changing environments, yet they continued for 500 million years and are still with us today.

The immediate issue is to question where the signs of evolutionary development are from these ancient fish if they are physically unchanged after 500 million years.

If new species appear as a result of mutations, there should be many examples of mutant fish showing the gradual progression from one species to another, both alive today and in the fossil record. With twenty one thousand different species and millions of small changes to each of these over the millennia, the fossil record should be overflowing with mutants. Looking at the actual fossil evidence that does exist, what it shows are continuous lines of fish with no evidence of mutants anywhere. There are many new, different sea creatures appearing in the fossil record at different times but no known link attributable to other sea creatures. The scientific evidence is that these new sea creatures appeared suddenly, complete and fully functioning.

Biologically, some creatures can adapt to different environments and change to a point where eventually they will not reproduce with the original group. Technically, by definition, these will become a new species. However, the type of creature is still the same (that is, it's still a barracuda fish not a catfish) as they have not changed into a new or different type of creature. Charles Darwin identified different species of finches on the Galapagos Islands that did not reproduce with the finch species from which they originated. But they were still finches. It would be clearer to classify them as finch species type A and type B, each being slightly different from the other but finches nonetheless. This is an example of the confusion caused by the definition of species. It creates the impression creatures have changed into something else whereas they have only adapted within a pre-set range of variables to cope with a different environment, yet they still remain the same type of creatures.

The available scientific evidence is that fish do not change into other unrelated species today and they never have in the past. Minor adaptations occur in many creatures, but these adaptions do not change the basic design components of any particular species. Mutations tend to be malformations and generally appear as a result of contamination or some other external factor. They also tend to die out quickly.

The types of changes proposed by evolution from one sea creature to another include the following examples which show an extraordinary range of variations. Evolutionary Theory generalizes that these changes occurred as a result of mutations, yet fails to provide any detail how the changes could occur biologically. Although the crab in Figure 109: Crustacean is not a fish by definition (as it has limbs), it is a clear example of the substantial number of mutational changes that would be required to change from a fish, particularly highlighted by crustaceans' 'skeletons' being their outsides.

Figure 109: Hard Shelled Fish

Figure 110: Squid

Figure 111: Colored Fish

Figure 112: Octopus

In summary, evolution's supposition is that one original fish ultimately mutated into twenty one thousand different types, shapes, sizes, structures, and components that in themselves are perfectly formed, symmetrical and in proportion. Yet not one example, from the billons of fish living and in the fossil record, is anything other than complete with no superfluous evolutionary parts or evidence of mutational development towards a different type of fish.

The scientific truth is this:

There is no actual evidence to validate the theory that aquatic creatures changed into other types or species.

Chapter 15 Unraveled

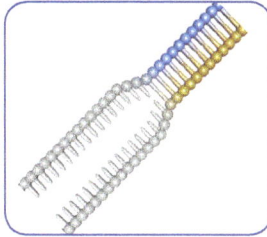

- *A species is a definition used by scientists to categorize creatures, both extinct and living.*

- *There is no single definition that covers the variability of living creatures.*

- *As the world of living things cannot be covered by one definition, there is confusion about what can and cannot occur.*

- *There is no scientific explanation that details the mechanics of how twenty one thousand species of fish developed from one original line.*

- *There is no known biological method for one fish to change into another different type of fish.*

- *There is no evidence in the fossil record that any fish has ever transformed into another type.*

- *The evidence in the fossil record shows fish remaining the same over hundreds of millions of years and new species appearing in full without any previous trace of development.*

16

LIFE MOVES TO LAND

When tracing life to its beginnings, evolution's foundation is that events happened in chronological order. As ocean creatures appeared before land life, evolution assumes life originally formed in water and progressed to land. Evolution does not cater for two separate groups of creatures, one in water and one on land, forming themselves by different methods with the same basic physical layouts and bodily functions. For biological evolution to be a viable theory, each species had to follow a previous species in the order they appear in the fossil record. This is the major premise on which all biological evolution is based. The fact is, Evolutionary Theory claims life came out of water and adapted to land simply because it appeared in water first, not because there are any known processes or evidence of self-invention or mutational development.

The generally accepted steps for evolution are:

- All life came from one original microbe that self-created in water. Later microbes moved to land.

- Ocean living microbes changed into nucleated cells that made a partial living creature, which in turn mutated into a full living creature, itself mutating into different species of aquatic creatures.

- The original nucleated cell was probably also responsible for the development of ocean plant life, (even though these cells differ from animal cells; see Chapter 18, Figure 139: Plant Cell and Figure 140: Animal Cell), then land-based flowers, trees and all plant life and all invertebrates (covered in Chapter 17, being predominantly insects).

- The millions of significantly different species that live outside water originated from one original ocean creature. Effectively, a fish mutated to become a snake, a mosquito, a dinosaur, an ant, a bird, a human, and every other living creature.

If Evolutionary Theory is correct, there should be one simple biological path to follow. However, what is actually presented is a variety of options, which are still only theories, not facts. There are three types of life on land; microbes, plants and creatures. Plants are reviewed in Chapter 18.

Microbes

Evolution proposes that microbes in water are the source of microbes on land. This is a reasonable view as microbes are extraordinarily adaptable without needing to change structure. Their ability to relocate to almost every part of earth is reliant on natural events such as wind and rain, as well as on creatures' movements.

Mobile Creatures

A number of theories about the appearance of nucleated cell land life have been proposed, although none of these have supporting evidence. As a background to understanding the challenges of exiting water to survive on land, the likely environment facing land-based development of life about 400 million years ago was:

- Microbes had existed for 1,600 million years, co-existing with ocean creatures in abundance for about 180—200 million years.

- Oxygen levels were reasonable, with probably more sulfur in the atmosphere than today from active volcanoes.

- There were many earthquakes caused by tectonic plate movements.

- The weather was fairly settled, with the ferocity of the sun's rays slowly easing as the ozone layer matured and thickened.

There are three basic theories how ocean life changed into land creatures.

Option 1. Microbes

Microbes adjusted to land life, changed into nucleated cells then formed themselves into a mobile air-breathing creature.

Despite many dismissing this scenario as impossible, there have been a number of theories that propose early bacteria were washed ashore on the beaches and adapted to the new environment on land much the same as they had in the oceans. The process by which ocean-going microbes may have formed nucleated cells has already been reviewed in Chapter 11, which we have assessed as biologically unfeasible. Any theories that consider repeating this process on dry land are scientifically unrealistic at best, particularly as there was no food source and the sun, wind and rain would have readily destroyed unprotected cells. Unlikely as it is, even if this scenario did happen it still brings us to the point where there is a group of nucleated cells beached on the shore, which is Option 2.

Option 2. Nucleated Cells

A group of nucleated cells adjusted to land life then formed into a mobile air-breathing creature.

A group of nucleated cells repeated the process of forming themselves into a living being, this time on land. There is no explanation for the existence of these independent nucleated cells as, for the previous 200 million years, the fossil record shows these types of cells only existed as part of living creatures in the oceans, not living independently by themselves. The concept that single cells would again combine into identical nucleated cells then make a living body with components similar to those already randomly mutated in ocean creatures is not considered realistic under mainstream evolution. The likelihood of these things accidentally happening in a group of cells lying on the beach in air and as a fish forming independently in water, hundreds of millions of years apart, is not plausible, considering every theoretical mutation is random and by chance.

Option 3. Fish

A fish accidentally mutated to breathe in air, developed legs, supported its own weight, changed its flesh type, reproduced and ate on land, as well as changed many other fundamental operations of its body such as linking both its eyes to operate in tandem.

This is the most common theory as it eliminates the need for evolution to explain the development of nucleated cells into a living being on land, even though there is no biological explanation for such development in water.

The first creatures believed to live both on land and in water were the eurypterids, which were arthropods with external skeletons and looked like scorpions.

Figure 113: Eurypterid

There is no record of eurypterids evolving from any other organism, they simply appeared around 460 million years ago then disappeared about 250 million years ago after populating most seas of the world. Eurypterids didn't come from another evolutionary line, and they had hard shells, which are not a feature of most land animals. Therefore the most commonly assumed theory is that fish became stranded and were forced to adapt to survive, possibly evidenced by the existence of lungs in lung fish. Why or how any fish could be 'forced' to adapt, rather than die of suffocation as they do now, is not explained by Evolutionary Theory. This possibility will be reviewed in more depth, but first a better understanding of what we know about the adaptability of fish is needed.

Adaptability Of Fish

Fish are vertebrates, having an internal skeleton, as are animals. Most fish swim, some crawl on the bottom, others can skip across mud or leap out of the water. Their abilities allow them to live in many different environments—freshwater, salt water, shallow brooks, deep abysses in cold oceans, close to shore, in warm tropics, or freezing river streams.

Amazingly, different processes operate in fresh and saltwater fish. Freshwater fish don't drink. If they did, the process of osmosis would swell their bodies because their internal fluids are saltier than freshwater, thus causing the freshwater to move across cellular membranes into the cells. Too much movement into the cells would cause them to swell and burst, causing cellular death. Conversely, saltwater fish would dehydrate if they did not drink large amounts of water, as their internal cellular environment is less salty than seawater. These two fluid-balancing systems are effectively the opposite of each other. As it happens, there are a few fish species such as salmon and eels that can live in both fresh and saltwater.

Every nook and cranny of the world's oceans seems to have a fish specifically aligned to live in it. So much so, many fish varieties will generally not venture too far outside their regular habitat or from their annual spawning journey, even to avoid regular predators. Few fish species are able to survive outside of a range of limited conditions. There are exceptions such as Great White sharks, which are known to hunt over thousands of kilometers.

The question raised here is this: With most of the twenty one thousand fish species only living in a limited environment, what scientific evidence exists that adaptation works as an evolutionary principle?

We have briefly mentioned adaption throughout the book. At this point, we can summarize the principle of adaptation as the process by which the best mutation of each species will adapt and flourish in any environment. Interestingly, it is not a process whereby a species will be limited to a tight environment once it has adapted to it. This distinction needs to be clear. According to evolution, a single original line of fish is theorized to have been broadcast into all types of difficult environments, the best mutation(s) of which has adapted to each of them and survived.

Yet what is actually observed is different from what evolution theorizes. That is, specific fish live in different and specific environments but most do not leave them as, generally, they do not adapt to other environments. This is highlighted by the recent over-fishing of the seas. In bays where the waters have been fished out, the fish don't return. The waters remain devoid of fish life because other fish do not exploit the vacancy, which is evidence that fish are not highly adaptive species nor do they fill empty niches by adaptation. Furthermore, for most fish species, there is a predetermined place of birth they return to just to reproduce, regardless of the hazards of the journey. Atlantic Bluefin tuna, for example, are fish that travel great distances for food but return to two main spawning grounds eight thousand kilometers apart. These are around the Balearic Islands in the Mediterranean Sea, east of the North Atlantic Ocean, and the Gulf of Mexico, which is west.

Aside from the Great White shark, salmon and several other species, fish tend to reveal themselves as creatures of habit and mostly incapable of adapting to different water environments. The concept of fish adapting to a totally foreign environment on land does not concur with their observed behaviors. When fish are stranded on shore they suffocate and die quickly; they do not start mutating various options for breathing air. There is no known evidence of fish in the process of developing lungs from gills, even lungfish.

Lungfish have two sets of lungs and can breathe air directly into a 'pocketed' lung or they can extract oxygen through their gills, as happens in normal fish. When using their lungs, various valves, arteries and gills change from open to closed or vice versa. The exception is the Australian lungfish that breathes through its gills (some other fish can breathe air through gas bladders). Fossil evidence dating back almost 400

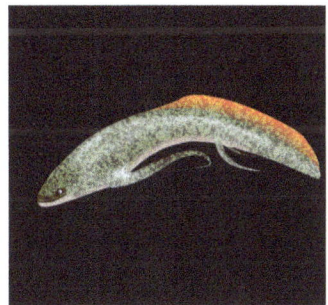

Figure 114: Lungfish

million years shows no evolutionary changes in lungfish. Being freshwater fish, they are not associated with saltwater development at this point in history.

First Fish On Land

Evolutionary Theory assumes that fish mutated until they could breathe air (that is, developed lungs) and then by chance were stranded on land rather than deliberately seeking it, as some theorists propose. It is assumed these fish were a form of 'proto-lungfish' as no other type of fish could live out of water for more than a few minutes.

This theorized process requires a fundamental change in breathing systems. When fish breathe through their gills they aren't separating the water's components into hydrogen and oxygen, they are extracting oxygen gas that is already dissolved in the water (roughly 5% of the oxygen that is present in air). This oxygen gas (O_2) is different from the oxygen in water (H_2O). Fish don't need as much oxygen as most land animals because they are cold-blooded. Whales and dolphins, on the other hand, are warm-blooded mammals that breathe air into their lungs then hold their breath under water. In order for fish to breathe on land, gills had to accidentally mutate into lungs without an awareness of:

a) What lungs were.

b) How lungs operated.

c) What air was.

d) The requirements of lungs for breathing air.

Lungs transfer oxygen into the blood stream, which is then used as an energy stimulant, and they remove carbon dioxide from the bloodstream, a waste product of cellular metabolism. Air travels from the mouth or other orifice (such as the nose) into the lungs via tubes (trachea, bronchi, bronchioles) that end in minute air sacs called alveoli, which are surrounded by blood vessels (arteries, capillaries, veins) and act like a filter capturing the oxygen from the air and transferring it into the blood stream. Oxygenated blood is then carried via the pulmonary vein into the heart that pumps it around the body where the oxygen is used for energy release. The returning blood carries carbon dioxide, which is transferred back into the lungs via the pulmonary artery and expelled from the body during exhalation. Lungs have a deliberate function without which land life would not be possible.

Across 400 million years, fish have not fundamentally changed. When 'proto-man' suddenly appeared on earth about 65 million years ago (a figure hotly disputed, which we will discuss in Chapter 23), fish such as mackerel, herring, catfish, and garfish also appeared, interestingly, with no trace of previous ancestry. In this interim period of 335 million years (between 400—65 million years ago), evolutionists theorize that mutations led to the full range of the living creatures that we know about; birds, reptiles, amphibians, insects, quadrupeds, mammals, and so forth, while ocean-based development slowed almost to a halt. Evolution is suggesting that the massive rate of mutation almost stopped in water but continued unabated on land, then just as suddenly restarted in water 335 million years later when 'proto-man' appeared.

Fish To Amphibians

This brings us to the lobe-finned fishes that are theorized as being the direct ancestors of amphibians. One definition of an amphibian is:

"A cold-blooded, smooth-skinned vertebrate of the class Amphibia, such as a frog or salamander, which characteristically hatches as an aquatic larva with gills. The larva then transforms into an adult having air-breathing lungs".

An amphibian can also be defined as: "An animal capable of living both on land and in water".

(Both quotes from http://www.thefreedictionary.com/amphibian)

The major changes from fish to amphibians include:

• Four legs that self-created.

• An eardrum that self-created.

• The skin invented glands to avoid dehydration.

• The gills changed into lungs.

• Eyelids that self-created to adapt to vision in air.

There are some scientists who believe there is a similarity between the two creatures in the figures below, one a fish, the other an amphibian.

Figure 115: Eusthenopteron

Figure 116: Ichthyostega

Some scientists propose the lobe-finned fish *Eusthenopteron*, from the Devonian period (420-370 million years ago), appears similar to *Ichthyostega*, one of the first amphibians. The conclusion is that these two creatures represent proof that life on land emerged from the water (some others attribute the exodus from water to the Coelacanths at 360 million years ago). The most obvious difference between the *Eusthenopteron* (fish) in Figure 115 and the *Ichthyostega* (amphibian) in Figure 116 is the legs. When a fossil of *Ichthyostega*, still embedded in sediment, was examined under three-dimensional computed tomography scans in 2012 by Stephanie Pierce and Professor John Hutchinson from the UK Royal Veterinary College and Professor Jennifer Clack from the University of Cambridge, they determined that *Ichthyostega* was only able to use its front limbs to drag itself along as all four limbs did not have the structure to rotate and allow walking on land, at best making it a transitional development to land life.

The development of just one leg, with all its components (hinge joints, feet, and so forth) represents an incredible matter of chance. It would also require the fish to lose a fin and develop an entirely new cell memory for a leg whilst it was developing as an embryo. Evolutionary Theory, however, provides no details about the processes required to develop a leg, therefore the challenge left to us is to work through the potential ways that:

1. One leg might form.

2. Three more matching legs developed.

3. How the creature could breathe air and live on land.

4. How these developments occurred simultaneously and for no particular reason other than the accidental drive of evolutionary adaptation.

To achieve a different, new development, such as an amphibious creature, would require millions of lobe-finned fish with mutant cells in the embryo stage to re-configure their bodies, such as different eardrums, skin and eyelids, as listed above, as well as new types of food processing and breathing systems and stronger muscles to support land-based movement. One, or maybe a few, would have to create the start of a leg structure. Do we assume that the millions of cells containing new layouts for muscles, skin, bone, joints, nerves, foot padding, blood, veins and so forth mutated in the same embryo at the same time? Are we also to assume that a complete functioning leg appeared in one single fish, the first land-based leg in history, despite there being no awareness of what legs were and that they might be required to walk on land?

Realistically, leg development could only have occurred over many thousands or millions of successive generations in linear progression. As the evolution of a leg is the result of accidental cell mutation, it is reasonable to theorize that a leg could form in fish that are swimming around in the ocean. However, whilst there are plenty of ocean creatures with legs, Evolutionary Theory avoids linking them with land-based life.

How Did A Leg Evolve?

Science offers no known biological process that allows a leg to self-create, so we begin our proposed process of leg evolution with the assumption (for illustrative purposes) that bone cells began at the front left fin. These left fin cells mutated and increased every generation. Matching cells for muscle, skin, nerves, veins, and arteries must have mutated in the same area, in the same ratio, in

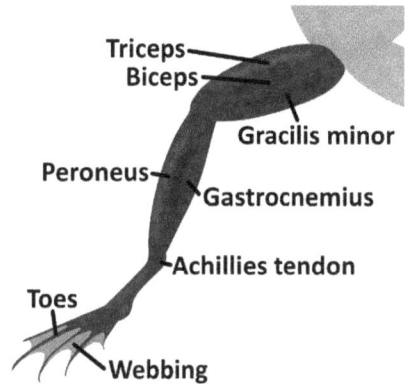

Figure 117: Frog Leg Components

successive generations, by pure chance. Why? Because there is no pre-determined design for the very first land-based leg. All of these additional, complementary and necessary parts needed for a leg to develop also had to form themselves by pure chance. The types and range of parts required are shown in Figure 117: Frog Leg Components:

There is no way for a cell to know where it has to go (that is, its anatomical position) if it is a mutation of another cell because a mutation is a random event without a pre-determined outcome. The chance that millions of muscle cells, accidentally formed from other cells in an embryo, maneuvered themselves to exactly the right spot where bone cells have grown is reliant on chance. The same applies to millions of skin, nerve and vein cells. Evolution assumes that, whilst still in the gamete cells in the embryo, the sequence and cell type program accidentally changed to instruct skin cells to form around an area where muscle has formed without actually knowing it had happened. In answer to this, we will investigate this process during human embryonic development in Chapter 25.

Similarly, the cell program for veins would have to reprogram their positions in order for blood to circulate properly in time with the gradual extension of the leg. Furthermore, while this is happening, the fish is also accidentally changing the cell programs for the position and structure of its eyes, mouth, teeth, gums, and digestive processes. It is as if it deletes the genetic program to make cells for fins and scales and replaces these with skin programs, all by accident. It forms new programs for stronger vertebrae, by accident. Its breathing process cells develop new programs for a different design, greater capacity and efficiency, by accident. The creature has to find some way of recording these changes as they occur in each cell in each embryo, each generation, by accident. That is, assuming just one family line of fish is responsible for the changes. It seems there is no option other than pre-knowledge of the outcomes before these numerous and major changes could be implemented.

Development now has reached the point of the first land-based fish family. This fish species has only one left front leg and has air breathing lungs. Assuming they have accidentally managed to connect the new leg muscles to their brains, which have accidentally learned to send, receive and interpret messages, we would find that they are easy to identify among the millions of different mutants on the shoreline. They are the ones scraping around in clockwise circles while the rest of their relatives flop around in the same spot.

Figure 118: The First Land-Based Leg In History

As unlikely as this scene appears, this scenario is as the Theory of Evolution sees the first movement on land from the sea. There is no other mobile life on land at this time except disputably some developing insects which are, so far, without any evolutionary explanation of their origin.

For evolution to continue, many one-legged creatures must exist. Of these, many mutate many different types of body parts. Yet only one line will form a second leg to enhance development to successful land life. Without laboring the point, the same questions arise:

• Will it be by the same process?

• Will bone cells start first or muscles or veins?

• Was it a matching front leg that formed next? Or was it a rear leg?

• Was it a leg on top of their back or right next to the first one?

Consider the complexity of an amphibian (frog) leg displayed in Figure 117: Frog Leg Components. The leg consists of ankle joints, muscles, foot padding, toes, nerves, blood vessels, tendons, skin, and various other components. According to Evolutionary Theory, the occurrence of these components on the first leg was totally accidental and unplanned. The question therefore is: How did these same features appear on four legs?

Legs are complex layouts requiring multiple coordinating parts within themselves and with each other. To mutate more than one leg is an incredible result of chance. Amphibians have four legs (with the exception of legless caecilians), evenly distributed, identical in proportion, in the correct places to allow balance, co-ordinate walking (or hopping), hold the animal off the ground and allow bending down to eat and drink. This would seem to be a highly contentious outcome from a series of chance events with millions of variables in cell development. Any new leg could face any direction in three hundred and sixty degrees. And that is just in the first creature on earth. Of the millions of subsequent creatures, most of them would have legs facing in different directions if formed by chance mutations. The reality is different. They all point to the ground and line up in the perfect dimensions, directions and locations.

According to Evolutionary Theory, this new family of 'proto-amphibians' is lying at the water's edge. There are possibly millions of one-legged creatures pulling at the sand trying to operate this new mutated extension. Many have other mutations, protrusions and weird shapes forming over their bodies. Some gasp for air as their feeble new lungs struggle to cope. Others have been born with stronger fins, scales and gills that are useless on land and die quickly or possibly head back to the water. Of the multitude of mutating fish probably only one line of them will survive and develop four identical leg structures that will allow them to adapt to their new environment. These surviving creatures will also do something that is extraordinary. They will accidentally develop a new reproductive method that is superior and safer on land.

Land Reproduction

The new semi-species, proto-amphibian, is now facing a dilemma—land reproduction. Most amphibians use water for reproduction. Predominantly their fertilized eggs initially develop into fish-like larva then change into a 'normal' amphibian[23]. The proto-amphibians would have maintained the

---

23. There are exceptions, such as one type of Brazilian tree frog that builds mud nests, the Surinam toad that hatches eggs on its back and the Alpine black salamander that develops its eggs in the female's body.

same sexual process as fish where sperm and eggs were released in the same location at the same time in water. It is the larvae that change into the recognizable amphibian. This does not happen as a result of cell mutation in the embryo; it is an intended outcome, not an accidental one.

Both male and female proto-amphibian, predominantly stuck on land, now breathing air, had to scrape their way back to water to release their sperm and eggs and then return to land again to live their daily lives. Their offspring were born in the water but left it to again struggle on land and then repeat the process of going back into water to produce their own offspring. This process needed to repeat itself millions of times until both male and female had four legs and became proper amphibians. These developments had to be coordinated with changes from gills to lungs or other breathing methods so that newly formed baby fish/amphibians did not drown once hatched in water. Then their entire reproductive processes needed to mutate to include internal fertilization – a major change in functionality for females and a requirement for males to implant their sperm in a previously unfamiliar location without receiving instruction or knowledge of the change. The resultant amphibian species have a variety of different processes including land-based egg-laying, which required both male and female to change independently of each other, at the same time, by accident. The changes to the new sexual organs and processes required are covered in land reproduction in reptiles in Chapter 19.

At this point, there is an overwhelming problem that challenges this part of Evolutionary Theory: amphibians are essentially freshwater creatures. Evolution's theory is that amphibians are descendants of ocean fish or some other saltwater creature and the ancestors to all land creatures having come out of salt water. Yet, amphibians can't live in salt water. Perhaps freshwater fish were the ancestors to land life. But again there is no evidence, leaving evolution in a quandary. Evolutionary Theory, however, is not silent. There are some claims that it was not always that way—amphibians evolved from the oceans but have since changed to live in freshwater—without any logic, evidence or known biological processes to validate these ideas.

Furthermore, what is it that these creatures are using for food? They are the first and only mobile living things on land. There is no food source other than possibly ocean-based plants such as seaweed or perhaps some type of land plant. Tadpoles mostly eat plants but adult amphibians predominantly eat insects, slugs and worms. How could they eat something that didn't exist and if they did exist, where did they come from, considering amphibians are assumed to be the first land creatures?

Many, many questions arise when examining the changes required to adjust the shape, function and layout of an aquatic, finned, scaled, water-breathing fish body to a land-based, skinned, walking, air-breathing animal. Many questions for which there are no answers to explain how ocean creatures might change most aspects of their bodies to adapt to land life.

Water-based life changing itself to land-based life leaves evolution with a huge gap, as there is a lack of evidence, there are no known biological processes and every proposal defies the normal functions of observed life.

Chapter 16 Unraveled

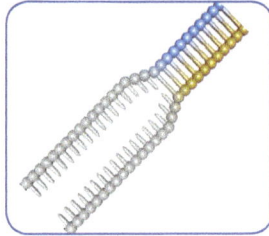

- Fish of today are limited in their ability to adapt to any significant environmental changes. They do not mutate into different species.

- The main proposal explaining the start of life on land is that fish were stranded on shore and they adapted to cope. This contradicts the observations of fish cast on shore these days as they die very quickly, as do air breathing mammals such as whales.

- The self-appearance of four separate, identical, coordinating legs by accident is a concept not supported by scientific evidence.

- There is no scientific explanation regarding how a 'proto-amphibian' thrashing around on the shoreline would get access to a land-based food source nor what the food could be.

- There is no scientific explanation how a completely different process for reproducing on land could self-create.

- The scientific evidence is that amphibians shun salt water and live in freshwater and are therefore not a valid candidate for the evolutionary link between oceanic life and the emergence of life on land.

17
INVERTEBRATES

Around this period of 400 million years ago, other animal life appears in the fossil record. Technically, animals are defined as organisms created from nucleated cells, which have life and voluntary motion. These early fossil recordings are predominantly insects, which are from the class of animals called invertebrates. Invertebrates breathe air but have no backbone or spinal cord, with most having an external skeleton. The main categories of invertebrates are:

1. Arthropods:

- Arachnids (for example, spiders)

- Insects (for example, ants)

- Crustaceans (for example, lobsters / crabs)

2. Protozoa: (meaning first animals)

Protozoa are microscopic life forms, such as amoebas. They are classified as single eukaryote cell animals without any skeletons or brains. They are neither plants, animals nor fungi and are considered to be microbes. They differ from other eukaryote cells in that they survive alone having internal processes that allow them to live. They live in wet environments and move with tails that thrash around. They consume nutrients and oxygen from water that they take in through their cell membranes. As they reproduce by cell duplication, they are genderless. Effectively they can

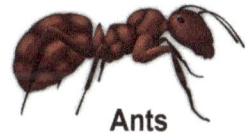

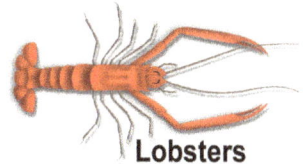

Figure 119: Arthropods

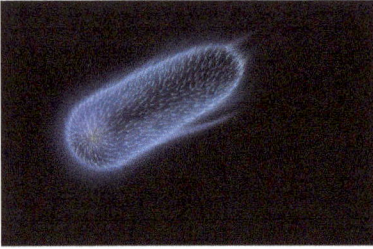

Figure 120: Protozoa

move, breathe and reproduce, take in oxygen and nutrients, consume energy and give off carbon dioxide similar to multi-celled animals, but they are just one cell doing all of these things. Some protozoans cause disease in humans whereas others are eaten by fish or animals as a food source. They are part of the taxonomic kingdom Protista which is a group of single-celled organisms that are in their own unique category as they don't comply with any others. There are a number of differing times put forward by science regarding the first appearance of Protozoa as they are almost impossible to find, so in essence most dates are guesswork. They are generally expected to have existed from about 750 million years ago.

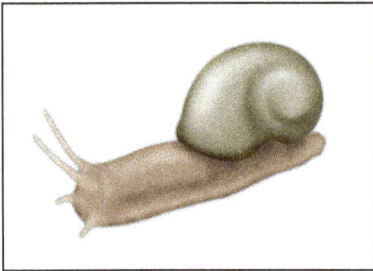

Figure 121: Mollusks

3. Mollusks: (Snails, scallops, octopus, squid)

Mollusks are either soft bodied, like slugs, or have a shell, like snails. They have one muscle that is classified as a foot. It moves by contracting and stretching and slides on its own secretions. Ocean mollusks breathe through a set of gills whereas land-based mollusks have basic lungs. Aquatic mollusks develop from larvae into adults while terrestrial mollusks start out as fully formed babies once released from their eggs.

Figure 122: Annelids

4. Annelids: (Worms)

Annelids are soft-bodied ringed worms that live near water or moist soils. Their bodies are laid out in segments that have the same sets of organs and have external body projections called parapodia that enable them to move. Although some can reproduce asexually, most species employ sexual reproduction. They are burrowing animals and help to aerate soils as well as keeping them moist at the same time. Their first appearance was about 500 million years ago.

5. Echinoderms

Echinoderms are essentially ocean animals, such as starfish. They only inhabit ocean waters and have been found in high pressure depths as far down as twenty thousand feet. Their skeletons are comprised of chalky plates. Their bodies mostly have five different segments that contain the

Figure 123: Echinoderms

same internal organs. These have suction pads at the ends through which they withdraw water, effectively sucking on a surface, thereby securing them. They move by pushing water into the pads, which releases the suction. Curiously, echinoderms don't have hearts, brains or eyes. They are able to detect different light intensities from eye spots on the ends of their 'arms'. They consume via a mouth on the bottom side of their bodies. They can regrow various body parts including tissues and limbs. Their origins date back 540 million years ago.

Invertebrates are thought to have developed in the oceans between 525 and 540 million years ago. Within these groups above, various invertebrates such as lobsters, crabs and starfish live predominantly in water, most others live on land. With species on both land and in water, Evolutionary Theory assumes that invertebrates must have originally formed in water then moved to land, making it the second time such a transition occurred, at around the same general time as fish are thought to have sourced land as a home. These modifications included examples such as mollusks changing both their breathing (gills into lungs) and reproductive (larvae into babies) processes at the same time as heading to shore to live.

Insects

Over one million different species of insects have been identified so far and there may even be up to several million. Insects numerically dominate life on earth. Over ninety percent of all the species in the world are insects. As such, it is reasonable to expect that any evolutionary changes occurring in insects should be representative of the class of invertebrates as a whole. This is where our attention will be focused.

Insect bodies are in three parts:

1. The head, which has two antennae generally located between the eyes.

2. The thorax, which has six legs, in pairs (and often two pairs of wings).

3. The abdomen, which has all the digestive and reproductive organs.

The antennae are the primary organs for smell but they also sense a variety of external factors such as motion, heat, sound, touch, air movement, taste, humidity and chemical detection but not all of these functions are common to all insect antennae. Their layouts include designs that are clubbed, beaded, hinged, pouched, saw-toothed, bristled, combed, feathered or threaded. The range of functions and designs is broad and each appears to be suited to the type of insect. The following ten figures display different insects. Each of these shows intricate formation in every aspect of their bodies despite their tiny sizes. They are beautiful, colorful and perfectly formed in their own ways.

Figure 124: Wasp

Figure 125: Butterfly

Figure 126: Assassin Bug

Figure 127: Beetle

Figure 128: Termite

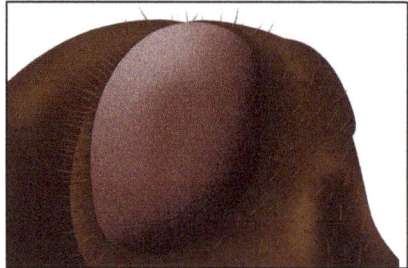

Figure 129: Fly Eye In Close Up

Figure 130: Dragonfly

Figure 131: Ants

Figure 132: Moth

Figure 133: Rose Chaffer Beetle

There have been various evolutionary proposals put forward regarding insects' appearance on earth. It is thought by many evolutionists that a possible ancestor to insects may have been a legless worm-like creature that came out of water. No evidence of such a being has ever been found nor is there any suggestion that this creature is also the ancestor to other invertebrates. As there are no known biological processes allowing an ocean-going creature to form into a land-based creature, nor is there any evidence supporting this theory, we will not pursue the detail about a supposed third exodus from water to land (following fish/amphibians and mollusks), leaving the appearance of insects as another mystery that evolution does not explain.

As previously noted, evolution considers it virtually impossible that, by chance, more than one living entity can independently form similar, if not identical, components such as hearts, blood, veins, muscles, sight, skeletons, heads, hearing, taste, touch, digestive systems and so forth. Consequently, Evolutionary Theory assumes that the million different species of insects sharing the common structures noted above (head, thorax, and abdomen) have all come from one original insect family line and then changed by mutation into each different, unique type. In line with this thinking, it seems reasonable that flying insects must also have all come from one, original insect that invented flight.

With over one million different species of insects, it would be expected that if cell mutation does occur, and is the cause of one type of creature changing into another, there will be billions of examples of insects that have been, or still are, in the intermediate stages of change. But these examples do not exist in the fossil record or in observable life. The fossil record shows that insects arrived on earth fully developed about 400 million years ago with no trace of any ancestral lines. They fed on vascular plants (vegetation with ducts that carry fluids or sap) whereas many of today's insects diet on other insects, some others eat plants, dead animals, garbage or wood, and some consume blood.

The first complete insect discovered is believed to be *Strudiella devonica*, dated in the Late Devonian period around 425 to 385 million years ago. Insect fossils then disappear for 50 million years and reappear suddenly in massive numbers in the Carboniferous period. Around 300 million years ago, the fossil record reveals a world teeming with beetles, dragonflies, termites, millipedes, cockroaches, moths, butterflies, and grasshoppers. All of these show a strong resemblance to their modern equivalents and remain the same throughout the fossil record. The majority of insect families that exist today appeared around 200-150 million years ago, with some appearing in the Cretaceous period from 145-65 million years ago. Insects thrived and their population expanded rapidly from around 90-80 million years ago, concurrent with the development of flowering plants. Most insect fossils show that they are fundamentally unchanged over these enormous timeframes.

The fossil record indicates that insects formed independently from other animals, yet they still possessed the same basic bodily functions necessary for life on earth. Any suggestion that they were part of the theorized amphibian exodus from water does not explain why or how they would completely change their skeletal structure, mobility, diet, sensory systems, and particularly their size. There is no biological explanation regarding how insects could either transfer an internal skeleton to external or accidentally form a skeletal structure similar to other animals whilst forming independently.

There are so many differences in the manner in which insects operate that it would be logistically inappropriate to detail them all in this book. For example, adult butterflies do not have lungs, they have tiny openings running down the sides of their bodies through which they breathe. Other butterflies have taste buds on their feet so they can test the surface of leaves before placing their eggs.

The oldest flying insects in the fossil record are dragonflies, around 300 million years ago. They had wings with veins that carried blood and they needed special muscles to move their wings. These muscles have no anatomical relationship to flying birds or dinosaurs, meaning they developed flight independently. Evolutionary Theory must therefore answer this question for which to date it remains silent: Are dragonflies therefore assumed to be the ancestors to all flying insects, or did insect flight develop itself in many different species unrelated to each other?

Following are a few fascinating aspects about selected insects that give an insight into the potential for self-development of a variety of specialized abilities.

Mosquitoes

Male mosquitoes flap their wings at an amazing rate of four hundred and fifty to six hundred beats per second. They do not glide. In an evolutionary development process, until the mosquito's wings and muscles had fully formed they had no purpose. Evolutionary Theory assumes that mosquitoes accidentally mutated two identical, mirror-image wings over countless generations without being able to fly, and without any intention of flying, as it is a chance outcome. Evolutionary Theory also assumes that the flapping of two wings at the same speed, in unison, by the first mosquito was accidental, as there was no plan for a mosquito with wings to ever appear from nowhere or to fly.

A mosquito is an accidental mutant, as are all of the proto-mosquitoes without wings, or with partial wings, that were its theoretical ancestors, although there is no evidence of these partial ancestors in the fossil record. Here is a thought experiment to ponder on the first ever mosquito: At the point where a partial mosquito had fully formed wings on both sides, what was the event or incident that caused it to accidentally fly? Did it start at one beat per second and accidentally keep building up until it reached four hundred and fifty beats and then took off? Why or how would muscles accidentally develop that could move so fast without any awareness that such a frantic speed was required for flight to occur? Did this occur with just one mosquito or over many generations? Why and how would a mosquito keep flapping at four hundred beats per second while still remaining grounded? Why would it keep flapping at all if it didn't go anywhere? When it reached four hundred and fifty beats and took off, how did it control direction, speed, altitude and hovering with such a fast beat rate, accidentally?

For Evolutionary Theory, however, these same questions apply to all flying creatures. They also apply to every activity such as breathing, walking, crawling, slithering, climbing, eating, and so forth because every function that exists in life forms, according to evolution, was completely accidental and required the formation of the appropriate body parts *prior* to the action being performed.

Ants

Ants have well organized social structures where they co-operate with one another for survival. Ants as a group are sophisticated and well organized with their activities targeted at maintaining the colony. Their colonies use distinctive chemical profiles to recognize members. These chemical communication processes instruct each ant to follow specific roles and actions. Scouts search for food and, on their return, leave chemical trails for others to follow directly to the food source. They farm their own crops using dung as fertilizer. They have defense personnel who specifically protect food sources and living quarters. They maintain the quality of their environments by removing parasites from host plants. They have disproportionate strength to their size and can carry up to fifty times their own weight.

There are probably more ants living on land than any other creature. One of their most important duties is to help clean up the mess that other creatures leave. These well-structured, strong, obedient workers numerically dominate life on earth with scientists estimating there are one and a half million ants for every human, yet their purpose appears to be as the lowest form, removing leftovers and assisting most other life forms to enjoy a relatively clean environment. Proportionately, the human brain is one to two million times the size of ants' brains. Yet these miniscule creatures display controlled organization of massive groups of many millions. It appears their mental abilities far exceed their physical dimensions. It's interesting to ponder what happened in the chain of evolution that enabled such a small brain to be so structured and to have such specific skills?

Myrmecologists, who study ants, estimate there are possibly twenty thousand different kinds of ant species inhabiting the earth. Of those studied, not one is showing any evidence of changing into anything else. The first fossil record of an ant is 92 million years ago. It appeared fully developed with no ancestral trace. It is the same size and structure as today's ants, showing no changes over this timeframe.

Termites

Termites first appeared about 100 million years ago. Termite mounds can be found in various parts of the world. On the savannas of southern Africa these can be up to thirty feet tall and are topped with earthen spires. These mounds are made by a termite called *Macrotermes michaelseni*, which cultivates a fungus called *Termitomyces*. The internal structure of the mound is both complicated and clever. There is a large chimney that runs up the middle. Branching out from this are multiple passageways that connect with tunnels that lie just beneath the external surface. Oddly, the termites live underground, not in the actual mound. The levels of oxygen and humidity in the nest are quite stable but differ from the outside environment. Oxygen levels are about two percent below the atmosphere (nineteen percent) while humidity is high at seventy percent (approximately twenty percent in the atmosphere).

The termites eat a range of cellulose-based foods as well as other animals' feces. These pass through their systems quickly, mostly undigested, along with the fungus. The termite feces are deposited inside the mound where the fungus breaks down the other materials. The mound is essentially the place where the fungus lives, but it also provides the termites with breathable air, housing for their nurseries and storage for food.

The mound acts as a large ventilator for both the termites and the fungus. Because the mounds are quite high, their tops are subject to stronger external winds, which power the internal air movements. The strength of the wind pushes air through the soil on one side while the lower pressure on the rear side provides a suction-like function, giving access to fresh air. As water erosion compacts the external soil and other factors, such as wind and heat, alter the fundamentals of the mound, the termites regularly build onto or excavate from the mound to maintain the internal atmospheric integrity. For example, if airflow is too low the termites will increase the mound height.

The spores of other bacterial components are deposited in the mound in the termite feces. If allowed to grow at their normal rates, these faster growing organisms would overcome the fungus, which is a slow grower. It appears that the higher level of carbon dioxide in the mound limits the ability of all but the *Termitomyces* to grow. As the fungus and the termites are interdependent, both would be under threat if the carbon dioxide levels changed. Effectively, this mound structure is a gas-control structure.

Termite mounds are complex, specifically designed, air conditioned, humidity- and gas-control ecosystems. They are so sophisticated that architects have used their unique ventilation system in the design of the Eastgate Complex in Zimbabwe, which is that country's largest shopping center.

Evolutionary Theory proposes that termites, their mounds, their ecosystems, and their collaborative liaison with fungi are all the result of unrelated chance events over history. Further questions to ponder about termites include:

- How did the first termites learn to build a mound of such enormous relative proportions with intricate tunnels, passageways, arches, chimneys, and nurseries in order to function at the correct temperature, humidity and gas levels?

- When the internal environment changes, how do termites measure the levels of heat, oxygen, carbon dioxide, and humidity?

- How do termites know that adjusting the mound will change these levels?

- How do termites know exactly how high to build their mound or where to add or remove materials so that it changes the conditions back to original 'design' levels'?

- How are instructions given and received?

Whilst pondering these questions, also remember:

- The termites had to build these massive structures one small piece at a time.

- Termites are blind: how do they know where to go or what to do or how to do it?

When humans build complex structures a plan is used, as well as measuring devices and tools at a minimum. Yet termites have none of these. They manage to control their building perfectly and, in this case, humans have copied the achievements of a blind creature one millionth their size.

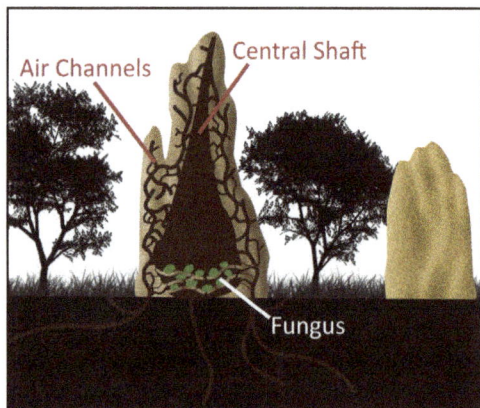

Figure 134: Termite Mound

The termites' role in the survival of animal life in desert environments cannot be understated. Without their collection of cellulose rubbish (trees, grass, leaves, and so forth) and dung, the ground surface would fill up with this matter. This would limit the ability of plants to live due to low levels of nutrients and little free space for seeds. The result would be a reduction in the food available for herbivorous animals, who would also not survive, thus also affecting the survival of carnivorous animals further along the food chain, including humans.

Chapter 17 Unraveled

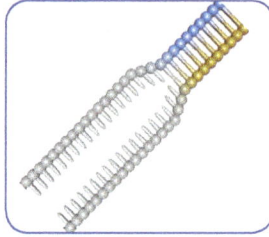

- *Science cannot explain the sudden existence of insects in the fossil record, their intricate body designs, their clever constructions, organization and their duties.*

- *The fossil record is confirmation that insects appeared from nowhere instantaneously and have remained the same ever since. There is no evidence of any mutations in any of the million different species of insects.*

- *Science cannot explain how insects such as mosquitoes could accidentally evolve with the ability to fly and maneuver, particularly as their wings flap at amazingly rapid rates.*

- *Various insects such as ants and termites live in highly organized social environments and can build structures with sophisticated features that science cannot explain.*

- *There is no scientific evidence that indicate insects have ever adapted from one species to another, or that they do so now. All scientific evidence points to the conclusion that they were formed as complete creatures, individually and independently.*

- *Insect reproductive methods are the same as other creatures but had to invent themselves independently.*

Size Of Insects

How fortunate it is that insects are not the same size as other animals? Why should they be so small? There is no reason in random evolution why tens of thousands of different insect species should not have grown to the general size of other animals and therefore dominated. Why shouldn't spiders be as large as dogs, or mosquitoes as large as birds?

Invertebrate Reproduction

Most invertebrates reproduce by sexually copulating. Others use a similar method to most fish by broadcasting male and female gametes in the same general area. There are a few hermaphrodites that have reproductive organs of both sexes. With no claimed connections to any other creatures, these processes had to invent themselves independently for the second time in history – an evolutionary impossibility.

18

PLANTS

As a source of food for insects and herbivorous animals, and the main producers of oxygen and absorbers of carbon dioxide, plants are critical to the development of land life. There are differing records of the first arrival of plants on earth. Some reports claim sea algae as far back as 3,500 million years ago. Other research indicates fungi on land at 1,300 million years ago with plants appearing around 700 million years ago. The most commonly proposed date that plants first appeared on land is between 450-425 million years ago, roughly about the time that insects first appeared.

What Is A Plant?

The Oxford Dictionary describes a plant as:

- "A living organism of the kind exemplified by trees, shrubs, herbs, grasses, ferns, and mosses, typically growing in a permanent site, absorbing water and inorganic substances through its roots, and synthesizing nutrients in its leaves by photosynthesis using the green pigment chlorophyll".

- "A small plant, as distinct from a shrub or tree: *garden plants*".

The Oxford Dictionary also states that plants differ from animals in several key ways:

- They lack specialized sense organs.

- They have no capacity for voluntary movement.

- They have cell walls that hold in the pressure when excess water is absorbed (refer Figure 140: Plant Cell).

- They grow to suit their surroundings rather than having a fixed body plan.

Of note in this description is that animals have a fixed plan for their bodies, namely the size, shape, placement and operation of components. Plants, however, will grow to suit the environment in which they live. Their cell growth programs have flexibility so they can adapt their own shape to their environment, presumably because they are fixed and cannot move of their own accord.

There are over three hundred thousand plant species on earth that have been identified so far. As general categories, these are roughly as follows:

1. Flowering plants 90%   2. Ferns 4-5%   3. Mosses 5-6%

The main classifications of plants are:

Gymnosperm

Gymnosperm plants are those in which their seeds are exposed. Examples of gymnosperm plants are pine, redwood and fir trees. They have defense mechanisms against predators to prevent their seeds being eaten, such as toxic juices and hard seed cases, in order to protect their colonies' futures.

Figure 135: Pine Trees

Figure 136: Tulips

Angiosperm

Angiosperm plants have their seeds enclosed in ovaries, such as fruit and bulbs. Examples of angiosperm plants are tulips, daffodils, apples, and avocados. Angiosperms have a variety of attractions such as scent and flavor, ensuring their attraction to predators. This is to enable them to be eaten, thus allowing their offspring to spread across wider areas when the animals move and excrete the seeds in their feces.

Herbaceous

Herbaceous plants (herbs) are those plants where the above ground section, such as the stem, is not woody. Examples of herbaceous plants are carrots, alfalfa and basil. These plants can have medicinal,

Figure 137: Alfalfa

flavor or scent properties. For example Aloe Vera leaves are used to heal various skin conditions and burns, parsley is added as a common flavoring and roses are used in perfumes.

Herbaceous Flowering

Herbaceous flowering plants have parts that last for only a short amount of time above ground. They sprout, flower, seed and die in a couple of months. A short reproductive cycle is the key to their adaptability. There are a variety of under-surface methods including

Figure 138: Cyclamen

bulbs, tubers and seeds. Examples of herbaceous flowering plants include azaleas and cyclamen.

The Cells Of Plants

Plants are made from nucleated cells, as are all living creatures. Plant cells are different, however, in that they have cell walls. These cell walls are made of cellulose, are rectangular in shape and have large vacuoles[24]. Animal cells, however, have no walls are round or irregular in shape and have much smaller vacuoles. These differences are displayed in the figures below.

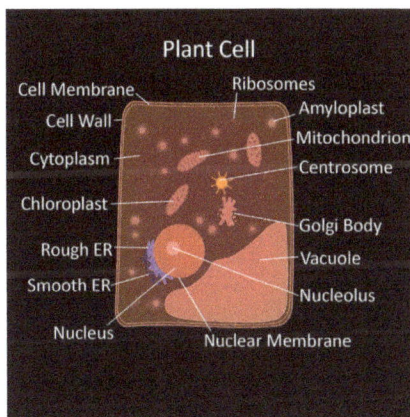
Figure 139: Plant Cell

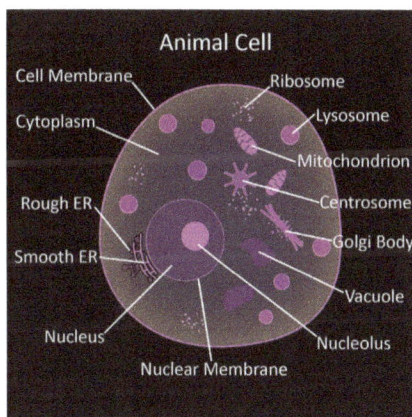
Figure 140: Animal Cell

The Evolutionary Theory explanation for the appearance of plants centers on the development of some form of algae. However, there is no detail about the processes required for algae to change into plant life.

---

24. A vacuole is a small internal compartment containing water and enzymes.

Algae lack the leafy structures (phyllids) in nonvascular plants and the roots, leaves and organs in vascular plants (tracheophytes). It appears the reason algae have been selected is that they already existed on earth prior to plants' arrival. Algae have been classified into different plant kingdoms (groups) with some kingdom terms now superseded. Some are in the Monera kingdom, which are true bacteria, namely single cells. Other algae, such as sea lettuce, are in the Protista kingdom, which can either have nucleated cells without specialized tissues or are single cellular, but neither are considered to be plants per se.

There are, therefore, three basic types of non-animal life cells:

1. Single cell algae.

2. Partial nucleated cell algae.

3. Nucleated cell plants.

Evolutionary Theory could argue a logical line of development from single cell algae to partial nucleated cell algae and then to nucleated cell plants, but the theory fails to address the detail regarding how any of these developments might have taken place. As we now know, there is no known scientific or biological method that allows single cells to build themselves into nucleated cells, or for nucleated cells to accidentally aggregate into a living entity, in this case a plant.

Evolutionary Theory fails to explain:

1. How single cells became nucleated plant cells, and how they came to be different from nucleated animal cells.

2. How these cells survived while not part of a plant (they will die if separated).

3. How they gather themselves together to make a functioning living plant.

Now for the second time, Evolutionary Theory requires these three events to have occurred, despite known biology and all the scientific evidence to the contrary. Furthermore, one of evolution's main tenets is that such dramatic change is highly unlikely to occur more than once because it is purely by chance.

There are, nonetheless, several common evolutionary theories about the start of plant life. Some parts of these theories have been blended together below. The information presented is generalized because Evolutionary Theory offers no details of any of the processes that are covered.

The Evolution Of Plant Life

As part of their root structure, most plants have a fungus called *mycorrhia* that is critical to their survival. It attaches to the tips of plant roots and grows into the root system, then transfers nutrients and water from the soil to the plant. Evolutionary Theory suggests that all plants originated from one initial plant that interacted with this fungus. Plants are predominantly fixed in the ground or in rock, ice, water and so forth, while some travel on water movements. Most have vascular tissue that dispenses nutrients and water throughout the plant.

In the oceans, their structures are thought to have been supported by water, from which they also took nutrients. Although the fossil record shows no evidence of ocean plants moving ashore, land plants appeared fully formed with no trace of origin, thrived, multiplied, and spread with great speed. The fossil record shows the instantaneous arrival of an incredible range and variety of features, colors, shapes, sizes, fruits, leaves, root systems, flowers, and stems with varying reproductive processes all fully developed. Around 130 million years ago, the arrival of new types of plants suddenly stopped, with the exception of grasses, which appeared without ancestral trace between 100 and 15 million years ago. Evolutionary Theory provides no reasons why this exciting and plentiful appearance of new plant species, claimed to be as the result of mutations, suddenly stopped across the entire world.

Seed Development

Water-based plants reproduce by a variety of methods. Runners or shoots that grow underwater can develop into new plants. Other plants have gametes, which are reproductive cells being either male (pollen, released into the air) or female (an ovule, inside the middle of the top part of the plant). Pollen can be distributed by wind, water or insects. If land-based plants did come from water-based plants and were to colonize widely, they had to change their reproductive processes to accommodate new limitations presented by air and lack of water. One theory claims that plants found the shortage of water on land as a problem for reproduction and accidentally invented a new way of ensuring their continuity. Seeds appeared. These were capable of lying dormant until the environment suited reproduction. This new invention needed to occur relatively quickly, however, despite evolution's tenet that progress is gradual over extended periods of time.

A plant has no brain or neurological cells and therefore no thoughts. A plant is not aware that the process of shedding a seed produces offspring.

Evolutionary Theory therefore fails to answer why seeds would develop at all, and why plants would not have simply perished on land.

Seed development is a huge evolutionary leap. A seed has to redesign the growth program used by a water-based plant and, at the same time, invent a new structure allowing it to survive indefinitely in a foreign and harsh environment without any nutrients or liquids. These abilities had to happen:

• Without any awareness of the new, drier environment.

• By pure chance from accidental mutations.

• Relatively quickly, in evolutionary terms.

A plant having offspring is of no benefit to the plant itself, only for other creatures on earth. Plants are effectively servants providing food, shade, shelter, nutrients, changing the mix of gases, and soil stability via their root network. They benefit other living creatures and the environment without an awareness of it.

Even if the first few generations of plants on land had been lucky enough to survive, their components that had worked well in water had none of the internal structures needed to support their own weight in air and on land. They would not have lived long without strong roots to hold them in place and to provide basic fluid and nutrient needs, as well as stability against the buffering winds. It would have been critical that these early plants had to produce a vast array of mutants, one line of which had to incorporate many of the features that we see in plants today. Furthermore, they had to start by evolving a line that completely changed reproduction into seeds within a relatively few generations. If that isn't enough, they also had to adapt their biochemical makeup to compensate for the change from seawater to freshwater.

Plants are claimed by Evolutionary Theory to have colonized vast amounts of land due to the impact of insects that spread their seeds and pollen (remembering that insect life existed before plant life on land). The mainstream thinking is that land-based plants stemmed from algae living in shallow water that dried up and became primitive mosses and liverworts without stems or leaves. The first plants are thought to be a type of fern bush that produced seeds. Some plants did not reproduce themselves and would have divided this process into male and female parts, releasing spores that relied on floating in air or water to fertilize. This astonishing development requires more consideration.

According to Evolutionary Theory, it was an ocean creature, such as a fish, that first accidentally split reproduction into two genders. As we have seen, this was a new, complicated, inefficient, and risky method, a method biologically unexplainable. Evolutionary Theory makes no claims that plants developed from ocean creatures, rightly so. They formed separately, as evidenced by their differing cells. This means, therefore, that plants invented the same sexual method of two sex genders independently from animals. In order to move forward, Evolutionary Theory thus requires two improbable chance occurrences of two different, but matching, genders to materialize in similar ways in two completely different, unconnected life forms.

The fossil record shows that plants spread widely from 350 to 220 million years ago when conifers appeared. These were much more hardy and more temperature resistant. They used water to protect against cell wall dehydration. They are thought to have mutated lignin, a complex chemical compound that is found today in most plants, interestingly as part of a secondary cell wall. They then mutated a variant that moved water up the plant by inventing a duct (vascular) system, as seen in the middle of the root tip in Figure 141: Plant Vascular System. This system also brought photosynthesized food down from branches and leaves to the rest of the plant.

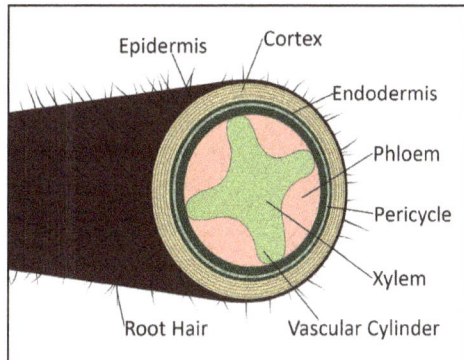

Figure 141: Plant Vascular System

Flowering plants arrived another 100 million years later, approximately 250 million years ago. These plants also relied on insects to pollinate them and their numbers expanded quickly, to the point where today flowering plants account for up to ninety percent of all green plants. These early flowering plants are claimed to have been eaten by animals, which spread their development, as birds did not arrive for at least another 100 million years. Evolutionary Theory does not identify the types of plant-eating animals, nor suggest what they might have been, but the overwhelming consensus regarding the first appearance of plant-eating animals has them present around 65 million years ago. Evolution therefore claims animals were supposedly eating plants for 185 million years before any animals actually appeared on earth.

This leaves us with the conclusion that either the theory is:

• Pure speculation, or

• The evidence has been misinterpreted, or

• The dating is incorrect.

Plants' Survival Systems

Following are four different processes detailing plants that repel animals and insects and others that attract them, all with the intended purpose of survival.

1. Gymnosperms have toxins that repel animals ensuring their own protection.

2. Angiosperms produce nuts and pits that protect their embryos by making them capable of surviving animal digestion when eaten. These plants spread their colonies through animal droppings as they graze and migrate.

3. Some fruit trees, which are asexual, such as bananas and oranges, have their seeds dispersed when their fruit is eaten and the seeds and skins discarded in another place.

4. Other plants have defenses such as thorns, spikes, poisons, and toxic juices that only allow certain animals and insects to help them without allowing over-abuse to hinder their colonization.

Mainstream evolution assumes that plants developed these systems by chance with some suggestions that they 'needed' to be aware of their circumstances, enabling continuity for their offspring when in fact there was no inherent understanding that:

• They were having offspring.

• Animals and insects were eating them and causing their demise.

• Animals and insects could spread their colonies.

• Animals' digestive systems existed or what was required to get through them unscathed.

• A spike or thorn would cause pain to an attacker.

• A poison could harm an attacker, how a poison was constructed, or what chemical components were required.

• The plant itself would need to resist the debilitating effects created by toxins and poisons.

These concepts of plants self-creating complex chemical formulas and components perfectly suited to repel and attract animals, birds and insects without affecting the plants themselves are not supported with any detail about how a plant would know what components to source, or how to manufacture then combine them into either poisons and toxins or attractive odors and tastes. What is observed today is that plants provide food for specific animals, which provide a spreading colonization for the plants. Nature is full of these types of inter-reliant species, much more so than the process of survival of the fittest.

Grasses

Recent evidence of grass pollen up to 60 million years ago contradicts the long held view that grasses were the last new plant form. Grasses are very different from traditional plants. They grow quickly, use relatively little soil nutrients, and grow after cropping. More or less, the whole grass plant is a food source. Many animals eat field grasses as their main food source (horses, cows, sheep, deer, and so forth). Those grass species consumed by humans include rice, wheat and maize. Grasses also adapt much better than other plants to changes in climate.

Photosynthesis

Plants are also blessed with a chemical known as chlorophyll. Chlorophyll contains a green pigment that absorbs all the colors of the light spectrum except green, which it reflects, giving rise to the predominant color of plants. More importantly, chlorophyll converts sunlight and carbon dioxide through the process of photosynthesis into energy that can be stored in its cells. This energy is used later by a range of animals, plants, insects, and humans. Evolutionary Theory proposes that chlorophyll was an accidental mutation that happened only in plants, not in animals.

Plants are the primary food base on earth and are the start of the food chain (refer Chapter 10 subheading: *The Pyramid Of Life*). Plants absorb carbon dioxide and produce oxygen and glucose, which is the base energy source for all living organisms. The process of photosynthesis is critical to life on earth.

Water is drawn by the plant's roots into its vascular system, some of which perspires onto its leaves and is held there by a wax covering (cuticle). The photons in sunlight strike the leaves and are absorbed by chlorophyll, which causes an enzyme to break water down into its constituent molecules, oxygen and hydrogen. Oxygen is released as a gas to the atmosphere and hydrogen ions are converted into nicotinamide adenine dinucleotide phosphate (NADPH), discussed below. Photosynthesis takes place in chloroplasts, which house the chlorophyll. The photosynthesis process comprises of two stages, called light and dark (the Calvin cycle).

In the light process, the plant transfers energy received from sunlight into two molecules:

ATP (Adenosine Triphosphate) is the energy component that allows cells to operate, reproduce and repair.

NADPH is a coenzyme that transfers ions from one reaction to another and is then used as a reducing (oxidizing) agent in the dark stage.

In the dark process, ATP, NADPH, carbon dioxide and various other chemicals found in the plant make sucrose, starch, cellulose, fats, and proteins. Essentially the process is: six molecules of carbon dioxide combined with six molecules of water result in one molecule of sugar and six molecules of oxygen. The basic equation for photosynthesis is:

$$6CO_2 + 6H_2O + \text{light photons} = C_6H_{12}O_6 + 6O_2$$

(Carbon dioxide + water + light = sugar + oxygen)

Animals consume the sugars as part of the food chain and breathe the oxygen from the air. Animals' respiration produces the opposite outcomes to photosynthesis, as displayed in the following equation.

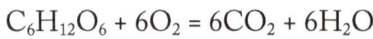

$$C_6H_{12}O_6 + 6O_2 = 6CO_2 + 6H_2O$$

(Sugar + oxygen = carbon dioxide + water)

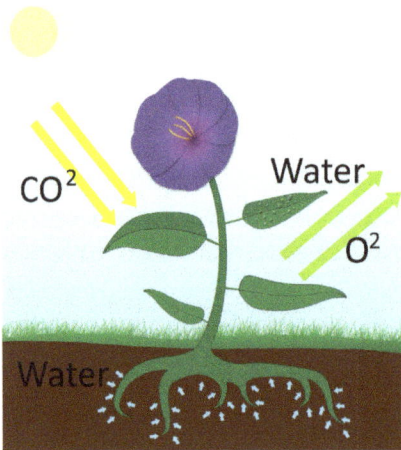

Oxygen is released by the plant through its stomates, which are holes in the leaves that can also draw in carbon dioxide. This is a similar process to animals breathing through their lungs, whereby plants breathe through their stomata. Photosynthesis is a daytime activity that gives off oxygen whereas at night the plant will release carbon dioxide by closing its stomates. Increased levels of carbon-dioxide are considered in some circles to be

Figure 142: Primary Food Source

detrimental to earth's environment[25] but a reduction may in fact result in lower temperatures because carbon dioxide is a fundamental component keeping the earth at a warm temperature[26].

Photosynthesis has many positive outcomes, helping maintain the balance of key atmospheric gases and the earth's temperature, as well as provision of the primary food source.

Summary

Today, plants adapt within a pre-set range of environmental limits and reproduce the same type of plant. There is no scientific evidence to suggest that plants mutate into other types of plants. They simply are not observed to do so. The flexibility of plants to procreate is no more evident than in their combination with other things. They utilize:

- Insects and birds: spreading pollen from plant to plant.

- Air: casting seeds to float on the wind.

- Water: for survival and colonization.

- Animals: eat their nuts and spread them through their excrement or cast off uneaten parts.

The functions of plants are critical to life on earth. Without them, there would be no regeneration of oxygen or removal of carbon dioxide–life would suffocate. They bind the soil and draw nutrients from it. They are the beginning of the food chain, harnessing the energy from sunlight and converting it into energy that can be used by living cells, ultimately providing energy for all other life forms. Without them, life would fail to survive.

---

25. In reference to global warming.
26. Acting like a gaseous blanket, carbon dioxide in the atmosphere keeps heat radiating away from earth into space.

Chapter 18 Unraveled

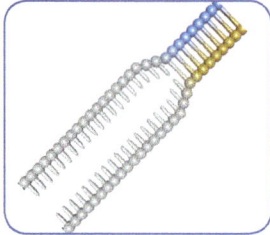

- *The appearance of plants is disputed, between 430-700 million years ago, with algae claimed to exist as far back as 3,500 million years ago.*

- *Somewhere around 400 million years ago, plant life suddenly appeared on land fully developed.*

- *There is no scientific evidence of plants migrating from water to land.*

- *At 250 million years ago, flowering plants appeared from nowhere.*

- *At 60-15 million years ago, grasses appeared from nowhere.*

- *With the exception of grasses, no new plant species have appeared for hundreds of millions of years.*

- *Plants are not observed today, or in the fossil record, to mutate into different species.*

Points To Ponder - Reproduction In Plants

To reproduce, plants release seeds which may or may not be encapsulated (for instance, in a fruit). Some are self-fertilized, others require male and female inputs. These produce miniature versions (saplings) that eventually grow into mature plants. This plant then repeats the process and makes more seeds for reproduction.

The most important question here is: Which came first, the adult plant or the seed?

- If the seed came first, where did it come from if not an adult plant, and why did that unknown thing stop making seeds and disappear from trace?

- If the adult plant came first, how did it become a plant without being a seed first?

This question is essentially the same as the 'chicken or the egg' conundrum. How can you get a plant without a seed and how can you get a seed without a plant? You can't. The answer for the chicken and egg question[27] is the same for all life forms.

---

27. A scientific answer for the chicken and egg question can be found in "Other Things You Should Know" - SECTION A at the rear of this book.

19

AMPHIBIANS CHANGE INTO REPTILES

Now that plants and insects are living on earth and providing a food base, the spread of life can expand. Mainstream Evolutionary Theory proposes that amphibians evolved into reptiles, simply because they are the next group to appear chronologically in the fossil record.

Reptiles

Reptiles are very different from other land animals. Below are some points that highlight the main differences:

- They are cold-blooded and need regular outside heat sources, such as sunlight.
- They have to warm up daily, losing heat quickly at night when they are fairly inactive.
- When catching food, reptiles trap their prey in wide open mouths and swallow.
- They spend a lot of time digesting food.
- They have seven-piece jaws with teeth already formed when born; these are regularly replaced with others of increasing size to fit the growing mouth.
- Reptiles hear through a simple eardrum.
- Most have offspring that develop in hard-shelled eggs and are born externally.
- Snakes and a couple of lizards have no legs.
- They act in the chain of life by eating vast quantities of insects.

There are four basic groups of reptiles:

• Lizards/Snakes

• Crocodiles/Alligators

• Turtles/Tortoises

• Tuatara[28]

The most famous reptiles were the dinosaurs although they are considered to be in a separate group of their own. The fossil record clearly shows reptile dinosaurs appeared on earth complete with complex and effective ball and socket hip joints, relatively advanced eyes and brains, and fully developed legs, hands and teeth. They had a range of members with thick 'armor plated' skin, spikes and fangs. They ate insects, plants and other animals. They were two and four legged, they lived on land and in water and presumably flew.

Considering their massive sizes, their evolutionary development from other creatures would be expected to be found in the fossil record. However, there is no evidence of any previous development of dinosaurs prior to their first appearance, only the complete fully functioning species that, quite literally, appeared out of nowhere.

Flying dinosaur wings were apparently made up of membranes, not feathers. Some evolutionists propose the flying dinosaurs developed independently because there is no link to land-based dinosaurs. Mainstream Evolutionary Theory considers this proposal a virtual impossibility and dismisses the notion that both land-based and flying dinosaurs could evolve by accident from two separate ancestral lines. Their existence remains an unresolved mystery for science.

Dinosaurs became extinct 65 million years ago due to a massive, worldwide phenomenon that was quick and devastating, leaving many dinosaur remains intact today. The cause of their demise is still unknown but was possibly related to an asteroid that collided with earth or a supermassive volcano sending suffocating dust clouds into the atmosphere, or both. An alternate possibility of overwhelming flash flooding caused by the gravity of a large body passing very close to earth also has some validity.

In general, the reptiles we see today are a fairly restrictive group of creatures. They tend to move slowly unless threatened, stay static for long periods and have limited movement during night hours. Unless sourcing food, they tend to keep away from other creatures and spend a great deal of

---

28. Tuatara are found in New Zealand and look similar to lizards.

time covered or hidden from view. They are masters of camouflage, patient, quite strong, and well protected with thick skin and 'armor'. They have remained virtually unchanged since first identified and mostly operate as individuals or in small groups and rarely display intimacy between themselves. They seem well suited for the environments in which they have lived for hundreds of millions of years and appear to thrive in generally warmer climates.

Images Of Reptiles

Figure 143: Various Reptiles – Snake, Lizard, Tortoise, Crocodile

It is a challenge for Evolutionary Theory to explain how these species of mostly slow, limited, individuals are a positive step in a process of evolutionary survival of the fittest. Evolution also fails to explain how the huge evolutionary leap from amphibians to reptiles occurred, or why the evolution of amphibians into other new species would then stop once reptiles appeared. The question therefore is: Why did adaptation continue only in reptiles and not amphibians, if amphibians were already good at evolving?

From here, Evolutionary Theory proposes that reptiles became ancestors to the rest of the creatures on land, excluding insects. These mutations are claimed to have resulted in the huge and diverse range of creatures today, including humans, as well as those that have become extinct.

Reproduction Shifts To Land

One significant barrier in explaining how land-based reptiles followed amphibians is that reptiles have completely different birth processes. To overcome this, some evolutionists argue that different fish species with different birth processes became reptiles that give birth the same way. In this instance, giving birth to live young or external eggs. This would require removing amphibians from the chain of life and finding some other fish-like creature that breathed air and laid shelled eggs. However, as these don't exist in the fossil record, amphibians still remain the preferred ancestors in most evolutionary proposals.

Adult amphibians typically live both on land and in freshwater with many species returning to water for mating season. The reproductive method whereby amphibians lay eggs that are fertilized in water, which then develop into larvae and grow into adult creatures, is unsuitable on land. For land survival, the process of shedding eggs would need to be different, notwithstanding there are some frogs that reproduce on land in wet rain forests. Rather than surrounded by membranous fluid, most reptile eggs have a covering surface, such as a shell, with some lizards, snakes and chameleons giving birth to live young. Because Evolutionary Theory fails to offer any explanations for the differing reproductive methods of fish and reptiles, our challenge is to come up with a process that shows how such fundamental reproductive changes could occur by themselves.

Most reptiles lay eggs that vary from soft to leathery to hard-shelled, which comprise mostly calcium carbonate, the main constituent of everyday chalk. The shell is made in a woven fabric style. This amazing structure allows the embryo access to various external gases and fluids that assist in its growth while providing a protective casing. The shape of a hard shelled egg provides enormous strength resisting pressure from all directions. Even though the components themselves are relatively weak, they resist compression because its dome shape spreads the load evenly across its structure, pushing the force towards the object's center. Conversely, a strike with a sharp object at one point, such as pecking from a hatching baby, will break the weak individual components. In controlled circumstances, a chicken eggshell has been shown to be able to hold up a 200-pound weight.

The magnified structure of the egg shell is shown in Figure 144: Structure Of Egg Shell.

Figure 144: Structure Of Egg Shell

With most amphibians, the females cast soft eggs into water that are then fertilized separately by males. To develop a hard shell (to calcify it) a mutation in the genes coding for the shell must therefore occur. Essentially, where it had not previously been able to, the mutating embryo would now require the capacity to make calcium carbonate from its own cells. If the shell formed around the outside early in the growing process, it would crack apart as the embryo inside kept growing. The shell would be useless, causing the egg to die, which would not be able to pass on the genetic information about the failure to the next generation of eggs, thus making the new line of shelled-egg creatures extinct before it started.

To be viable, the hard shell can only form once the embryo is fully grown. Modern reptilian embryos send a hormonal signal to the outer cells causing them to change and solidify, sealing the shell. Whenever this occurred in the very first egg, the process required a detailed knowledge of when the embryo had reached maximum development; something that had never happened before and is not explained by evolutionists.

The first shelled egg was not going to be instantly perfect in shape, size thickness and strength by chance alone. The process of evolution requires it to develop by constant improvements over time, a concept that is challenged in this case. If the shell built up in stages over successive generations it would be redundant because babies were born and survived successfully

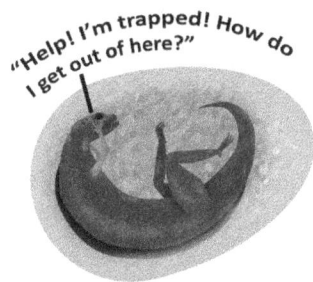

Figure 145: The First Hard Shell Egg

without a complete shell. If it was a full covering from its first evolutionary trial, it contradicts evolutionary principles in that it would have been an instant and fundamental change in one generation[29]. If the first shell was too weak, it would break and the embryo would be lost. If it was too strong, the matured baby could not escape. As it happens, the construction, shape and strength of the eggshell is hardy enough to stay intact and protect the embryo from most elements such as sun, rain and wind, but soft enough to allow the weak infant animal to force its way out once it has reached a level of maturity, whatever that might be. Most creatures born in shells have some form of tooth or horn that allows them to break the egg surface from within. Generally these will disappear or drop off as the animal grows.

---

29. Evolution can't occur in 'leaps' such as described by the formation of the instantaneous egg shell because it would imply a knowledge of what the end result is before it happens, and would therefore not be the result of millions of random, chance mutations but of one directed or planned event, which evolution strictly forbids.

There are only two ways the first shelled egg could have occurred:

1. It developed externally in water, or

2. It developed internally in the mother amphibian.

If fertilized externally in water, it would have to self-create its own shell from the nutrients within. However, the genetic coding hasn't been altered. Therefore, when the egg hatched and grew into an adult, the female offspring would still produce soft eggs and release them into water to be fertilized by a male, which would then have to reinvent the hard shelled egg again. Possibly, this process could have continued forever but it is not observed to happen like that today. At some point, fertilization had to change where the shelled egg formed internally inside the mother, as in today's reptiles. This process requires sexual intercourse[30].

Sexual Intercourse

If shelled eggs developed inside the mother proto-reptile, they had to be internally fertilized by a male through sexual intercourse. Some fish use intercourse for reproduction, so this process self-invented somewhere in the history of evolution. Whilst some amphibians such as some caecilians and salamanders use intercourse for reproduction, most amphibians do not have intercourse. The most likely scenario for evolution is that male and female proto-reptiles had to invent sexual intercourse on land at the same time as each other. The male had to insert his sperm into the female rather than just cast it into water. The coordinated release of sperm and eggs in water is a challenging task as it is, but now this process that has worked successfully for up to a billion years suddenly changed.

How did the male sperm enter the female?

For the reproductive process to continue on land, males would now be required to deliberately implant their sperm in the female reproductive canal at the same time that females were producing eggs. The evolutionary assumption is, at some earlier time, cells have mutated to form an external penis that would perform this function. As this appendage first formed, this new invention was initially without any purpose. It provided no additional benefit for surviving in any particular environment and its development over successive generations continued despite its inability to do anything until fully operational. A system making the process pleasurable would also need to be self-created.

---

30. There are rare exceptions, for example, lizards can reproduce asexually, and new females are also capable of reproducing females. The same applies to rotifers.

Similarly, for females, a vagina, uterus and fallopian tubes (connecting the ovaries to the uterus) would have to self-create at the same time. A uterus is a highly specialized organ. Before it appeared it had to be structured and fully functioning so that it could be a place where sperm could move and meet with an ovum to fertilize it.

1. For externally delivered hard shelled eggs, once fertilized they develop rapidly often being expelled in a day or so. The embryo within develops an umbilical cord that attaches itself to a surrounding sac containing all the fluids necessary for its growth. This required pre-knowledge of the components and their quantities that the embryo would need or it was doomed to failure.

2. For live births where the embryo grows internally, the mother had to have suitable cell walls that allowed the fertilized ovum to attach itself. These walls had to enable the zygote, embryo and fetus to obtain oxygen and nutrients while excreting gases and other fluid wastes without harming it or the mother—requiring an umbilical cord to self-create. The uterus also had to invent a system to deliver antibodies to protect the embryo from diseases. The uterus, in conjunction with the rest of the female body, had to determine when the fetus had reached a certain point of growth that then required expulsion from the mother. It also needed to develop a program for recognizing this, starting labor and it had to invent the muscles to enact the birthing process. A birth canal (vagina) was necessary as was a structure that allowed the release of the umbilical cord.

There are several theories about the self-creation of these female reproductive body parts. It is commonly suggested that the expelling systems for urine and feces are intermingled in the same area of the body as the reproductive parts because they are all basically expelling functions. Such thinking somewhat undermines the fact that a developing fetus is not a waste product. Nonetheless, we are presented with historical examples of previous creatures with combined systems releasing in the same body area. For instance, tetrapods (four legged animals) are claimed to have evolved into today's mammalian reproductive organs.

The concept of intercourse requires the willingness of an animal to reproduce because the process, through necessity, has to be pleasurable in order for it to occur. Otherwise they wouldn't do it. Single and nucleated cells reproduce by splitting in two. We can reasonably assume this is functional, not pleasurable, otherwise our bodies would be in a constant state of delirious pleasure as cells reproduce in such vast numbers every moment[31]. Evolutionary Theory supposes that once a full living body had self-created and two genders had appeared from nowhere, there was a change in the process of reproduction that required both desire and pleasure to be invented. If it didn't, nucleated-cell life would have failed to continue. Evolutionary Theory assumes that sexual pleasure was an accidental mutation, evolving for no reason, and yet it could be argued sexual pleasure is one of the most important influences for life to continue. Without it there would be no reason for reproduction in creatures other than humans, who are the only ones to understand the process. But at what point this understanding may have occurred in an evolutionary world from apes to humans is unidentified by science.

It is just as well this level of pleasure wasn't a mutation that happened when an animal opened its mouth or turned its head to the right, or for humans for that matter. Under evolution, these options are just as valid as intercourse because it is random, accidental and unintentional. Here are some questions that Evolutionary Theory also fails to answer:

- How did the first male proto-reptile know how to insert his penis into a female's vagina, where previously the penis was used only to release sperm into water?

- How did he know what it was, where it was or why he should do such a thing?

- What was the reason the first couple in history had sexual intercourse when they had no instruction or concept to initiate the act?

- Considering the size of the earth, how did the first reptilian couple happen to be in the same place at the same time with the same new process?

- As amphibians can lay up to several thousand eggs at a time, the female reptile that invented sexual intercourse and hard-shelled eggs also had to be responsible for dramatically reducing egg production, all at the same time.

Figure 146: The First Ever Act Of Land-Based Sexual Intercourse – A Coincidental Accident

---

31. For example in humans, one million red blood cells are released from the bone marrow into the bloodstream every second.

270

Summary

Summarizing the concepts of reptilian evolution from amphibians so far, a set of unique events had to occur:

1. A fully functioning penis self-creates by accidental mutation.

2. A vagina and uterus also self-created without any intended purpose for the same reproductive process that was not planned.

3. Both of these are coincidentally the perfect size, shape and location for each other.

4. Females retain eggs internally for fertilization rather than cast them out from their bodies.

5. A male accidentally inserts his penis into a female in the right spot and ejaculates sperm, allowed by the female.

6. Sexual intercourse, the most important process for continuing any land-based species, becomes pleasurable by physical contact, by accident, after a billion years of aquatic reproduction.

In an overwhelming evolutionary coincidence, land-based sexual intercourse just happened to invent itself in the same timeframe that eggs started making hard shells. We know this because hard-shelled, egg-laying reptiles that reproduce sexually were among the first land-based animals to exist in the fossil record and, chronologically, they followed amphibians that have soft eggs that were fertilized externally in water.

At the present time, the majority of amphibians still release soft eggs into water, having successfully reproduced this way for almost 400 million years since the Devonian period. For tens of millions of generations there has not been any known change to their reproductive processes. Evolutionary Theory claims these creatures are ancestors to reptiles, birds, quadrupeds, mammals, and humans. They are meant to be mutating with every generation, every day. Yet the observations today show:

- Amphibians still reproduce the same way even though this process has changed often in successive mutations on land.

- Amphibians do not mutate.

- Amphibians still exist even though they are a temporary evolutionary step between life in water and on land.

Chapter 19 Unraveled

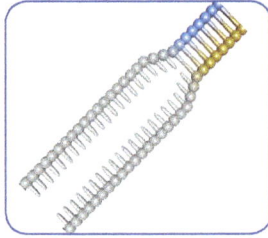

- Amphibians mostly reproduce by sperm and egg fertilizing in water.

- Reptiles predominantly reproduce by internally fertilized eggs.

- The Theory of Evolution assumes that the amphibian system for reproduction changed to the reptilian system.

- For the reproductive process to change, new sexual organs and processes had to independently and coincidentally develop in male and female at the same time.

- Then, land-based, sexual intercourse had to occur, by accident and, at the same time, the female had to grow a new shelled egg that was perfect in the first few attempts.

20

REPTILES CHANGE INTO BIRDS

At this stage in history around 65 million years ago, the oceans, rivers and streams abound with fish. Plants are in the oceans and spreading widely on land. Insects are everywhere. Amphibians live around water edges and more reptile species are appearing. The next step in the fossil record brings us to the birds.

The main proposal of Evolutionary Theory is that birds mutated from reptiles, possibly a small dinosaur, as there is no other natural option to explain their existence. At 270 million years ago, the first early reptiles appeared. There are some disputed records of mammals shortly thereafter at 260 million years ago and there is evidence of a precursor to mammals, the Therapsids, at 225 million years ago. The next distinct arrival is the birds, which are first identified at 140 million years ago, although this timing is in dispute because fossil datings are not always based on actual finds or have agreed, or standard, measurement processes. Timings, such as these, are often tenuous as they are proposed from a complex variety of various estimates of things like theories about the reconstruction of bone fragments in fossil finds, or molecular clock rates, or probabilities based on mathematically calculated sea levels, or soil deposit rates, or variables observed on different continents. Birds did appear in abundance about at 65 million years ago, around the same time as mammals with this timing commonly accepted as more reliable.

Reptiles are claimed to be the ancestors to all later life forms, including birds, mammals and quadrupeds. This implies reptiles changed into three entirely new forms whilst continuing as unchanged species and never disappearing from the fossil records. The evolutionary leap to birds is enormous in biological terms as there are a vast amount of fundamental changes needed. The only basic similarities between the reptiles and birds are the actual body organs and operational processes, such as blood circulation and the abilities to eat, breathe, see, hear, move, and excrete.

To analyze how this transition could possibly occur we will review the most common theory for the appearance of birds. But before we do, we will first look at some of their more interesting characteristics.

Birds

There is a certain majesty about a bird in flight. Man has been fascinated by the ability of birds to fly, hover, swoop on prey, and to see great distances. Yet, it is not flight that distinguishes birds from other creatures, as insects, bats and possibly dinosaurs have all flown. What separates birds from all other animals is their complex feathers.

Feathers

Made of a protein called keratin, feathers are extremely lightweight, yet they are also very strong and waterproof. Feathers allow lift, protect the bird from water, and maintain body temperature through the insulating process of air trapping. Which is why oil spills are disastrous to birds—oil is heavy and prevents the bird's ability to take flight, as well as inhibiting air trapping in their

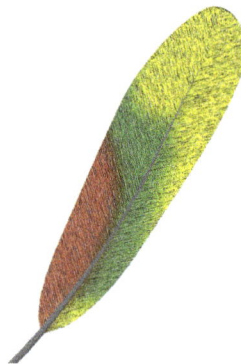

Figure 147: Parrot Feather

feathers; they can die of hypothermia, literally freezing to death.

The range, size, number, and positioning of feathers required to enable each type of bird to fly is extremely important. Robin feathers on an eagle, for instance, would not work because each type of bird needs a unique range of feathers suited to its body shape and size to fly. Pigments mostly determine a feather's colors. However, some greens and violets are based on the reflection of light, making the feather appear colored when it is not. For example, a jaybird's feathers appear as bright blues due to their inner structure, not pigments, and are dark when examined in the laboratory.

A central quill running up the feather's shaft or rachis has protruding linear clusters of barbs that secure the feather's structure. These have miniature interlocking hooklets that give the feather both stiffness and flexibility. In one feather alone, there can be up to three hundred million tiny hooklets. This is where the majority of the feather's pigments are located. The complex features of a feather are shown in Figures 148: Feather Layout and 149: Feather Hooklets. Evolutionary Theory proposes that each of these hundreds of millions of hooklets, on each feather, independently formed as mutations in the sex cells of a 'proto reptile/bird' embryo. Although these mutations were by pure chance, mathematically speaking, billions of hooklets have self-created on each type of bird[32].

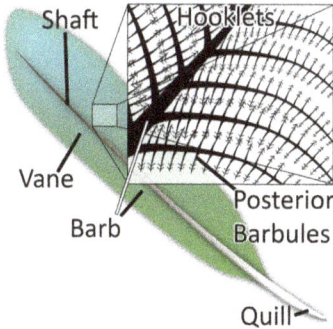

Figure 148: Feather Layout

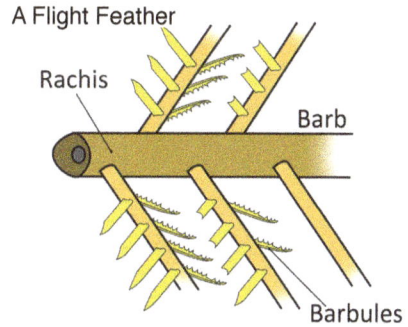

Figure 149: Feather Hooklets

Feathers get their water repulsion from oils secreted from preen glands located at the base of the bird's tail where it is able to access them with its beak. These oils are constantly applied to its plumage as a deliberate and necessary function. The oils consist of various waxes, fatty acids and wax-alcohols, which can vary from bird to bird. Evolutionary Theory, however, fails to explain:

• Why a bird would have a preen gland.

• Why it would fortuitously be at the base of the tail.

• Why this gland would store oil.

• Where the oil came from.

• How the preen oil was perfectly suited as a necessary waterproofing agent.

• How the bird would know that its application was vital to its long term survival.

Not all feathers on a bird are the same. The wings and backs have contour feathers, under these are down feathers. The wings and tail have

---

32. This level of intricacy in hooklets was also observed by George De Mestral in 1948 when he examined cockleburs under a microscope after they had stuck to his pants while out walking. These burs had tiny hooks on their ends which had embedded themselves into the woven loops of his pants. Realizing this was a brilliant way to get things to stick together, but would also allow them to be separated again, George copied the design and marketed his product across the world under the name of Velcro.

flight feathers. All of these feathers are different, have their own design, size, shape, position, function, and are perfect for the specific task they perform. The wing feathers even expand during flight to give a bigger surface area and more lift, as shown in Figure 150: Wing Feathers. Different types of feathers are shown in Figure 151: Flight, Contour, and Down Feathers.

Figure 150: Wing Feathers

Figure 151: Flight, Contour And Down Feathers

Body Structure

In the flesh and bones underneath their covering of feathers, birds generally don't appear to carry excess weight. Flying birds have porous bones resulting in lighter weight compared with other animals. Their bones also contain a series of internal supports similar to struts or braces that improve strength, like flying buttresses on baroque church buildings. Birds that spend more time gliding have more porous bones whereas birds like ostriches, which

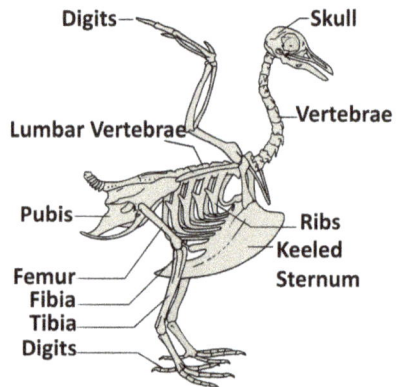

Figure 152: Typical Bird Skeleton

essentially don't fly, don't have porous bones which raises the question about whether dinosaurs with heavy bones actually flew. Birds have greater neck flexibility due to a larger number of neck vertebrae than most other animals. They are unique in having fused collarbones and a keel-shaped sternum where the flight muscles are attached. Figure 152: Typical Bird Skeleton displays the minimalist bone structure of a typical bird skeleton.

Beaks

There are over eight and a half thousand bird species worldwide with most having a preferred habitat. Some birds will migrate large distances to seek their preferred foods, while many live in a wide range of habitats within a small distance. Birds operate with an array of different abilities using their beaks as tools for building, feeding, hunting, protection, and to assist with some movement. Rather than jaws with teeth, birds have beaks which are lighter in weight. A few species have notches on their mandibles that look like teeth but these are not used in food mastication. Each beak of each species is specifically suited to that bird's skills. The range of birds' beaks is vast. Some examples can be seen in the images below.

Figure 153: An Eagle's Beak

Vultures and eagles have sharp strong beaks that are used for tearing flesh apart.

Figure 154: A Pelican's Beak

The pelican can scoop and sift water.

Figure 155: A Parrot's Beak

Parrots have hooked beaks that are used to dig into foods such as berries, vegetables or fruit.

Figure 156: A Heron's Beak

Herons have a long thin beak that is used to spear fish and insects.

Figure 157: A Sparrow's Beak

Sparrows have tough strong beaks used for cracking hard shells and grabbing insects and berries.

Figure 158: A Robin's Beak

Robins have beaks like tweezers that are used for pulling worms from the ground.

Figure 159: A Hummingbird's Beak

Hummingbirds have long thin beaks that allow them to suck nectar from flowers.
Their wings flap at enormous speeds enabling them to hover in one spot while sucking.

Figure 160: A Duck's Beak

Ducks beaks are designed to strain and sift mud and water so they can collect plants, seeds and other small foods.

Other birds use their beaks for a wide range of activities, including probing, raking, stabbing, and clinging to trees.

Other Bird Features

Birds that feed in shallow waters have long skinny legs; those in deeper waters have webbed feet. Birds are very territorial and will defend their territory while, in the main, respecting others' territories. Their breeding habits, courtship rituals, grooming, feeding and movements are different in each species.

Birds show no interest in any type other than their own. There is no natural cross-breeding. Courtship rituals vary from dancing and preening to fencing with bills. Birds generally prepare nests for their eggs. Some build elaborate constructions requiring the use of special materials, such as the Mistletoe Bird that uses down feathers and spider webs. Birds vary in size, ranging from the Bee Hummingbird, which is about two to three centimeters, up to the Ostrich at two meters, approximately one hundred times taller. Some types, such as the Owl, have many different varieties, each with a specific skill. Oddly, Owls' heads are a different shape from most other birds, perhaps to assist their incredible sight and hearing skills

In summary, birds are beautiful creatures with an enormous array of colors, shapes, sizes, and varieties. They fit perfectly in specific environments with unique structures, sizes, feet, wings, eyes, heads, legs, and beaks. Almost every component of a bird is lightweight with great strength. Yet science does not corroborate the Evolutionary Theory that birds developed from reptiles by evolutionary processes. There no evidence today or in the fossil records across millions of years of failed mutations or graduated changes. Furthermore, with such observable diversity, science would expect plentiful evidence of mutating birds around the planet, now and in the past. However, no such evidence exists.

Evolution's Main Theory Regarding Birds

As far as flying creatures are concerned, fossil records show insects appearing between 400 and 300 million years ago. These were followed by dinosaurs (the *pterosaurs*) at 200 million years, then birds at 65 million years, about the (disputed) time of proto-ape/man (for details refer to Chapter 23). Insects are therefore the first flyers on earth but are not claimed to be related to birds or dinosaurs. Flying dinosaurs are believed to have had 85 flying species, one of which had a wingspan of twelve meters and this type is thought to have existed for 135 million years.

There is one famous creature that is claimed to be a step or link between reptiles and birds. It had feathers and a reptile tail and is called *Archaeopteryx Lithographica* (see Figure 161: *Archaeopteryx Lithographica*). There are eleven known fossils discovered so far in sizes up to five hundred millimeters long, all found in Germany and dated around 150 million years ago. There are variations in these fossils that still have experts debating their classification as a single species. They appear to be much more dinosaur-like with common features such as:

• The long, bony tail.

• Three clawed fingers.

• A jaw with teeth rather than a bird's beak.

• Feathers that appear more like fur rather than any flight feathers.

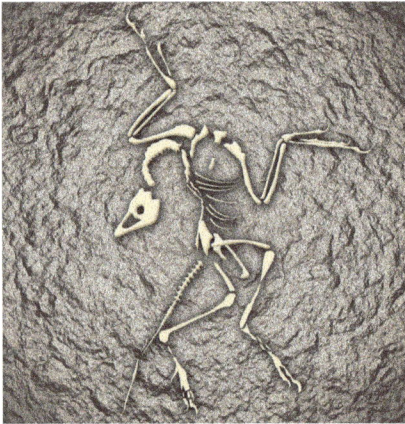

Figure 161: *Archaeopteryx Lithographica*

One small patch of feathers was found on the back of the Berlin sample, but these lacked the stiffness required for flight. With a flattish sternum less likely to support strong flight muscles compared with flying birds, and few, if any, proper flight feathers, this creature would probably not have been able to elevate for more than very small distances.

Thomas Huxley, an English biologist in the 1800's, proposed that small carnivorous dinosaurs evolved into birds. This theory is now fairly widely accepted by evolutionists. Thus, the most popular theory is that a slow, heavy, four legged, cold-blooded, scaled reptile/dinosaur with teeth accidentally altered almost every single aspect of its structure to change into a quick, light, two legged, alert, warm-blooded, winged flying bird with a beak.

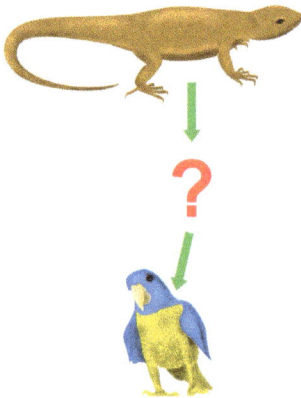

Figure 162: A Reptile Becomes A Bird

The most common version of this theory is that reptiles were struggling to survive and began climbing trees in search of safety and food. They then developed flight as a way to avoid predators and increase their source of available food. However, unlike most previous theories that require developments or improvements to occur in an accidental manner, this proposal now suggests the evolution of flight happened in response to a deliberate need, which contradicts the previous evolutionary rules of randomness. A living entity cannot will itself to form new body parts or structures, particularly something such as a feathered wing, which did not exist previously.

The evolutionary assumption is that some type of reptile adapted to trees and started to develop feathers by mutation. It is assumed to be a tree climber with possibly a strong, lightweight bone structure needed for agile movement. Seemingly, it would have had strong muscles in its arms and legs, but there are no specific fossil records of it, making it a theoretical concept. For flight, it needs wings, feathers, new muscles, and to be extremely lightweight. As flying requires much higher levels of energy than reptiles generally use, this new creature, the 'proto-bird', also needs to be warm-blooded rather than cold-blooded, a complete change from all previous creatures.

Warm Blood

The differences between warm- and cold-blooded animals are significant. Essentially, cold-blooded animals have a low rate of metabolism with their internal temperature dependent on the surrounding environment. For example, reptiles are more active during the day when it is warmer but are fairly inactive at night when it is cold. The low metabolism in cold-blooded animals generates very little internal heat, whereas warm-blooded animals have higher metabolism, higher blood temperatures and more evenly controlled heat levels.

Metabolism refers to chemical reactions in cells, which separate glucose into water and carbon dioxide. This process results in about sixty percent heat creation with the balance becoming energy, Adenosine Triphosphate (ATP), which is the fuel source that powers the body. Excess energy is stored in body tissues as fat. Cold-blooded animals generally lose their internal heat to the external environment. Warm-blooded animals such as mammals and birds have external coverings like feathers, fur or layers of fat (such as blubber in whales) that insulate against external heat transfer. They can also shiver when cold, resulting in rapid muscle contractions that use ATP and increase metabolism, thus making more heat. When too hot, mammals release heat by the cooling effects of panting, or, as in the case of humans and horses, by sweating. The nervous system is mostly responsible for controlling the rate of metabolism.

Evolutionary Theory, however, does not explain the processes which caused these differences in metabolism, namely the warmer blood temperatures and insulating mediums that appeared with the arrival of birds. Science cannot explain the changes in the chemical, biological and physical structures from a reptile to a bird.

The First Feathers

For a reptile to begin evolving into a bird, at least one feather had to start to form. The 'repto-bird' would not have mutated a perfect feather instantaneously, the evolutionary process for which would require many thousands, if not millions, of generations of new mutations followed by failure. A single feather, with hundreds of millions of components formed over thousands of generations, would be useless by itself because flight requires a full complement of feathers. All of the 'one feathered' creatures would never be able to leave the ground. With just a single feather, this line of creatures is not better adapted to its environment, because the first feather by itself, is not a progression, nor would be five, ten, fifteen, or even twenty feathers. It is not until the bird has all its feathers in the perfect sizes, shapes, number and locations that feathers are a progressive adaptation. If five, ten, fifteen, or even twenty feathers provided no positive benefit to a line of creatures, there is no scientific reason why that creature would continue to progress over and above other non-feathered creatures over millions of years. These partially feathered birds would not have flight until all types of feathers had developed in exactly the right numbers in exactly the right areas of the bird.

A further curiosity is how a feather recognized that it was fully constructed, or that its shape had to be so specific and different from other feathers on the same body. Despite forming by millions of mutations over millions of generations, feathers stopped mutating into other shapes, sizes or forms once they were perfect for the body they were attached to. They are an end product and there is no observable scientific evidence to suggest they are mutating further. Through evolutionary process, the new accidental body attachment of feathers would most likely have been redundant very early after its arrival, as it served no purpose for possibly millions of years until a full bird developed.

Even if a creature was born with front and/or rear legs covered in feathers, it could not know that it would be able to fly and only an accidental fall might allow it to glide. It would not know how many feathers were needed or how a 'wing' should be shaped to get uplift. For flight to become natural, even with the appropriate feathers in place, birds would need to randomly mutate new muscles that could operate the wings and tail via new instructions from the brain.

The main theory for this part of the evolutionary story tells us that many adaptations were occurring in tree dwelling reptiles. One family line was to develop many crude feathers over many generations. Despite these being useless during these generations, possibly only providing extra warmth, they would surely have been severely and regularly damaged as these creatures scurried among the trees. Nonetheless, at some stage, one of these feathered tree dwellers may have fallen from the trees. Having landed safely without any serious injury, it somehow became the most adaptable mutation of its type. This first feathered creature's family, generation after generation, then continued to accidentally develop changes to their bodies as well as specialized feathers for different areas until they became the perfect design for their environment. Over millions of generations the following, totally random, events had to occur to one family line of proto-birds:

- They mutated the perfect shape, strength and number of feathers, perfectly placed on two mirror image wings of perfect shape for flight.

- Their bones mutated to become porous and stronger.

- Their jaws and mouths mutated to a reshaped format of a beak to access different food bases during which time teeth disappeared.

- Their eyes mutated to a smaller size and their eyesight accidentally mutated to improve dramatically.

- They mutated a process to secret preen oil that they instinctively knew to spread on their wings.

- Their muscles mutated to much smaller sizes and mutated actions that would flap new wings.

- At some point, during all of these changes, the new family of birds also did something totally amazing—they mutated changes to their physical and chemical body functions and became warm-blooded.

In other words, they performed millions of mutations, in direct sequence, from a cold-blooded, land-based reptile with scales to a warm-blooded flying animal with feathers. Evolution assumes these changes occurred despite only a single claimed transitional species in the fossil record and no evidence of any changes occurring to reptiles living today.

Bird Species

Notwithstanding the emergence of the first proto-bird, it then reached a point whereby it began to mutate and change into other bird species, implying the knowledge that it somehow knew it was a 'completed' end product. Evolutionary Theory then suggests this one creature commenced the process of mutating into every type of bird on earth, all the while maintaining each different type as a separate and unique species that will not normally associate with other bird types/species. Yet over millions of years, despite the available evidence of different, individual bird species, none of these show any transformation from any of the eight and a half thousand bird species to any others.

The picture that is painted by Evolutionary Theory is developmental carnage: billions of repto-birds mutating over millions of years, forming all sorts of useless body parts and developments that then became redundant and died out. However, there is no fossil evidence to support such a picture:

• There is no evidence of redundant mutations of repto-birds.

• There are no fossils of the countless, stepped changes of feathered tree dwellers.

• There is no evidence today of non-flying tree dwellers—chipmunks, insects, monkeys, apes, possums, various reptiles, including snakes—that have any mutations that are suitable for flight, despite millions of years of evolutionary adaption.

• As both male and female birds are needed to create more bird eggs, there is no evidence that both sexes develop all these changes step-by-step simultaneously over history.

In a world of evolutionary birds, science would expect a vast and diverse range of mixed birds with many variable characteristics. Science does not predict thousands of different independent species, regardless of how any single species might adapt to an environment, yet that's exactly what is observed today. Furthermore, birds should be mutating into millions of strange variants. But there is no indication this is happening at present, nor in the past.

In any large park or forest, we can see many different and independent bird species sharing the same environment, food sources, threats to their safety and survival, and so forth.

Yet, according to the tenets of Evolutionary Theory, we should see either:

- Just one or a few types of bird that are best suited to that environment, or
- An intricate mixture of non-specific birds with many common features but not one necessarily identifiable species, as every bird would be a unique mix of a complicated ancestral history of differing adaptations to that environment.

What is scientifically observable is very different. There are very specific, independent species (groups) of birds living separately from all other species in the same environment. Each of these groups has their own uniqueness, purpose, function, and beauty. They do not mutate to any other types of bird species and they do not reproduce or mix with other bird species. There is simple, clear, scientific evidence of independent species, each having a different and unique functionality to that group, yet sharing the same environment.

In such a type of park or forest, there are many other creatures and plants sharing the same environment. There are thousands of insects, animals and plants, all completely different, all with their own separate functions—foxes, rabbits, worms, beetles, ferns, bushes, trees, ants, flies, mosquitoes, and so forth. There is no observable evidence to suggest that these species intermix outside their own species, or that they mutate into any other species.

The scientific observations are such:

- Each species has their own individual functionality and performs unique, specific roles.
- Each species shares a common environment with other life forms but is different from all of them.
- Each species is not just one entity that is specifically matched for the environment; there are many differing types that successfully co-exist.
- Rather than fighting for survival of the fittest, all species are mostly inter-reliant on each other.
- Each species only reproduces their own kind.
- None of the species mutates into anything else.

Each of these life forms is separate and different from all the others and has a different but deliberate function within that environment. There is no single best adaptation dominating all others[33].

---

33. In the natural world, excluding humans.

In summary, birds are extremely sophisticated flying animals with specific functionality for their purpose. Their weight is specifically low in all areas of their body, from porous bones with great strength to super lightweight feathers and beaks. Feathers have intricate and complex individual designs, depending on their location throughout their body, specifically to achieve flight. Birds are remarkable creatures without scientific support that they evolved from reptiles.

Although there is no scientific evidence that creatures can, or do now, change from one species or class to another, we will continue the scientific dissection of Evolutionary Theory by looking at the next arrival on earth—mammals.

Chapter 20 Unraveled

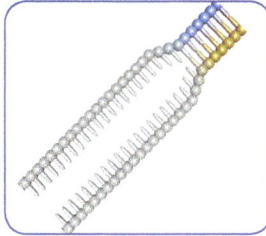

- *Evolutionary Theory does not adequately provide scientific evidence for the notion that reptiles, or anything else for that matter, developed into birds.*

- *Scientific evidence does not support the theory that reptiles changed every aspect of their body, including:*

- *Bone structure and porosity.*

- *Cold- to warm-blooded systems.*

- *Skin covering from individual scales to a single skin organ.*

- *The development of one of the most sophisticated and complicated body components in history—the feather—then adapt the shape and size of this component for a range of varying locations and functions on the body before they collaborated to allow flight.*

- *Science predicts that an evolutionary world would be flooded with partly formed and constantly mutating birds (and all other creatures) rather than individual and separate species.*

- *In the natural environment, there is no single best adapted species dominating all others.*

21

REPTILES CHANGE INTO MAMMALS

The next group of animals that appear in significant numbers is mammals. There are fossil recordings around 260 million years ago, but this timing is highly disputed among scientists. A four-legged mammal called Eozostrodon has been recorded at living 210 million years ago. Various other mammalian species appear over the following 150 million years but none in any significant volume. A large quantity of mammal fossils is recorded at 65 million years ago, which is the same general timeframe that dinosaurs disappeared and birds first appeared.

Evolutionary Theory, as a rule, proposes that reptiles formed into both mammals and birds separately at or around the same time. There don't appear to be any theories that mammals evolved into birds or vice versa, even though this would seem more logical due to both being warm-blooded, as opposed to reptiles which are cold-blooded. In keeping with Evolutionary Theory, reptiles are therefore required to change their metabolic systems from cold- to warm-blooded on two separate and unrelated occasions while continuing to remain cold-blooded themselves.

There are over four and a half thousand species of mammals. Some of the more common types are: humans, apes, cats, dogs, elephants, horses, whales, and dolphins.

The following are some of the main characteristics common to mammals:

• Mammals are warm-blooded, have high body temperatures and generally have an insulating medium of hair or fat resulting in a more even temperature.

• They give birth to live offspring (excepting the duckbill platypus and echidnas of Australia and New Guinea that are egg layers).

- Mammals can participate in more physical activities than reptiles over longer periods but need more food and fuel; they therefore need a more efficient metabolism of oxygen and nutrients.

- They have larger aerobic and circulatory systems and have four-chambered hearts.

- Nasal and food passages in the cranium are separate and allow breathing and eating at the same time.

- Their excreting functions (feces and urine) are separate, whereas reptiles are combined.

- Mammals chew whereas reptiles swallow – mammals' jaws are one piece with a strong fixed joint that provides better leverage. Mammals' jaws are shorter than reptiles with muscle attachments allowing sideways movement. The upper and lower jaws fit together and provide a cutting action.

- Mammals' teeth are complex with four types that are only replaced once in a lifetime.

- The majority of mammals are placental and birth their young at a relatively advanced stage. While in gestation, the young receive direct nutrients and oxygen via a placenta.

- Some mammals, such as marsupials, have pouches for their young[34], which are born very immature and develop in the pouch.

- The parental period after birth requires the mother's mammary glands to provide an exact blend of the correct nutrients for the infant.

- Mammals are generally protective of their young.

Evolutionary Theory assumes several processes in the development of reptiles to mammals:

1. All reptiles came from one original creature that displayed all of the reptilian body functions, being the first full reptile.

2. That first reptilian creature was the ancestor to all following species of reptiles. Creatures as different as snakes (with no legs), crocodiles (having four legs and thick leathery skin) and turtles (with hard shells) all came from one unidentified single ancestor.

---

34. Such as kangaroos, koalas and wombats.

3. A later reptile 'family' evolved into a complete mammal, becoming the first full mammal.

4. That first mammalian creature became the ancestor to all following species of mammals.

It is at this point that much conjecture and difference of opinion exists from evolutionists, particularly how these processes could have occurred. One such concern is the mammalian reproductive processes.

Mammalian Reproduction

There are two types of reproductive processes found in both reptiles and mammals: egg laying and placental. The likelihood of a reptile changing into a mammal more than once from separate, unconnected sources is considered essentially zero by evolution, as previously noted with other creatures. This therefore nullifies the prospect that placental reptiles evolved into placental mammals at the same time that egg laying reptiles evolved into egg laying mammals. The consequence is that either:

1. Placental reptiles mutated into placental mammals that then mutated into egg laying mammals.

2. Egg laying reptiles mutated into egg laying mammals that then mutated into placental mammals.

Evolutionary Theory unfortunately offers no specific direction to follow in this particular instance, so we will overview both options.

Because mammals share common functions, irrespective of which of the two just mentioned processes that might have occurred, the first identifiable mammal would have to develop warm-blooded circulation and metabolism, separate ingestion and excretion systems, single-piece jaws with muscles, and all of the other differences that mammals exhibit from reptiles, before they changed into other types of egg-laying and/ or placental mammals. These series of events are now examined in the following two options.

Option 1 – Placental reptiles changed into placental mammals then into egg laying mammals

Some placental-replicating reptiles that give birth to live young include some skinks, some snakes and chameleons. If a placental reptile evolved into a placental mammal then, to become an egg laying mammal, this creature would have to completely change its reproductive process and re-invent the process of making a hard-shelled egg for the second time in history (as discussed previously in Chapter 19). This chain of events is displayed in Figures 163-167.

1. The original placental reptile appears somewhere around 220 million years ago.

Figure 163: Cold Blooded Placental Reptile e.g. a Lizard

2. An unspecified type of reptile starts an accidental process of gradually developing into a placental mammal by mutating to new metabolic systems and blood temperature, split excretion systems, bigger lungs, split breathing and split eating systems.

3. As time passes and multiple mutations occur, its appearance gradually changes from a reptilian body into a mammalian body.

Figure 165: A Repto-Mammal[35]

4. Eventually after 100 million years, and millions of mutational changes in multiple directions, one placental proto-mammal changes into an identifiable mammal. Around this time, the millions of other creatures involved in the same mutational group from reptiles that are partially developing other forms, disappear from earth forever.

Figure 164: Developing Placental Mammal

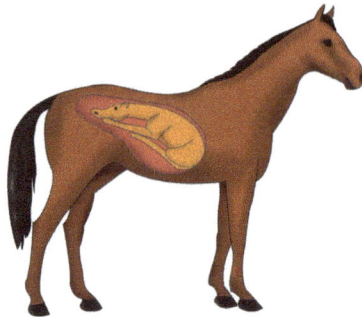

Figure 166: Placental Mammal: e.g. a Horse

35. For example, a lizard mutating into a horse.

5. By a process that is considered biologically unlikely, and impossible as a repeated evolutionary process, a placental creature changes its reproductive process to egg laying, limits these types to a couple of differing species in restricted environments, and then ceases any further mammalian evolution. All traces of these transitional changes disappear from earth once the new reproductive process is complete.

Figure 167: Egg Laying Mammal e.g. a Platypus

Option 2 – Egg laying reptiles changed into egg laying mammals then into placental mammals

Some egg laying reptiles include turtles, crocodilians, some lizards and some snakes. Under this scenario, the newly evolved mammal is most likely to be a platypus or an echidna, as these are the only egg laying mammals. This creature then became the forefather to all mammals. Evolution therefore requires a platypus to evolve into animals as different as elephants, humans, cats, and whales. This chain of events is displayed in Figures 168-172.

1. The original egg laying reptile appears somewhere around 220 million years ago.

Figure 168: Cold Blooded Egg Laying Reptile e.g. *Trachylepis ivensi* A Lizard

Figure 169: Developing Egg Laying Mammal

2. An unspecified type of reptile starts an accidental process of gradually developing into an egg laying mammal by mutating to new metabolic systems and blood temperature, split excretion systems, bigger lungs, split breathing and split eating systems. (displayed in Figure 169).

3. As time passes and multiple mutations occur, its appearance gradually changes from a reptilian body into a mammalian body.

Figure 170: A Repto-Mammal[36]

---

36. For example, a lizard mutating into a platypus.

4. Eventually after 100 million years, and millions of mutational changes going in multiple directions, one egg laying proto-mammal changes into an identifiable mammal. Around this time, the millions of other creatures involved in the same mutational group from reptiles that are partially developing other forms, disappear from earth forever.

5. By a process that is considered biologically unlikely and impossible as a repeated evolutionary process, an egg laying creature changes its reproductive process to placental. All traces of the above transitional changes disappear from earth once the new reproductive process is complete.

Figure 171: Egg Laying Mammal e.g. a Platypus

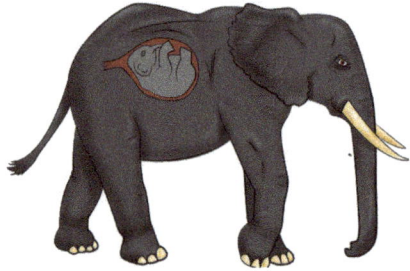

Figure 172: Placental Mammal[37]

Observation about the above exercise

What we learn from trying to track the necessary steps in reptilian to mammalian evolution is that there are no clear or obvious directions to follow. The result is a multitude of options, all of which fail the test of reasoning and proof. When generalized theories are examined in detail, the processes required to occur are exposed as impossibilities, from a biological perspective, as well as contradicting the founding principles of Evolutionary Theory. None of the events is known to be possible and the absence of evidence is reasonable confirmation that the events never actually occurred. The obvious failing here is that no one has any realistic idea how any reptile could change into anything else, and so evolutionists avoid giving any detailed explanation. When the logical steps are followed, they simply highlight the lack of evidence, logic or known process for such evolutionary claims to actually occur.

---

37. For example, an elephant.

Marsupial Reproduction

Regardless of which process occurred, at some stage marsupial mammals had to appear. Marsupials such as kangaroos, koala bears and possums are categorized as being an infraclass of mammals, a tier below a subclass (secondary tier) that undergo a third, different reproductive process. Most marsupials are found in the southern hemisphere, seventy percent in Australia or neighboring islands. They carry their young in a pouch on the mother where she feeds them from milk teats. The young leave the womb after 2-5 weeks when they are very small, immature, blind, and lack insulating hair. They climb into the pouch after birth and attach themselves to the teats. They spend about 6-8 months in the pouch while developing.

In evolutionary terms, this process would appear to have resulted from a placental reproductive process gone wrong. Even so, the mother's pouch and teats had to exist and be fully functional before the first young could leave her uterus, otherwise they would die from exposure or lack of nutrition. This implies that the anatomy and operation of a feminine pouch and teat(s) needed to have been fully developed before they were actually required to nurture and protect a young fetus. Other than the luck of accidental mutations, Evolutionary Theory does not provide an explanation why pouches and teats developed prior to their requirement to exist, rather than after, as would be expected by evolutionary processes.

Aquatic Mammals

Despite the issues concerning reproduction and lack of fossil evidence, we come to the point where mammals exist in the fossil record with not two but three different reproductive processes. Rather than re-examining various ways that different species and reproductive processes might have happened by mutation, we will highlight one substantial obstacle facing Evolutionary Theory. As it turns out, mammals don't just live on land. Whales, dolphins/porpoises and sea cows are air-breathing mammals that live and reproduce in the oceans. Whales, porpoises and dolphins are identified as appearing around the same time about 50 million years ago. Sea cows appeared about 25 million years ago. In contrast, polar bears, which are mammals that swim and eat from the oceans but reproduce on land, are dated at just 130,000 years ago.

This brings us to a point where an explanation is required how a fish could completely change every aspect of its body and functionality to eventually live on land as a mammal and then go back to live in water again, where it came from originally, after hundreds of millions of years.

Effectively, the evolutionary story for ocean-based mammals follows as such: a water-breathing fish with gills and fins left the water, mutated into an amphibian, ate land-based food (insects), breathed air, mutated into a four-legged reptile body suitable for land, changed its diet again, mutated into a four-legged warm blooded mammal, mutated back into a legless fish shape, returned to water but kept breathing air, ate seafood or sea plants while retaining mammals' reproductive, breathing and functioning methods, mutated into a few other species of aquatic mammals, then stopped mutating over 25 million years ago. This series of changes is displayed in Figure 173: Evolution's Proposal About Whales.

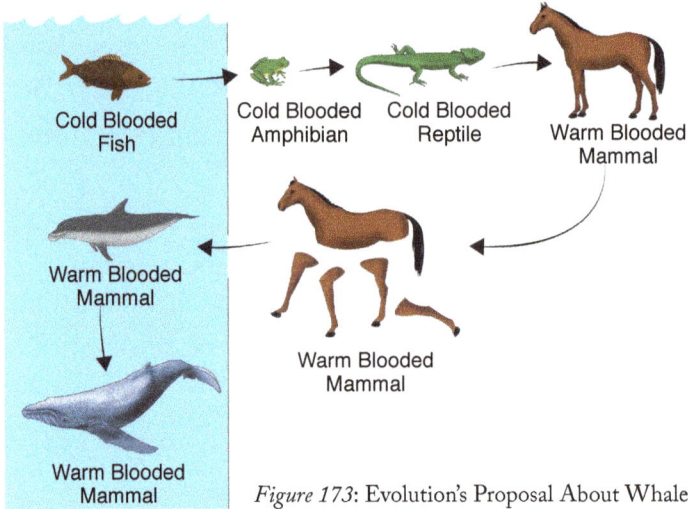

Figure 173: Evolution's Proposal About Whales

For the mammals that live both in water and on land (sea lions/walruses), the evolutionary story is similar to above except they were able to live in both environments by not losing their walking aids, as displayed in Figure 174: Evolution's Proposal About Sea Lions And Walruses.

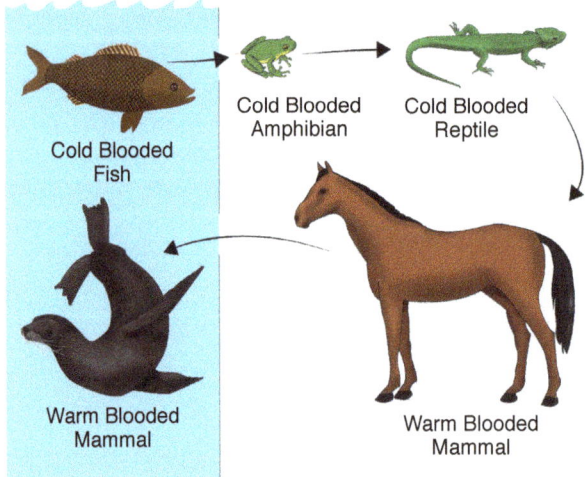

Figure 174: Evolution's Proposal About Sea Lions And Walruses

However, the story for polar bears is different. Evolutionary Theory assumes an established mammal changed into a complete brown bear that mutated into a polar bear, because only this adaptation could live in a period of cold freeze that occurred on earth tens of millions of years ago, as displayed in figure 175: Evolution's Proposal about Bears.

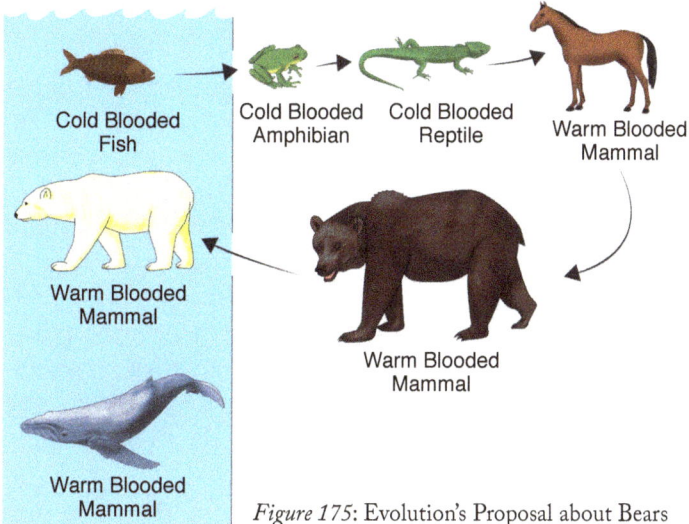

Cold Blooded Fish

Cold Blooded Amphibian

Cold Blooded Reptile

Warm Blooded Mammal

Warm Blooded Mammal

Warm Blooded Mammal

Warm Blooded Mammal

Figure 175: Evolution's Proposal about Bears

Evolution now has three different theories to explain how three different groups of land-based mammals came to live in the oceans (wholly or partially). Yet the creatures in these groups vary in many ways, including their ability to hold their breath under water. Some examples of this are otters: 5 minutes, dolphins: 10 minutes, polar bears: 30 minutes, sperm whales: 90 minutes[38].

An Alternate Theory

There are some recent evolutionary suggestions that the 'ocean-to-land-to-ocean' process was simpler than suggested above. Fish that tried to adapt to land, it is thought, changed into mammalian whales or dolphins while still on the shoreline, prior to amphibians. If this were so, the scenario would go something like this:

• Fish made it to shore, mutated to breathe air, but kept going back to the ocean daily to eat sea food, then back to shore to change to amphibians while a separate group changed to be warm blooded and have mammals' reproductive and functioning methods, without

---

38. Whales are able to remain submerged for so long because their lungs extract ninety percent of the oxygen in air whereas human lungs, for example, can only extract fifteen percent.

any connection to land-based mammals who would appear later, then they went back to the ocean to live where they mutated into the other ocean-based mammals.

However, this means that mammals had to develop identical functioning processes from two separate lines, which is one of evolution's own impossibilities. Furthermore, the proto-whales or dolphins had to hide any evidence of their existence for 310 million years – from the first amphibian appearance at 360 million years ago up to ocean mammals' appearance at 50 million years ago. If creatures as small as amphibians are found in the fossil record from 360 million years ago, there is no valid reason why these larger mammals, or their claimed transitional steps, are not also recorded as they were assumed to be living on the same shorelines across the same timeframes. This thinking also excludes any explanation for land- and ocean-based mammals (such as walruses) possessing walking limbs/aids as a separate development from amphibians. In other words, walking is said to have evolved in two different and unconnected groups (amphibians and land- and ocean-based mammals). It is common for different scientists to propose contradictory evolutionary theories such as these because there is no evidence on which to base any theory.

There is a significant difference in the functionality of ocean living mammals compared with those that live both in oceans and on land, assuming both came from one original ancestor. Walruses, sea lions and polar bears survive in both environments due to their physical construction. However, aquatic mammals' bodies (whales/dolphins) will either collapse under the pressure of gravity or drown when their blowholes are covered in water.

Aquatic mammals have had 50 million years to change their breathing processes back to gills, which they originally used according to evolution. But this has not happened, effectively eliminating the theoretical claims of mammalian return to the oceans, especially considering the evolutionary assumption that all creatures are undergoing constant, progressive major changes to every component of their bodies to adapt to differing environments. The concept of completely reshaping their bodies, re-growing fins and tails, changing their skin covering to live in water, while still holding their breath (rather than adapting to the more efficient system of gills again), is a regressive process in part and does not conform to evolution's own principles of progressive adaption.

Quadrupeds

Within the species of mammals, there are three basic types of four-legged animals, known as quadrupeds:

1. Herbivores (vegetarians)
2. Carnivores (meat eaters)
3. Insectivores (insect eaters)

Each of these types has a different diet which, naturally, requires different chewing and digestive systems.

Herbivores

Herbivores have wider teeth with denser enamel that suit chewing vegetable material, such as leaves, grass and tree bark. Among these are two different types of hooved animals: those with odd and even numbers of toes. For example, horses and rabbits have odd-numbered toes, while cows, goats, camels, and sheep have even-numbered toes (see the discussion below, *Hooved Animals*).

There is a difference in the manner these two groups digest grass, which has a high level of cellulose.

- The odd-number toed group (non-ruminants) pass grass through the large intestine, then excrete it anally with some re-ingesting it as feces, where it then passes through the small intestine.

- The even-numbered toed group (ruminants) have four stomachs. Anatomically, it is more correctly described as one stomach with four compartments. One compartment, the rumen, takes the food first and partially processes it, passes it to the rectilium for further digestion, then returns this partially processed food (called the cud) to the mouth for reprocessing. The cud is then swallowed and passes into the third (omasum) and fourth (abomasum) parts of the stomach where digestion is completed.

The non-ruminants have very long intestines to cope with the difficulty of processing cellulose feed. Some of these re-ingest the excreted waste to extract more nutrients whereas the ruminants use their four compartments for processing.

In evolutionary terms, it is nigh impossible for both of these groups to form hooves independently of each other. They therefore would have had the same evolutionary ancestor who developed two different mutants with either three or four toes on each foot. Most likely, the hooves formed before the groups split, then one group either dropped off or grew a toe on each foot and the hooves sizes adapted. Consequently, cows, goats, camels, and sheep would by necessity have originated from one common ancestor, while horses and rabbits came from another common ancestor.

Whilst this theory provides an explanatory link for these two groups, there is no supporting scientific evidence that this is true.

Carnivores

A carnivore's main diet is animal flesh. Ocean fish are considered to be the first carnivores. Land-based carnivores are believed to originate from insectivores (such as moles and hedgehogs) and are known as creodonts. They are separated into two groups: the felids (cats, including the big cats, such as lions and tigers) and canids (dogs). There are differing timelines proposed for the first appearance of cats and dogs. Some scientists believe the first feline appearance was only 9,500 years ago and canines anywhere from 14,000 to 31,700 years ago. Other records claim their ancestors were carnivorous mammals[39] appearing around 65 million years ago and disappearing within about another 10 million years. Their disappearance so long ago does not account for their relatively recently reappearance.

There are about two hundred and thirty breeds of dogs and thirty-nine breeds of cats, which include domestic cats as one species (with over eighty different domestic breeds). Cats have been suggested as having the perfect design for function because their skeletal structure and muscle construction allow them more mobility and flexibility than most creatures. Their basic skeleton shape suits their wide range of activities – hunting, killing and climbing – and is without any development from first appearance. Cats suddenly appeared in the fossil record from nowhere, showing no signs of having evolved or adapted from any other creature, yet they are anatomically perfect for their purpose and have shown no changes since.

Fossil records have identified that some types of animals have appeared independently of each other on different continents. Animals such as dogs, cats, rhinoceroses, and pigs have appeared without any common ancestors in different parts of the world. In evolutionary terms, this suggests that many unconnected reptiles on several different continents formed into the same creatures following a series of millions of random mutations over millions of years. Yet Evolutionary Theory considers this scientific finding to be impossible.

---

39. *Cimolestes* – originally classified as marsupials.

Insectivores

Insectivores are classified as carnivores that eat insects. Amphibians were chronologically the first insect consuming animals on land, followed by reptiles then mammalian insectivores, a small group perfectly structured for devouring tiny insects. Members of this family include moles, hedgehogs and shrews. Their natural body shape brings them closer to the ground where their main food source exists. These creatures are generally slow and cumbersome with in-built natural defenses, such as spikes or venomous spurs[40]. Their origins, as discussed under Option 1 and Option 2 in mammalian reproduction, are not clear. Other non-mammalian insectivores include lizards, marmosets, carp, frogs, lizards, and bats.

These three groups of mammalian quadrupeds – Herbivores, Carnivores and Insectivores – all share four-legged anatomical layouts that Evolutionary Theory suggests developed from one original source. For example, a hedgehog and a giraffe originated from the same ancestor. Yet no scientific evidence of any lines of changes exists.

Hooved Animals

A hoof is made by the protein keratin covering the tip of a toe with thick nail rolled around it. There are differing versions of hooves with some animals having one, two or three hooves per foot.

The most common evolutionary explanation for the existence of hooves refers to horses, which have one hoof. It is generally claimed that a creature started out with five toes which 'dropped off' each foot causing the foot to draw closer together. The same thing is assumed to have happened on all four feet, as each foot independently developed a hoof.

Biologically, however, any animal that loses a body part after it is born, will not have offspring who are missing that part because, genetically, the DNA of the creature has not been altered. The horses' offspring would have all of their toes, regardless of whether the parent lost one toe, a leg or a tail, as the design for the next generation in already preset in the ovum and sperm of the parents. It could only come about by DNA mutations in embryos. But this process is not the process that is proposed by evolution.

---

40. For example, the platypus.

Chapter 21 Unraveled

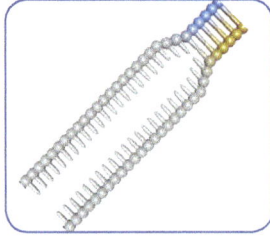

- *Mammalian functionality and basic design are significantly different from reptiles. Some of the fundamental changes that are necessary to evolve from one type of reptilian creature to mammal include:*

  - *Hard-shelled egg laying to internal soft egg.*
  - *Cold- to warm-blooded circulatory systems.*
  - *Single to twin channel excretion systems.*
  - *Simple to complex eating systems.*
  - *Care of the young.*

- *The fossils of many animals are common across different continents without any connection between them and no apparent way for them to migrate to different locations.*

- *Evolutionary Theory suggests that quadrupeds originated from one original creature. Hooved quadrupeds have either three or four toes per foot, originating from one original creature.*

- *The period of change from reptiles to mammals is estimated to have taken 140 million years, while the dinosaurs were living on earth. There are numerous fossil examples of the dinosaurs during that period but none of the changes from reptiles to mammals despite existing in the same conditions during the same time period.*

- *The development of marsupial pouches and teats, before the need for them existed, is unexplained by evolution.*

- *Three different groups of mammals predominantly live in ocean environments. The group that lives exclusively in the ocean dispelled of their land-based walking aids that they had possessed for hundreds of millions of years and returned to water. Other groups only partially did the same.*

Points To Ponder - Instantaneous Intelligence

There is no doubt that dogs (for example) show more intellect than reptiles, from which evolution suggests they originated. Evolution proposes that creatures have become increasingly more intelligent as time has elapsed through gradual intellectual development, presumably through evolution of the brain. This occurred by new, smarter animals suddenly appearing. However, the animals we see that have been with us for hundreds of millions of years haven't, in themselves, become any smarter or more developed. As far as we can determine, lizards, cockroaches, sharks, and crocodiles still have the same intellect as their ancestors, still inhabit the same environments and still have the same basic body shapes and attachments. They haven't developed any observable skills other than those required for survival, which we can assume always existed.

What is claimed to happen with evolution is that a new creature suddenly appears and it is either immediately more intelligent or it develops more intelligence very quickly. In other words, creatures that have been around for hundreds of millions of years don't evolve intellectually but the new ones do, evolving immediately without any sign that they have changed from being less intelligent. Scientifically, therefore, it would be expected that at least one creature exists today that is noticeably more intelligent now than it was a hundred million years ago, but there is no evidence of such a creature.

22

INSTINCT AND REASON

Living creatures possess physical attributes that change and develop over time. Bodies grow, mature, then age and die. Most creatures have mental abilities to make assessments about their immediate environment, direction of travel, food collection, and many other aspects about the world they live in. Creatures can generally see, hear, feel, taste, and smell in some form or another. All of these occur as a result of the layout, interconnections and collective operating programs of the billions, or trillions, of cells in creatures' bodies.

Beyond these attributes, there are characteristics of behavior, specifically instinct and reason, that significantly help creatures to survive. In this chapter, the origins of these attributes, their abilities to change from one creature to another, and the programs that control them are examined.

Instinct

Animals possess a unique function that makes them act in a specific, pre-determined manner. Animals have instincts.

According to *The Macquarie Dictionary*, instinct is:

"An inborn pattern of activity and response common to a given biological stock".

Any complex behavior is considered to be instinctive if it occurs without being based on a previous experience. Each member of a particular species, or type, will act automatically in a similar way to similar circumstances the first time they are encountered. These are different from reflex actions, which are mostly involuntary muscle responses, such as the pupil in an eye contracting when exposed to light or the lower leg jerking when the knee is tapped by a patella hammer. Reflex actions are related to responses from actions on the nervous system.

We can observe instinctive types of behaviors when watching any household cat hunt down a mouse or bird. Most household cats have never been exposed to hunting routines by their parents but instinctively know what to do. Newly hatched turtles instinctively know that they must immediately head to the ocean and they know in which direction it lies. Sea creatures instinctively know how to swim when they are born. All species instinctively know how to reproduce. Most species seem to know many unique actions without training, but not all actions are instinctive. There are many things that still require learning. For instance, human babies will instinctively use their voice by crying when they need attention but they need to learn how to speak a language. This means that only certain actions are 'hard coded' into creatures' response patterns. Instincts are not learned by each new generation, they are passed on as part of that creature's makeup while it is developing as an embryo. Thus, not only do physical characteristics get passed on but so do instincts and their accompanying actions.

When did instincts first occur?

This is a most interesting question for evolution. Where did they come from and how does a creature's body instinctively know how to react when it has never encountered a particular scenario before? Where are the instructions for instincts located in living bodies?

Although not offering comprehensive explanations about this topic, Evolutionary Theory by default assumes instincts are accidental mutations that most likely occurred in just one type of creature and were subsequently changed when passed on to later types, as mutations. Instinct, therefore, had to adjust from its original source to each new type of creature. As there is no scientific evidence to guide us, we can assume instincts started when there were only a few living creatures, perhaps the trilobites or some sort of early fish, and those that developed instincts had better skills to survive, thus becoming the forerunners to all other living creatures.

Under the premise of Evolutionary Theory, instincts are not deliberate; they are random in every species and can only appear, or change, as chance mutations in embryonic development. This brings us to the major questions surrounding instincts. The following information is primarily related to humans, as this is where the scientific research has been performed.

1. Where Are The Instructions For Instincts Located In The Body?

Sensory inputs such as sights, sounds, tastes, touches, and smells are sent to an area of the brain called the thalamus, which sorts and directs a basic level of information to another area called the amygdala. Here the inputs are assessed and a response is raised within an astonishingly quick 12 milliseconds. These become pure instinctive reactions.

The thalamus also sends more detailed information to the neocortex, an area of the brain which responds with a more considered reaction that includes instincts, emotions and self-status or ego. This is also very quick, taking about 25 milliseconds. These areas are pinpointed in Figure 176: The Human Brain.

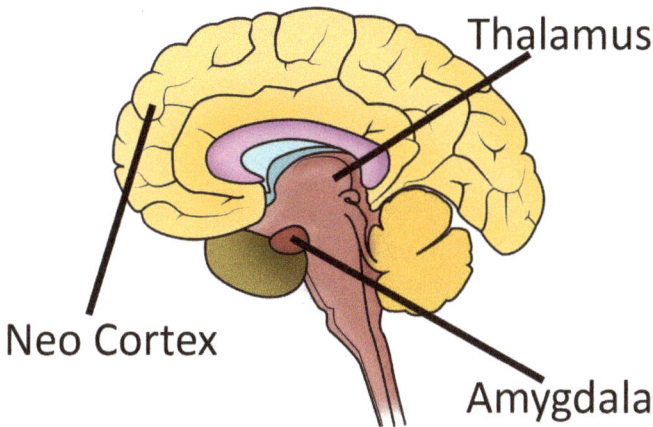

Figure 176: The Human Brain

The thalamus acts as a sorting and distribution center that is able to understand, filter and direct information instantaneously. It knows which data to send to each of the areas of the brain for any event or situation, utilizing a complex inbuilt program to enable neuro-electrical signals to be recognized, sorted, prioritized, and redistributed. Its primary activity is to identify and separate 'danger-based' data, such as fire and loud noises, and to direct it to the amygdala for an immediate and appropriate response.

With regard to instinct, there are three areas (thalamus, amygdala and neocortex), three programs and three response times for information being input to the brain every second. Events are assessed at computer-like speed, requiring perfect programming that complements the creature's physical abilities to respond. They require complementary electrical pathways that can transfer data at great speeds as well as muscle programs that can respond just as quickly. Evolutionary Theory suggests that these programs, connections and response times invented themselves by chance for no reason.

2. Where Did Instincts Come From Originally?

When considering the first ever creature to have instincts, the only way to know if an action or reaction is the best, in evolutionary survival terms, is for the activity to actually happen. The creature's brain can then record and store the best behavior or reaction and remember it for the next time a similar event or situation arises. This makes the reaction a memory and not an instinct. The memory-based response will be sent via the neurological pathways to a different part of the brain resulting in a different response and timeframe. As noted previously, any complex behavior is considered to be instinctive if it occurs without being based on a previous experience. Instincts are not related to previous events that occurred to a creature, otherwise they would be processed against a memory and a different, considered reaction would occur, rather than an instinctive one.

Consequently, under an evolutionary process, an instinct program would require the self-creation of that instinct without prior knowledge of threatening events or situations that could occur in the future or the prior knowledge that the reactions to be taken would be suitable to avoid these dangers. These instincts would by necessity have to be perfect from the first moment, as there is no possibility to retract those instinctive reactions and change the program following an event due to the fact that it has been coded as a memory. Similarly, if an event is encountered and recorded after a creature is born, there is no process by which it can change any of its DNA program in order for its offspring to benefit from the lesson of survival. There is also no biological method for these memories to be passed on to the next generation as they are not coded into that creature's DNA programs.

There are some suggestions by scientists that 'skills' learned during a lifetime can be passed from one generation to the next. There are examples of people with mental disabilities (savants) who have extraordinary skills in areas in which they have no training, such as music, mathematics or art. The proposition is that these skills were learned by their ancestors and passed down to them by 'genetic memory', although few are able to identify which ancestors, if any, may have learned these skills. This would mean that such skills, learned through the body's somatic cells, have a process for re-coding themselves back into the gamete cells of that person despite no known biological method for this to happen. Genetic memory is only a concept currently being studied by biologists and has not been proven nor does it have any known links to instinctive behaviors.

An alternate consideration is that every person is born with the knowledge about every possible skill already coded into their DNA, which is stored in their brain during embryo development. By practicing the skill, this information can be downloaded from its storage area into the temporal lobes (refer Chapter 25) that build working memory banks for activities such as music or art or riding a bike. This may explain why it is relatively easy to learn new skills, as the coding already exists in the brain's storehouse and it is simply a matter of transferring it to an accessible area in the brain by learning. When a person's mental abilities do not form normally, the different neurological connections in their brains may allow some of the knowledge to be downloaded automatically into temporal lobes for use in daily living. This may be restricted to the few who are observed as being savants. Of course, there is no evolutionary explanation about where these vast levels of knowledge could come from, particularly as no creatures in the evolutionary line of life had prior knowledge about music, art, balance on land, or any other skill yet to be invented.

Another suggestion, from British Biologist Rupert Sheldrake, is that all organisms are surrounded by undetectable supernatural 'morphic fields' that pass information to subsequent generations. It is not yet known how to prove or disprove this idea.

3. Changing Instinct Programs Between Different Creatures

Biologists assume that the program for instincts is coded in the DNA of all creatures. As each new cell appears in the developing embryo, instincts are programmed to react in line with the instructions from various other programs in the brain. However, this raises several questions:

- Does the instinct program alter itself marginally each time there is a gradual change in the physical makeup of an evolutionary mutant cell or creature?

- How could, and why would, an instinct program reprogram itself without prior knowledge of what new instincts will be required for the new creature?

As an example, we have previously looked at the evolutionary processes required for a lizard to change into a bird. Along with these physical changes, the instinct program would also need to alter so that the bird has a different set of instructions appropriate for its anatomical capabilities and the new circumstances that it will encounter. For instance, the act of flying presents new and vastly different activities from land mobilization, requiring instant movements to avoid damage at far greater speeds and heights.

As every creature has the appropriate instincts for its design, environment and survival, two evolutionary possibilities are raised. Firstly, while each of the millions of steps of evolutionary change occurs, the instinct programs would be rewritten to reflect each minor change in body variation. The second possibility is that the program remains the same throughout the evolutionary mutations until the new creature is fully formed, then and only then changing in one major rewriting of its code. The changes to the newly coded instincts would be accordingly random and, as such, more likely than not unsuited to that creature. Only the perfect mutation to recode the instinct program would benefit survival.

Insects also exhibit instincts. Evolutionary Theory, as we have discussed, claims insects appeared independently of all other creatures. This means that individualized survival programs, in this case instincts, evolved in two separate classes of creature. This, however, again contradicts the tenet of Evolutionary Theory that things happen by chance mutations and therefore can probably only happen once.

Instincts are observed in ants, bees, termites, and wasps, all of which instinctively know their roles and duties in life. For example, ant colonies are divided into groups that perform specific tasks. There is often only one queen, being the largest, who lays millions of fertilized eggs that develop through metamorphosis into females. Winged males are called drones whose role it is to fertilize the eggs. They have lifespans limited to several weeks only. The queens and their eggs are tended to by worker ants. Other workers also perform a range of specific tasks such as guards (having bigger heads and jaws), storage of honeydew (bigger abdomens), gathering food, making tunnels and rooms, and scouting for new colony locations. The colony is structured with each ant having a predetermined role to play in support of the whole colony. A queen ant can sacrifice the whole colony if defending herself from threats such as other queens, instinctively knowing she can start again as she is the source of reproduction.

Whilst human instincts differ from insects and other animals, we can observe an interesting instinctive reaction when we use a lie detector (polygraph). It will detect minor variations in blood pressure, heart rate, respiratory rate, and sweat levels when a human is under duress (usually from a probing question from the examiner). Research at the University of Pennsylvania using a Magnetic Resonance Imaging (MRI) machine shows that sections of the brain 'glow' when telling a lie. This is thought to be because the brain has to process more with a lie than the truth. This suggests that all of these human body patterns instinctively change from 'normal' when we give a misleading response to a question, even though the response may be meant in kindness.

4. Fight Or Flight Response

Vertebrate animals, including humans, possess an inbuilt reactive process, generally referred to as the fight or flight response, also known as hyper-arousal or acute stress response. It is an immediate physiological reaction to a potential confrontation or threat. An animal's awareness of immediate danger will cause part of the autonomic nervous system (ANS)[41] known as the sympathetic nervous system to swing into action, the purpose of which is to prime the body for an increase in normal strength levels should physical defense or speed of flight be required.

The types of physiological changes necessary to prepare the human body include redirection of blood from less important body areas into muscles, increase in heart rate and blood pressure, heightened muscle tension, improved vision through pupil dilation, and increased rate of perspiration. This response protocol starts in the spinal column and initially causes norepinephrine (also known as noradrenaline, a compound that acts as a hormone and neurotransmitter) to be released into the bloodstream, signaling many of the fight or flight responses to begin.

Without analyzing each activity, there is a sequence of actions that includes the following body components: amygdala, hypothalamus and pituitary gland, which are located in the brain, and the adrenal gland and liver, which are located in the abdomen. These initiate a series of events that includes specific and measured neural responses, hormone secretions, neurotransmitter release, constriction and dilation of blood vessels, and the release of energy from stored sugars and fats. These do not randomly burst out from their normal operations; they each appear in a specific sequence and quantity causing reactions in organs such as the heart, lungs, stomach, and sphincters[42]. Hearing and vision can also be restricted allowing focus on the event by avoiding external distractions.

Once the threat is over, another part of the autonomic nervous system known as the parasympathetic nervous system comes into operation to reverse the above effects by allowing the release of acetylcholine (a neural transmitter) which, through another series of actions, returns the body to its normal homeostatic state.

For these processes to be effective, the appropriate anatomical components had to be connected to the sympathetic and parasympathetic nervous systems and had to know their actions would cause specific and

---

41. Refer to discussion on the human nervous system in Chapter 24.
42. Such as muscle constrictors that reduce blood flow to non-critical areas like the stomach.

immediate responses in specific organs. The chemicals, fats, sugars, muscle constrictors, and neurotransmitters had to be instantaneously accessible and effective. The chemicals and compounds had to be complementary so the fight or flight response could be reversed and did not cause unwanted debilitating results, such as permanently closing down any body systems. Further, there had to be automatic recognition processes that the threatening event had started and then ceased, triggers to initiate both the reaction and calming processes, as well as knowledge that norepinephrine would initiate the responses and acetylcholine would reverse the process, and that both were available and accessible.

The detailed processes involved in the fight or flight response are claimed by evolutionists to have resulted from a random series of unrelated reactions that created the immediate preparation of the body for a possible trauma from the very first time it was required. Mathematically, if such actions were random, there would be many tens of thousands of other combinations of responses that a body could undertake as a reaction to threats. To strike any perfect mix of responses would at best be a most fortunate chance happening, particularly as the process is different in different animals and would need to change every time a new type of creature appeared. For example, observations of domestic animals show that their hairs rise as part of this process, something that only occurs in some animals.

Because there is no biological method for doing so, the issue again arises how this, or any series of responses developed during the lives of creatures, could be passed from one generation to another. Effectively, a program of operation has to invent then refine itself until it works over multiple generations, even though each step is more or less stuck in somatic cell memory of an individual creature and the new knowledge can't be back-coded into their gamete cells' DNA.

The observable science we do know shows us that the instinctive fight or flight response is a series of highly specific and detailed processes that are finely tuned to each creature in which this response is present. There is no scientific evidence or proof that the fight of flight response could be implemented as part of any evolutionary process.

Reason

As well as instincts, humans, in particular, have another unique ability that they can sometimes use to override natural instinct – reason.

There is no single agreed definition of reason, however. Generally, it can be described as:

The mental ability to draw conclusions based on facts, evidence or proof.

We observe many types of reasoning in animals.

- Birds will choose a location for their nests based on a variety of variables including safety, available materials and access to food.

- While deciding to leave alone or attack, sharks may use their noses or bodies to push, prod or rub against objects if uncertain about them.

- Scout ants will review a variety of locations for new colonies and then consult with other scouts to make a communal decision.

- Household pets work out many things from observing humans, and can even operate door handles, faucets or move furniture to gain access to elevated items.

Humans have significantly advanced reasoning skills compared with other animals. We are able to use reasoning across a vast range of skill areas such as the sciences, arts, construction, languages, and in most aspects of our daily lives. We can also use reasoning in areas based on our imaginative skills, such as theoretical reasoning (what a six-dimensional quantum string might look like) or creative reasoning (a new invention like the telephone).

Reasoning takes place in the cerebral cortex (the gray matter of our brain). Actions, reactions and stimuli are evaluated and assessed before response instructions are sent to muscles and organs. These are assessed against a hard-coded set of operating instructions in the brain that exist when we are born. The brain will pick up and copy actions (such as speech) while other things are automatic (such as heartbeats, breathing and digestion). Almost every person has the same basic ability to reason[43].

Conclusions are reached by following a line of thinking to an end point. Yet people can reach different conclusions from the same apparent thinking processes. This is particularly evident between males and females, and children and adults. Conversely, the same conclusion can be reached from different lines of thinking. Therefore, a conclusion is not absolute; it is only one of a number of options from various lines of thinking.

Humans are unique in that they have the mental ability to draw together a range of variables and to make a forecast or prediction about what might

---

43. With some exceptions of people born with abnormal or low mental capacities.

happen or what has happened and why. In other words, rather than just observing what happens around us and simply reacting, we are constantly working out how and why things occur. We have understanding.

Humans have a basic level of reason coded into their makeup at birth. Various experiences allow us to mature our reasoning skills throughout life but these experiences cannot be re-coded into our offspring's programs before they are born. We can teach them what we know once they are born but we can't rewrite the base program. The question, therefore, is: how did the reasoning program that is so advanced from all other animals embed itself into our brains? If apes were our direct ancestors, where is the evolutionary explanation about our staggeringly superior reasoning programs?

The abilities of humans are so far advanced from other creatures that it would only be reasonable for Evolutionary Theory to use reasoning to explain where these came from.

Chapter 22 Unraveled

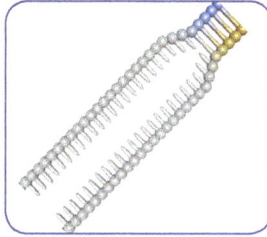

• *Instincts are automatically coded reactions to certain events.*

• *Individual species have instincts that are suited to their design and longevity.*

• *There is no scientific or biological explanation for instincts to exist or be changed in an evolutionary mutation process.*

• *Reasoning is based on predetermined processing of events in the brain against an assumed series of events.*

• *Reasoning allows understanding, predictability and flexibility.*

• *There is no scientific explanation for the existence or development of reasoning or the change of reasoning levels from one creature to another.*

PART FIVE

HUMAN EVOLUTION

23

APES AND HUMANS

The next group to appear in the fossil record are the apes who were able to walk on two legs, an attribute possessed by a limited number of creatures. Called bipedalism, it requires the body to have specific skeletal and muscle structures and the ability to balance and to correct imbalance[44]. Evolutionary Theory cannot explain this ability in apes as their appearance on earth was instant, with no previous ancestry and they could walk on two legs from the very moment of their sudden arrival.

Evolutionary Theory proposes a number of scenarios about apes' ancestors. Most of these are centered on squirrel-like creatures with hands that could grasp things, which may have become monkeys. Most scientific opinion is swayed against monkeys having a direct linear relationship to apes, leaving a void without any formal theories. The oldest bipedal fossil is a 290 million year old reptile called *Eudibamus cursoris* found in Germany that appears to have the structure required for two-legged walking. There are various dinosaurs, mammals and, of course, birds that have been able to walk on two legs, but these particular animals have no ancestral relationship to apes, leaving no explanation on the origins of bipedalism in apes.

Figure 177: Squirrels To Monkeys To Apes

---

44. For example, humans spring over a stiff leg, push from ankle to toe and rotate their hip around its horizontal and vertical axes. These actions require an appropriate anatomical layout, coordination, structure, and measuring processes to maintain balance and movement.

Apes are part of the group called primates (as are humans) and are generally forest dwellers with some living in trees. The skills and physical attributes required for climbing trees are vastly different from ground-based animals. Tree climbers need higher levels of agility, strength, eyesight, hearing, and coordination. Among their attributes, they have different teeth[45], bigger skulls, their eyes are nearer the front for better forward vision, and their faces are not as long.

Primates are different from other mammals in that they have:

• Four types of teeth.
• The ability to hold objects in their fingers and thumb.
• Two mammary glands on the front of their chests.
• Large brains.
• Front facing eyes.
• All but one has body hair - humans, who are considered to be hairless.

The evolutionary development of primates is thought to have started about 60 million years ago in the tropical rainforests of Africa, which was then part of the super-continent called Gondwanaland. Even though they are the most recent arrivals, there are very few fossil remains or scientific evidence to verify the evolutionary lineage of primates. As such, most of the theories on the origins of primates are derived through observations of modern day animals.

About 45 million years ago, 'more intelligent' primates are thought to have split into two distinct groups – the prosimians (lower primates) and the anthropoids (higher primates, who had thin noses). The lower primates are assumed to have developed to a point where they could cope with their needs and environment and then stopped development. This assumption, however, contradicts the basic tenet of 'adaptation by mutation' because it infers a limit to mutations, which Evolutionary Theory suggests are natural, random and unstoppable, without limits.

Even so, the higher primates are thought to have split into two groups somewhere around 20–30 million years ago. Yet there is little, if any, evidence of the primate split. The only tangible evidence is the two primate lines that followed. One group that was herbivorous became known as the Old World monkeys, and it includes the baboons and rhesus monkeys of today. The New World monkeys were a type of lower primate that developed in South America. They were omnivorous (eating plants and

45. More general teeth rather than incisors.

meat) and are claimed to have become the hominoids, of which humans are a member. These differ from other primates in that they have no tail. Gorillas, chimpanzees and orangutans are also part of this group.

In trying to reconstruct the evolutionary leaps from ape into man, science has identified what are referred to as the 'dawn apes' (*Aegyptopithecus Zeuxis*) as a possible first step. Found near Cairo in Egypt, dawn apes were smallish primates that climbed trees and had relatively large brains, believed to exist around 30 million years ago. However, the detail on evolutionary changes from this point is blurred and at times contradictory.

The primary issue at hand is that there is no valid scientific explanation for the appearance of apes on earth. The only scientific information that actually does exist about apes is:

- Apes appeared instantly in the fossil record.
- There is no record of their ancestry.
- They appeared fully developed with the ability to walk on two legs.
- They have distinctly different skills, abilities and design for the functions they perform.

The question now is whether there is any scientific evidence to substantiate an ancestral link between apes and humans.

Apes To Man

Although there is no universal agreement on the specific date that mankind first appeared on earth, from an anatomical perspective, humans' assumed ancestors (*Homo sapiens*) appeared around 200,000 years ago in Africa and are considered to have behaved as modern humans since about 75,000 years ago. That is, they displayed abilities of language and communication in various forms.

Where did humans actually come from?

Evolutionary Theory says humans are the result of a series of bizarre and unlikely events, mutations and chance happenings, which occurred over billions of years, the path of which we have followed from the first chapter in this book. Yet, the most challenging topic for humanity is the assertion that we have descended from the apes. This is a major aspect of evolution that features heavily in the media and our educational institutions, but there is little, if any, detail regarding the process of these changes. Typically, we see the following type of drawing, which is meant to depict the changes over time.

Figure 178: Example Of Ape-To-Human Diagram

Many of these types of ape-to-human drawings are based on an illustration called 'The March of Progress' from the 1965 *Early Man* volume of the Life Nature Library from Time-Life Books. The original was not intended to represent transitional changes from ape to man, with neither the illustrator nor the text supporting such a concept. Many people believe these images are confirmation of the facts that science has accumulated, in this instance proving the origins of man, but nothing could be further from the truth[46].

Despite apparent, general physical similarities, the biological gap between apes and humans is enormous. Apes have limited ability and do little more than just survive. Humans are very different beings from all other creatures on earth. Humans are far and away significantly more advanced than any other living being.

> "A human being should be able to change a diaper, plan an invasion, butcher a hog, conn a ship, design a building, write a sonnet, balance accounts, build a wall, set a bone, comfort the dying, take orders, give orders, cooperate, act alone, solve equations, analyze a new problem, pitch manure, program a computer, cook a tasty meal, fight efficiently, die gallantly. Specialization is for insects". – Robert A Heinlein, *Time Enough for Love* (Ace Books, 1973)

There have been many attempts to link humans' heritage to the apes. This thinking is quite reasonable as they are the creatures that display the most similar physical features and characteristics in common with humans. One of the more generally accepted views is that humankind originated from a line of beings traced back to Africa around 2-3 million years ago. However, there is also popular support for *Homo sapiens*, considered more anatomically close to modern humans, arising out of Africa around 75,000 to 200,000 years ago, as discussed before. This is quite a significant

---

46. The truth about the original image and text is covered in *Other Things You Should Know, Section F: False Interpretation Of The March Of Progress Illustration* at the rear of this book.

discrepancy. Despite extensive research and many assumptions about fossil 'finds', there is actually no specific evidence of any nature – biological, physical, geological, genetic, chemical, historical – that demonstrates a transition from ape to human. None of the various claims presented by media, evolutionists and scientific articles is based on any evidence[47]. They are assumed to have occurred solely based on evolutionary theories, not facts.

From fossil research and DNA analysis across the world, there is a growing belief that modern humans did not arise from a single African ancestor (Replacement Hypothesis), as evolution has previously assumed. The new theories suggest humans arose from a blend of different, unconnected groups living worldwide (Multiregional Hypothesis).

Replacement Hypothesis, also known as the 'Out of Africa' model, purports that a new species, *Homo sapiens*[48], developed in Africa and spread throughout the world without mixing with other existing species. They effectively took over the primary role in the development towards humans[49], while other lines died out. This concept requires extended periods where there are no fundamental changes to anatomical structures, contrary to the basic tenet of biological Evolutionary Theory, followed by quick identifiable changes. In other words, nothing happens for long periods then suddenly a new species appears.

The Out of Africa concept was predominantly based on DNA similarities. Within cells there are smaller components called organelles that perform specific functions, such as chloroplasts which take in sunlight for photosynthesis. The DNA in organelles are called mitochondria. Testing of mitochondrial DNA has been undertaken on a broad range of organisms including fossils dated back 2 million years ago to investigate whether there are any connections. The suggestion from this testing is that all females on earth have DNA links to one original African female from around 200,000 years ago, referred to as Mitochondrial Eve.

Contradicting this, DNA testing of Y-chromosomes, which come from only one parent, suggests that there has been interbreeding between various *Homo sapiens* types, including Neanderthals (*Homo sapiens neanderthalensis*) and *Denisova hominins* (specifically mixing with Australians and Melanesians). The conclusion being that interbreeding was occurring in various regions among different Homo sapiens species and

---

47. As explained in the next sub-chapter, *History of finds theorized as being 'evolution' from apes to human.*
48. Basically humans, including Neanderthals which are now extinct.
49. *Homo sapiens sapiens*, which is the term used to describe the modern human subspecies.

that at later times Asian species had bred back into Africa. This effectively makes human breeding multi-directional, not solely out of Africa.

The Multiregional Hypothesis was proposed by American Paleoanthropologist Milford H. Wolpoff and some associates. The underlying concept is that the basic genetics of a species remained essentially the same in the four major regions of human development (Africa, Asia, Europe, and Australia) whereas, over time, each of these groups experienced similar evolutionary changes to their physical traits. So there is parallel, if not identical, development that is unrelated to genetic mixing within these groups.

The result of many competing scientific views on human ancestry is that at any point in time, many theories exist, many are disputed and most are contradicted by other information. Furthermore, there exists an enormous level of acrimony, emotion and personal insults, often from professionals within the scientific community. The vast amount of information available results in an endless stream of contradictory, disputed, untested, and personalized opinions. To reinforce this point, an internet search for 'human's origins' will result in over twenty million references.

What evidence we actually know for sure is really only two things:

- There is no provable scientific evidence that apes evolved into humans – only theory and conjecture.
- There is scientific evidence that humans (*Homo sapiens sapiens*) appeared in different locations across the globe at the same general time and there is no absolute proof these were related.

In the absence of any proof, and no matter how many different opinions, tests, analyses, or theories are presented, there is only one conclusion to be drawn from the scientific evidence we actually know at this point in time. We can say this:

Humans suddenly appeared on earth in several unrelated locations, independently of apes.

Nonetheless, the most commonly proposed line of discoveries that Evolutionary Theory relates to humans tracks along the following line of creatures:

History Of Finds Theorized As Being Evolution From Apes To Humans

There are claimed to be ten major steps of named creatures in the evolutionary development from apes to humans starting with *Ardipithicus ramidus*. The figures below are interpretations of how such creatures might have appeared from a physical perspective. There are no official images endorsed by science because only minor, partial finds have been made, such as a part of a jaw. There is no common agreement on these species[50]. The dates are shown as million years ago (mya).

4.4 – 5.6 mya

Figure 179:
Ardipithicus ramidus

4 – 4.5 mya

Figure 180:
Australopithecus anamensis

2.9 – 3.9 mya

Figure 181:
Australopithecus afarensis

2 – 3 mya

Figure 182:
Australopithecus africanus

1 – 2 mya

Figure 183:
Australopithecus robustus

1.4 – 2.3 mya

Figure 184:
Homo habilis

0.4 – 2 mya

Figure 185:
Homo rectus

0.2 – 0.5 mya

Figure 186:
Homo sapiens archaic

50. The images proposed for these creatures are all displayed as walking on two legs whereas this may not have been the case.

The Missing Link

Next came Neanderthal Man. This was considered by Thomas Huxley, (British biologist, anthropologist and philosopher, 1825-1895), who supported Darwin's Theory of Evolution, to represent a strain of humans rather than a step up from the apes. He held the opinion that a 'missing link[51]' of ape to human still needed to be found.

0.04 – 0.25mya   Present – 0.2 mya

Figure 187:
Homo sapiens neanderthalensis

Figure 188:
Homo sapiens sapiens (human)

There appears to be progress in development towards humankind in this line of creatures. The assumption is that each of these developments was a step ahead of the previous one: less hair, a more upright posture, a smaller skull, and progressively smoother facial features. There is also an assumption of a higher level of intelligence with each step. All of these notions are quite reasonable when looking at these images.

However, there are two very important issues to regard about this proposal of our ancestry.

1. <u>Do these images represent actual creatures that existed?</u>

Apes have been on earth for several million years and there is evidence of them in the fossil record. They are still with us today, so we know what they look like. Consequently, we can identify their fossil remains. However, none of the nine 'species' that are claimed to lead up to humans is alive today, so we do not know if they were real or what they may have looked like. These images are only conceptual, an idea of how that creature might appear, if it had existed.

---

51. The missing link is essentially a creature 'between' ape and human that had the ability to walk on two legs, a larger brain, small front teeth, and larger back teeth.

A detailed study of the above finds has been undertaken by David N. Menton Ph.D. titled *The Scientific Evidence For the Origin of Man*. In this analysis, Dr. Menton reveals that most of the classifications of these assumed steps of evolution were based on small fragments of fossil evidence, which were then reconstructed by someone to represent a human-like creature that never actually existed. These were based on various fossil finds of teeth, jaws and odd bits of bone. The material in these finds is generally small and is often reconstructed to represent a possible body part, such as a skull. From these, theories were often developed or enhanced. The process isn't scientific, as it involves speculation and is open to inaccuracy. For instance, it suggests that a small part of a jawbone can be proof of an entire skull that then represents a body that is then regarded as a step of evolution. Naturally, the scientific reliability of this method is not sound. In fact, as science has improved its research and identification skills over the decades, almost all of these partial 'evolutionary finds' have been dismissed as likely to only be part of other primate lines, such as orangutans or pigs. Worse, some fossils were also deliberately tampered with to represent differing species and later found, under the scrutiny of higher levels of scientific analyses, to be fictitious.

Whilst these images above appear to show progress, they have for the most part been totally disproved, as proper scientific analysis has determined that they were not real and they did not exist[52]. Despite this, they are still quoted and displayed in text books, websites and the media as if they are in fact true scientific evidence of evolution.

Even Richard Leakey, famed evolutionist and author of *Origins and Origins Reconsidered*, admitted the failings of paleoanthropology in an interview in American Scientist (May-June 1978), "My reservations concern... the whole subject and methodology of paleoanthropology... that perhaps generations of students of human evolution, including myself, have been flailing about in the dark: that our data base is too sparse, too slippery, for it to be able to mold our theories. Rather the theories are more statements about us and ideology than about the past. Paleoanthropology reveals more about how humans view themselves than it does about how humans came about. But that is heresy".

---

52. For more information, please visit www.evolutionunraveled.com

The following table summarizes the findings about most of these assumed evolutionary steps.

Name	Age m.y.a	Component	Location	Claim Date	Disproved/ contradicted	Reason
PILTDOWN MAN (Eanthropus dawsoni)	0.5	Part skull, mandible	England	1912	c 1965	Proven as hoax
NEBRASKA MAN (Hesperoptiecus haroldcookii)	non specific	Tooth	U.S.A.	1922	1927	Pig's tooth
RAMAPITHECUS (1)	14	Teeth, jaws, bone fragments	Nepal (1932)	1932, 1961, 1977	1696, 1972, 1982	Chimpanzees /Apes
AUSTRALOPITH-ECINES (Africanus & Robustus)	2 - 4	Face, lower jaw, brain cast	South Africa	1924	1954	Combined parts from different finds
AUSTRALOPITH-ECUS AFARENSIS	3.5	Partial skeleton	Ethiopia	1974	Various	Based on strata dating assumptions
HOMO HABILIS	1.5 - 2.3	Skull fragments	Various	1959	Various	Based on tools found near these fossils
HOMO HABILIS KNMER1470	1.9 - 2.4	Skull fragments in dated strata	Kenya	1972	Various	Typical of human remains
HOMO ERECTUS 1. (Pithecanthropus erectus)	10	Skull cap, molar tooth, femur	Java / Sumatra	1891	late 1930s	Original proposer Eugene Dubois stated his original claim was incorrect
HOMO ERECTUS 2. (Peking Man)	1	Partial skeleton	China	1929	Original not widely accepted	Combined partial finds from different depths
NEANDERTHAL MAN	Various	Skull cap, femurs, humei	Europe & Asia	1856 - 1908	1950s - now	Some resembled human remains displaying rickets or arthritis

Figure 189: Table Of Ape To Human Disproven Finds

2. Are there any living examples?

Far more significant, however, is that none of these assumed species is alive today. Let's remember that evolution is a theory that says there has been progress by mutation and, if this is true for these nine species, then each is an improved, smarter and superior adaptation than its predecessor. Assuming each progressive step was better able to adapt and live successfully, there should be increasing evidence of each species as we head towards humankind. More so, if the apes were the first and most primitive of all of these species, and they still exist in the same form millions of years later, what possible reason can there be that none of these progressions has any living descendants today?[53] It would be reasonable to expect to find at least one of them, if not all, still in existence if each was a better adaptation than the previous.

The Changes From Ape To Human

Even though these finds have been disproved, the media and our educational institutions continue to propagate the concept that these creatures were found, in linear order, in the fossil record. The chain of life from ape to humankind, shown in Figure 178: Example of Ape-To-Human Diagram, and similar images, are familiar and repetitive, leading many of us to believe that these changes actually occurred over millions of years.

As this is the major and most familiar evolutionary proposal for our existence, we should examine what would actually happen, and what would be the status of these developments today, for the gradual change of apes into humans. We will use an example following just one family line of beings to determine what we would expect to see today as a result of these changes. The number of developments needed in this process is unknown, but we do know that the major physical differences between ape and human include:

• Muscle positioning and abilities

• Posture

• Bone structure (such as the ability to lock the knee)

• Bone length (enabling man to stand)

---

53. Some theories suggest that humans were the cause of their extinction, killing them out as is believed to have happened with mammoths, the giant marsupials of Australia and dodos. Although the giant marsupials and dodo were limited to single locations, Australia and Mauritius respectively, human 'descendants', like Neanderthals, are assumed to have existed all over the world, so their culling by humans is not a fact, just a theory.

- Spinal column flexibility (in humans)
- Skin
- Skull shape and size
- Brain capability
- Pelvic dimensions
- Design and use of fingers
- Facial structures
- Teeth

Conservatively, we could estimate there would be at least one million mutations required to the cellular development programs in order for apes to evolve to humans. This estimate allows for each change to the sequence in the embryo development program that makes up the fifty trillion cells of a human compared with a similar number for a similar size ape. The changes required to reorganize the specific sequences, locations and types of cells for the human brain alone, containing eighty-six billion cells, would seem to cover this figure of one million changes. If this number of changes occurred in direct linear sequence, we also need to consider how many changes per generation would be required. The estimations below represent a possible scenario.

Proconsul is thought to be one of the first very early apes, existing 20 million years ago. *Ardipithicus Ramidus*, at 5 million years ago, may be the earliest human primate. The most popular first possible descendant of man is *Australopithecus*, at about 2 million years ago. Modern humans are generally recognized as only being around for about 50,000 years. This leaves us with a period of up to 20 million years to change from early apes to humans in one family line.

Today's apes average around 10 years per generation. Humans average around 25 years. Using an average of 10 years per generation to get from ape to human, simplistically, over 15 million years there would be about one and a half million generations, meaning at least one of our one million progressive linear changes to the cell reproduction program would have to occur every one and a half generations. If these did not happen in the correct sequence or result in the perfect shape and construction first up, then the number of adaptations per generation would be magnified and consequently extrapolated.

With this frequency of change, in most generations, the offspring should have a new development that is hard coded into the DNA of the family tree. With perhaps four or five generations alive at the same time (40-50 years), any differences may be noticeable over a short history, particularly in the recorded history of humans over the last 6,000 years. One would expect that some history of these changes would be recorded – such as cave or rock drawings, or evidence in the fossil record. Evolutionary Theory also assumes that this massive number of changes would be accompanied by millions of other failed mutant attempts. Over this period, with hundreds of generations resulting in new progressive changes accompanied by millions of failures, the evidence should be available. But it isn't.

How can we better visualize the scope of changes that we should see around us?

Charting The Changes

The easiest way to understand the scale of changes proposed by Evolutionary Theory, where one species of creature mutates into another, is by making a chart of the type of steps involved, in this case, developing from ape into human. This should give us an indication of the real level of mutations that we should find everywhere – in every type of creature, not just the ape-to-human scenario.

We will look at a very simple model that will allow us to see how changes might be passed down the generations across developing species of creatures. It is claimed to take many millions of mutations for any type of creature to change, however, it is impractical to display such a vast number of steps in one chart. There is no assumption that this is in fact how it happened, as no one actually knows, as there is no evidence. It is merely an attempt to show the scale of changes based on what is only a theory.

Gorillas existed prior to man and are probably the closest creature to humans. By this, we mean that human structure has more in common with gorillas than any other animal. Do we assume that gorillas changed into humans or that, like humans, they are one of the offshoots of some other mutation? If they developed into humans, their huge skulls would need to reduce in size, which could be argued is a reversal of the evolutionary assumption that human brains and skulls had to increase in size to allow for better thinking capacity. Obviously, this is a difficult place to start, so we will go back to a smaller, less developed level that we speculate to be a non-specific ape in the absence of a specific creature agreed by evolutionists. For the sake of this exercise, we will assume that a small group of ape families has been responsible for developing into man.

For our chart, we will show just four original families (pairs) of apes – that is, eight apes (families 1, 2, 3, 4). We will assume that each of these families averages four offspring each generation, effectively doubling the previous generation. As far as this exercise is concerned, it is irrelevant if this high level of offspring actually occurs, as it is the principle of whether mutations exist or not that is at issue. Rather than drill down twenty or thirty generations, it is easier to display the numbers by increasing the birth rate over fewer generations. Either way, the results become clear.

We will assume that only one of the four offspring of each original couple has a mutation that develops into a progressive variant[54] and we will call these A, B, C, and D. Each family will have three normal and one variant offspring in the first generation and that will be the only mutant for each normal family in this exercise. Our four families will therefore give us a total of four different variants that form in the embryos of Generation 1.

(continued on opposite page)

Generations	Family 1 Larger Skull			Family 2 Walk Upright			Family 3 Loss Of Hair		
Original parent		2N			2N			2N	
Gen 1	A	3N		B	3N		C	3N	
Gen 2	2A	6N		B	6N	AB	2C	6N	
Gen 3	4A	12N		2B	12N	2AB	3C	12N	ABC
Gen 4	8A	24N		4B	24N	4AB	6C	24N	2ABC
Gen 5	16A	48N		8B	48N	8AB	12C	48N	4ABC
Gen 6	32A	69N		16B	69N	16AB	24C	69N	8ABC
Gen 7	64A	192N		32B	192N	32AB	48C	192N	16ABC
Gen 8	128A	384N		64B	384N	64AB	69C	384N	32ABC
Gen 9	256A	768N		128B	768N	128AB	192C	768N	64ABC
Gen10	512A	1536N		256B	1536N	256AB	384C	1536N	128ABC

Figure 190: Ape To Man Chart Across Ten Generations

54. Remember, this is claimed to be at least a 'one in a million' chance, with many higher estimates, so to actually get a mutant we may need many millions of offspring.

For argument's sake, we can say that each of the following variants occurs:

A: Larger skull   B: Ability to walk upright

C: Loss of body hair   D: Development of hooded noses.

The normal apes are shown as 'N'. These are the apes that still exist unchanged today, whereas the new developments A, B, C, and D will eventually blend to become humans, resulting in ABCD representing a human. So that we can chart the changes quickly, we will assume that these ape family variants only mix or interbreed one per generation, so that:

• Variant A mixes with variant B in Generation 2 = AB

• AB mixes with C in Generation 3 = ABC

• ABC mixes with D in Generation 4 = ABCD

Generations	Family 4 Hooded nose			Total normal apes per generation	Total variant apes per generation	Grand total per generation
Original parent		2N		8N	0	8
Gen 1	D	3N		12N	4	16
Gen 2	2D	6N		24N	8	32
Gen 3	4D	12N		48N	16	64
Gen 4	7D	24N	ABCD	96N	32	128
Gen 5	14D	48N	2ABCD	192	64	256
Gen 6	28D	69N	4ABCD	384N	128	512
Gen 7	56D	192N	8ABCD	768N	256	1024
Gen 8	112D	384N	16ABCD	1536N	512	2048
Gen 9	244D	384N	32ABCD	3072N	1024	4096
Gen10	448D	1536N	64ABCD	6144N	2048	8192

Figure 190: Ape To Man Chart Across Ten Generations (continued)

The result is eight different creatures, as shown in the chart:

- The original normal ape: shown as N
- Four original mutants: A, B, C, D, which are non-apes.
- Two combined mutants: AB, ABC, which are non-apes/non-humans.
- One final blend with all variants: ABCD, which is a human.

Under normal circumstances, all of these variants would mix randomly with every other variant in subsequent generations. The result would be the following seven non-ape/non-human variants:

- AC: larger skull, loss of body hair
- AD: larger skull hooded nose
- ACD: larger skull, loss of body hair, larger hooded nose
- BC: walk upright, loss of hair
- BD: walk upright, hooded nose
- BCD: walk upright, loss of hair, hooded nose
- CD: loss of hair, hooded nose

From this, we can see that the total number of different ape mutants totals fifteen (including N), however our chart only uses the previous eight, as the example would become too complicated otherwise.

At this point, there is a very significant issue to remember, as it is fundamental when assessing whether anything living has changed into anything else:

Our four families of normal apes have managed to survive and reproduce. They have managed to have normal offspring that also survived. We know this can happen for the ten generations in our example as we still have normal apes making the same normal apes today as they did theoretically 15-20 million years ago, which would represent over one million generations.

Each of these four families had one mutant that was an improvement towards the development of humans. Each improvement also had to be able to survive and reproduce the same improved offspring that were also able to survive like their parents – and so on until Generation 10.

Each combination of improvements from different families obviously did the same, as they were all better adapted than their ancestors and they also had offspring that survived. We know that all of these different types of ape/human combinations can lead normal lives and have offspring, essentially forever, as ultimately they changed into humans who survived and reproduced in abundance. Logically, it is reasonable to expect each different stage of the ape to human process will still be alive and producing offspring today.

If we follow the first four generations, we discover the following numbers of new differing creatures:

Generation 1: 12 Normal + 1A, 1B, 1C, 1D   = 16

Generation 2: 24 Normal + 2A, 2C, 2D, 1B, 1AB   = 32

Generation 3: 48 Normal + 4A, 4D, 2B, 3C, 2AB, 1ABC   = 64

Generation 4: 96 Normal + 8A, 4B, 6C, 7D, 4AB, 2ABC, 1ABCD = 128

As we know in this exercise, ABCD is human, so after four generations we have a new creature: human. If we extrapolate these to Generation 10 we find that there are numerous mutants in all manner of change. We can see that there are five hundred and twelve type A part-apes that have continued to produce type A mutants with a larger skull. This is because type A was able to survive, as its larger skull was a positive mutation. Similarly for each of the 256B, 384C and 448D types who would have survived based on the evolutionary principle that their new features are improvements that give them superior ability to adapt and the expectation is that each would have produced their own types in reproduction.

Those mutants that combined with other mutants (for instance, AB, ABC, ABD) were also capable of surviving and continuing to produce those types of offspring as they all had to survive for the mutation to get passed on. In total, in Generation 10, there are eight thousand, one hundred and ninety-two creatures developed from the original eight. The total numbers of the different variants are:

6,144N + 512A + 256B + 384C + 448D + 256AB + 128ABC + 64ABCD = 8,192

After just ten generations, there are many different types of part apes able to satisfactorily survive and reproduce. In our example, after thousands (and millions) of generations there will now be masses of these mutants alive today, as with the original normal apes. Even allowing for the 'one in a million' rate of mutations, billions of evolving mutant apes are required to be able to develop into just one human. Thus, with billions of apes and billions of mutants there would be the expected masses of intermediary fossils.

Evolutionary Theory fails to explain why there are no part ape/human creatures still living today and why the fossils of their ancestors are not readily found in the fossil record. The claim is that every step of mutation, including the millions of successful ones, died off without trace leaving only the full and complete identifiable, unique species.

In summary, we can make the following observations:

- Apes are still in the same environment as 15-20 million years ago and show no signs of mutating to anything else.

- There are no fossil records of any changes from apes to other creatures, no changes in any living apes now, and they still live how they always have.

- They are, to all intents and purposes, exactly as they were when they first appeared in the fossil record.

- There is no scientific evidence to suggest that they formed from anything else prior to them, nor is there any evidence to prove they changed into any other species, including humans.

Chapter 23 Unraveled

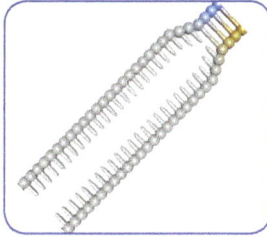

- *Apes appeared instantaneously in the fossil record fully developed with the ability to walk on two legs.*

- *Humans suddenly appeared on earth in multiple locations independently of apes, fully developed.*

- *There is no scientific evidence of links between the apes and humans – any supposed relationship is theoretical and not factual.*

- *There is no scientific evidence that apes evolved from any other creature or that humans evolved from apes or any other creature.*

- *Humans are significantly different from all other creatures on earth.*

- *Human skills, abilities and mental aptitude are inexplicable developments under evolution.*

24

HOW A HUMAN EMBRYO FORMS: THE KEYSTONE FOR BIOLOGICAL EVOLUTION

For life to exist and flourish, reproduction of living entities has to occur. From one fertilized egg, each different creature with millions, billions or trillions of cells is formed from a complex, timed and specific series of processes. All organisms require these types of programs – plants, insects, animals – but they vary in complexity depending upon their individual components.

As we have more knowledge about human embryo development than any other creature, we will look at this process in some detail. Similar types of development happen in all other creatures but in entirely different sequences, sizes, structures, timings, and outcomes unique to that creature. Within this close examination, we will investigate the key element in Evolutionary Theory regarding the emergence of the different life forms on earth.

Human Reproduction

To create a new human being, the basic requirement is that a male's reproductive cell, a sperm or male gamete, unites and combines with a female's reproductive cell, an ovum or female gamete, both of which are initially inside their respective bodies. The male's sperm leaves his body, enters the female reproductive tract, locates the ovum, combines with it, and then blends information from its twenty three chromosomes with the twenty three chromosomes of the ovum. The fertilized ovum is called a zygote. The initial zygote cell will divide and multiply to become multicellular eukaryotic cell called an embryo, then after approximately eight weeks it will become a fetus and eventually develop to a point where it can survive outside the mother. After approximately forty weeks, or nine months, the fetus will be expelled from the mother as a newly born infant (baby). We will now approach these developments from the beginning where it becomes apparent that several barriers impede the progress of biological evolution.

A human male ejects semen, containing forty to six hundred million sperm cells, from an erect penis into the female vagina. The optimal time for a human female to become pregnant is midway into her menstrual cycle, which is normally twenty-eight days[55]. After entering a female vagina, sperm will swim through the cervix up into the uterus and finally into the fallopian tube where they may meet a released ovum passing down from the ovary. The sperm's journey can take up to ten hours.

The female ovum is released from the ovary in a process known as ovulation. Involved in this process are hormones, such as luteinizing hormone, which help to control the rate of a female's menstrual cycles. They also stimulate production of testosterone in males. Normally only one ovum or egg is released per monthly menstrual cycle, usually alternating between the left and right ovaries. For a female ovum to mature prior to ovulation, luteinizing hormones and progesterone prepare the uterus for an embryo to attach or embed itself into the lining of the uterus. If the ovum is not fertilized, progesterone production ceases causing the uterus lining to disintegrate, which is then expelled from the body as menstrual blood. Female ova are finite in number. When the last ovum has ovulated, typically around the age of fifty, the female enters menopause and can no longer reproduce[56].

Another hormone, testosterone, is predominantly a male hormone that aids male sperm production. Females' ovaries, however, do produce some testosterone, which aids in sexual desire. Both genders get other benefits including bone and muscle development.

Sperm can live for about two days but only a few percent will reach the fallopian tube where the ovum is fertilized. Initially, just one of the male's sperm will penetrate the surface of the female's ovum and fertilize it. The ovum then undergoes a chemical process that seals itself to prevent other sperm from entering and flooding it with excess amounts of DNA. This process involves complex chemical and electrical activities.

To ensure that human and other species' reproduction remains protected and avoids contamination, there are various processes, barriers and tests that are engrained in the processes.

---

55. The menstrual cycle can vary between twenty-five to thirty-five days.
56. Without medical intervention and donor eggs from another female.

Sex Cell Security Codes

When approached by sperm, the female ovum releases a gelatinous substance called fertilizin, which is composed of carbohydrate and protein. This discharge attracts the sperm and ensures quicker bonding with the egg membrane. However, when the first sperm comes into contact with the ovum surface, its outer membrane binds to receptors on the ovum surface, then it casts off its external cover (acrosome) and it releases 'anti-fertlizin', which counters the attraction of fertilizin and stops other sperm from approaching. If the receptors on the sperm and the egg do not match neither will accept the other and fertilization will not occur, and therefore no embryo will form (explained below). When the sperm enters the ovum, it drops its tail and the membrane on the ovum surface immediately re-seals itself, allowing only one sperm to enter. The sperm head expands as it approaches the center of the ovum and within thirty minutes has completely dissolved and released its genetic information to blend with the genetic information in the ovum.

A second activity is also in play during the binding process. Sperm carry a positive electrical charge, which is attracted to the negative charge of the ovum and causes the sperm to head towards the ovum. When the first sperm enters the ovum, it triggers the ovum to instantaneously change its charge to positive, which then repels all other sperm.

This series of events are displayed in Figures 191 to 196:

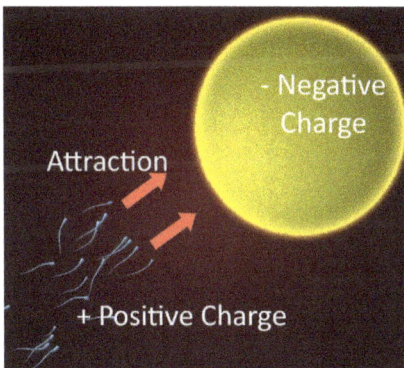

Figure 191: Sperm Approach The Egg

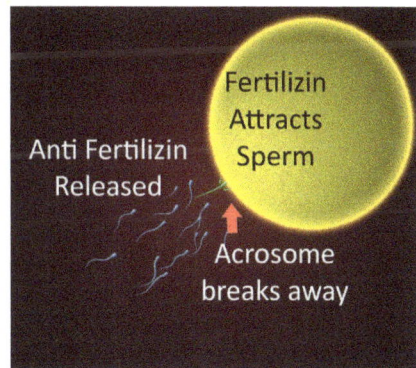

Figure 192: Sperm And Egg Contact

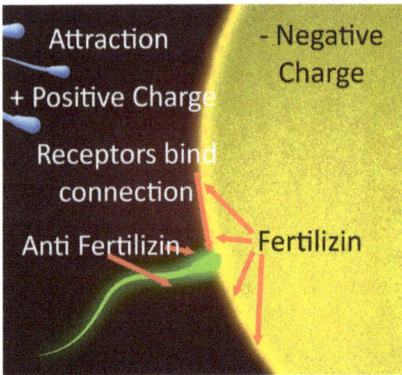

Figure 193: Sperm And Egg Surfaces Bind

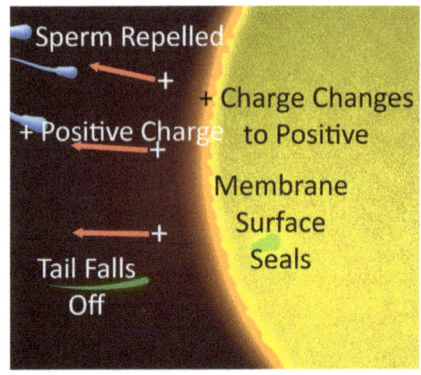

Figure 194: Sperm Enters Egg Which Changes To Positive Charge

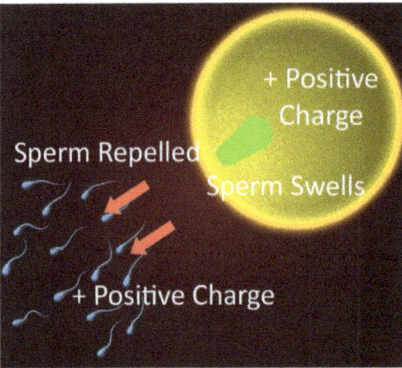

Figure 195: Sperm And Egg Unite Other Sperm Repelled

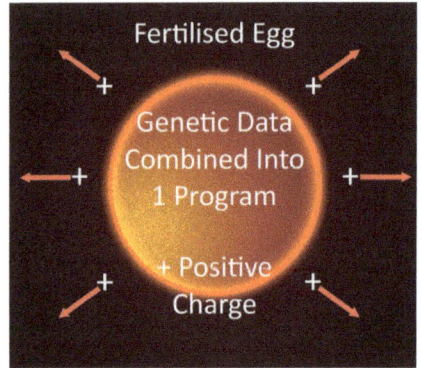

Figure 196: Union Complete After 30 Minutes

All of these processes, chemical reactions and information blending guarantee the embryonic process begins without contamination of multiple or non-human sperm. One of these fundamental processes, however, is worth highlighting in that it is structured to ensure every different creature on earth *fertilizes and creates its own kind*, and not anything else. In respect to the human fertilization process, female ova secrete a unique blend of fertilizin that only binds with matching receptors in human sperm. It will not match with the sperm codes from other creatures. This specific matching of codes occurs prior to binding in order to prevent mixing of different reproductive cells from different creatures. Added to this, once the ovum allows a sperm to penetrate inside, it changes its own negative electrical charge to positive in order to repel other positively charged sperm. These fail-safe mechanisms have established two different but effective levels of protection.

For the above protection to work, every creature's sperm and eggs must have their own matching and unique anti-fertilizin composition that prohibits contamination from other sperm/egg combinations. Effectively, each type of creature has its own personal security system, each with its own unique code specific to that creature which its reproductive cells must be able to test and analyze to ensure matching.

For the evolutionary development of all life forms, this code would be required, through necessity, to change constantly from one creature to another. If it changed every time there was a mutation along the evolutionary path, trillions of trillions of reproductive sperm/egg codes would be created, resulting in an ever-increasing challenge that any two would match and allow reproduction. However, if the reproductive code only changed when a totally different creature had evolved, the question Evolutionary Theory must answer is: How did randomly mutating cells collectively determine what is, and what isn't, a complete change?

Furthermore, the last mutation to form the newly evolved creature will be either a male or female embryo, not both. Evolutionary Theory therefore suggests that a matching reproductive code appeared in two different genders emergent from separate origins, with neither aware of the other's changed reproductive code. Even if this were the actual evolutionary pathway, it contradicts Evolutionary Theory's basic premise that life results from random chance events, because the reproductive codes would have to deliberately match each other.

With a couple of rare exceptions, these codes do not allow creatures to reproduce with other species and thus create a new type of evolutionary line. Although there are rare examples of blendings of very similar creatures, these usually produce sterile offspring which cannot reproduce and continue the species line. The two most well-known cases are donkeys and horses (producing mules and hinnies) and lions and tigers (covered in Chapter 15). There is documented evidence of the same type of creature adapting to different environments and not reproducing with the original group, such as Darwin's finches on the Galapagos Islands, but these were still finches, not some other type of bird.

Human Embryo Development

Following the union of the sperm and egg, the human embryo starts its amazing journey:

---

57. Trophoblastic tissue coats the developing cells and supplies nourishment, eventually forming into the placenta.

58. Without attachment to the endometrium, the early embryo dies.

Week 1

Within eleven hours of fertilization, the sperm and ovum have fused together to be one cell – a zygote. Within another twelve hours or so, it starts to duplicate. Cell division will continue for several days until there are eight or sixteen cells. Then it makes its way via the fallopian tube to the uterus, arriving in about four days. The zygotic cells then attach themselves to the wall of the uterus, known as the endometrium, in order to receive nutrients and oxygen and thus the ability to continue development. The zygote is now an embryo of about one hundred cells and at this stage is called a blastocyst. It is enclosed in an outer layer of cells called trophoblasts[57], the first cells to differentiate from the developing fertilized ovum[58]. The embryo will form from the internal cells while the placenta develops from the external trophoblasts. This placenta (or sac) surrounds the developing embryo and will fill up with a protective liquid called amniotic fluid that consists mostly of water and, later, excreted urine from the developing kidneys.

At around six days, the outer trophoblasts automatically secrete an enzyme which erodes the endometrial lining of the uterus so that it can attach itself in a process of implantation. The cells produce various hormones that cause the uterus to engorge in that area, forming new capillaries essential for maintaining the pregnancy.

Week 2

At this very early stage, the future development of the embryo is defined by three layers – the outer ectoderm, middle mesoderm and inner endoderm.

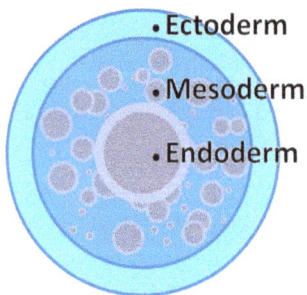

Figure 197: The Three Cellular Layers Of The Embryo (Week 2)

Although there are less than two hundred new cells, they have already organized themselves into distinct positions and layers:

• The outer ectoderm, which is relatively thin, becomes the brain, nervous system, skin, nails, hair, eyes, ears, nose, mouth, pituitary, and mammary glands.

• The middle mesoderm, which is thicker, becomes the muscles, bones, spleen, blood cells, heart, lungs, and components needed for excreting waste, such as liver and kidneys, as well as the sexual organs.

• The inner endoderm core develops into organs supporting digestion and respiration.

Even at this stage of just a few days old, the cell is operating on a clear and specific genetic program that is manipulating the cells for future changes.

Weeks 3 – 4

The beginnings of arms and legs appear as smallish extensions or protrusions just before the end of the fourth week. The middle cells produce a chemical that makes the outer ectodermal cells become larger. A range of cells appear that will shape into the blood vessels. The beginning of heart cells appear and a primitive heartbeat is present before blood vessels appear, indicating that a heart is an immediate priority in the development of a life. The beginnings of the brain, liver, lungs, pancreas, and stomach are identifiable. A very basic blood flow develops with cells travelling along the central nervous system then into the mother via the uterine capillary network.

Weeks 5 – 6

The umbilical cord starts to form from the material that connects the embryo to the uterus. There is a jelly-like substance (Wharton's jelly) that forms around these tubes that protects them from impacts. The umbilical cells have a strong enough connection to prevent it from shaking free or becoming damaged in normal circumstances during pregnancy but supple enough to be removed after birth without damage to the mother or baby.

Next, the umbilical cord develops a central vein that carries blood, oxygen and nutrients to the embryo. Two arteries also develop that remove waste products from the embryo into the mother's blood stream where they are filtered out by her kidneys. If these tubes formed in reverse for any reason, the embryo would be fed the mother's waste, such as urea and carbon dioxide, and quickly die.

At this time, the head begins to take shape in preparation for future growths. It starts to form the sockets for the eyes and shaping for the ears. Various parts of the eyes and ears then start to develop. The backbone and sex organs also start to form as does the digestive system. The heart now fills with blood and plasma. Plasma consists mostly of water and dissolved salts and is critically important to life due to its ability to carry blood cells around the body.

Among various developments in this period:

• The brain develops into the three main areas (some biologists claim five areas).

• A mouth and tongue start.

- Both sides of the lungs becoming visible.

- The esophagus and four sections of the heart can be seen.

- Blood enters the liver through the umbilical cord.

- The spinal cord wall has developed into three distinct areas that each perform a different range of functions.

- A thin layer of skin covers the embryo.

Weeks 7 – 9

The general sequence of events is:

- A basic heart is in place and beating.

- Fingers, ears, lungs, and eyes start forming although the eyes have a membrane over them.

- Muscles and nerves appear to be working while the throat and abdomen are formed and in place.

- As the skull gradually forms, the face, nasal cavities and parts of the jaws start appearing.

- Nerve distribution starts.

- The feet are just beginning to form.

- Kidneys are almost fully formed.

- Hands, arms and shoulders are visible.

- The tubes that carry urine between the kidneys and bladder, the ureters, grow longer.

- Jaw and facial muscles commence.

- The brain forms the olfactory bulb (for detecting and identifying smells, and is reviewed in more detail in Chapter 25).

- The thigh, legs and feet become more developed.

- The major organs like the heart, brain, intestines and lungs, although not fully formed, are in place. The face and jaw are already formed, but now the facial muscles and teeth buds begin to form.

- Up to this point, the skeletal structure is mainly cartilage, which now starts to be replaced by bone cells. As this happens, the joints start to form.

- Genitalia begin to develop and nipples appear on the chest.

- The kidneys start working and secrete urine.
- Legs, knees, ankles, toes, and nails are still developing.
- Muscles develop and get stronger.
- The brain is connected to tiny muscles and nerves that enable the embryo to make spontaneous movements.
- The nose forms.
- Up to this point the embryo is sexless but now, testes or ovaries are distinguishable.
- The clitoris or penis starts forming.
- The fingers fully separate.
- The feet lengthen and become more defined.
- Taste buds begin to form on the surface of the tongue. Bones of the palate begin to fuse.
- Hands and feet develop.
- The areas around the throat also begin to form.

Weeks 10 – 11

The general sequence of events is:

- Basic brain structure is complete and the brain mass grows quickly. The brain will finish up with eighty billion connections between neurons.
- Sockets for the teeth are formed in the gums.
- Separate folds of the mouth fuse to form the palate.
- Early facial hair follicles begin to develop.
- Vocal cords form in the larynx and the fetus can make sounds.
- Digestive tract muscles are functional.
- The liver starts to secrete bile.
- Development of the thyroid, pancreas and gall bladder is complete. The pancreatic islet cells start to produce insulin.
- Fingernails begin to grow.
- The abdomen and parts of the ears form.
- Amniotic fluid is swallowed by the fetus which it excretes through

urination. Effectively, the processes of ingestion and excretion have started and function satisfactorily.

Weeks 12 – 13

The general sequence of events is:

- The sucking muscles in the mouth form, tooth buds continue to develop and saliva glands begin to work.
- The lungs develop further as the fetus inhales and exhales amniotic fluid.
- Breathing, swallowing and sucking develop more.
- Arms have almost reached their final proportion and length.
- Hands, particularly the thumbs, become more functional.
- Muscles and nervous systems continue to advance.
- Sweat glands appear and body hair begins to grow.
- Muscles start working as seen by the movement of arms and hands.
- Facial movements are apparent.

Weeks 14 – 21

This is a period of growth as cells replicate to increase the size and strength of all body parts. Milk teeth have formed and, underneath these, the base for permanent teeth is starting. By the end of this period the baby is covered in soft hairs. These protect it from the increasing urine content in the amniotic fluid.

Other developments in this period include:

- Eyes and ears are in place.
- After about week sixteen, most body parts have formed and are growing.
- The nerves get sheathed in a coating called myelin that insulates them and allows greater transmission speed.
- Blood circulation is operating quite strongly.
- The head develops wispy hair (that falls out after birth).
- Females develop a uterus and ova.
- Vernix covers the whole body. This white substance is a mixture of epithelial cells, hairs and skin secretions thought to provide resistance against bacterial infection.

Week 22

Developments in this period include:

- The baby starts reacting to the mother's voice or external touch.
- Eyebrows and eyelashes start to form.
- The baby will start to make white blood cells to resist infections.

Week 23

Developments in this period include:

- Fingerprints start to develop.
- The enamel required for the milk teeth starts to appear.
- Nerve cells commence joining up to make the nervous system.

Week 24

Developments in this period include:

- The inner ear is formed and won't change size from this point.
- Nails start to grow.
- The lungs start to grow blood vessels. These are essential for breathing air and taking oxygen throughout the body. Along with these, air sacs in the lungs called alveoli start to make surfactant, a lipoprotein that prevents the minute lung sacs sticking together (which would make them unable to expand during inspiration) and enhance the movement of oxygen and carbon dioxide across the alveoli membrane.
- We can identify the parts of the spine, which comprises thirty-three rings, one hundred and fifty joints and about one thousand ligaments.

Week 25

Developments in this period include:

- Bones continue to harden.
- The skin is covered with more vernix, which assists in retaining an even temperature and protects against the mild acidity of the amniotic fluid.

Week 28

Developments in this period include:

- The eyelids, which cover the eyes fully, now split to allow sight.
- The brain's operation is evidenced by its ability to control body temperature.
- Lanugo hairs now only appear on the head and shoulders. These fall out 3-4 months after birth.
- Red blood cells start being made by the bone marrow and will continue this function for life.
- Fat, which acts as an insulating medium and energy storehouse, develops in the skin.

Weeks 30 – 31

Developments in this period include:

- Nerve cell connections increase.
- The toenails are fully formed.
- Alveoli now cover themselves with a new cell layer. This in turn secretes a liquid that allows the lungs to retract completely after the first intake of air at birth.

Week 32

Developments in this period include:

- The eyes are still blue, as pigmentation only occurs after birth when exposed to light.
- The immune system develops.

Week 35

- After this period of staggering growth, suddenly, the protective hairs and tissues disappear.

Week 38

The skull is comprised of five separate plates. At birth, these plates move together and overlap at their edges to enable the baby to fit through the birth canal, minimizing stress to the mother while protecting the baby's brain. Because of the size of the human brain, the head is so large, proportionally, that it would not fit through the female birth canal if the skull was solid and not divided into plates. Effectively, the head stretches and squeezes into a shape almost like a chicken egg, then, after birth, reshapes into a rounder form as these plates return to their normal pre-birth position and begin to fuse together over time. Easing of the birth process also happens elsewhere throughout the body, not just with the skull. When a baby is born it has a total of three hundred bones, which later fuse together to become just over two hundred bones.

While all cells are replicating, not only are they making body parts, they are also sculpting cavities, such as the mouth, throat, lungs, nasal passages, ears, and brains. A significant degree of intricacy is required to make complex cavities in a developing embryo that maintains perfectly symmetry with its surroundings during growth in the uterus, then after birth, then as an adult. During the embryo stage, many cavities are being formed in advance of future components. For example, the sockets for the eyes are made in the skull before the eyes start forming. The development of cavities requires prior cellular memory of the later body parts that are yet to appear, meaning that cells are working to a specific predetermined genetic blueprint.

Now, after this brief overview of the developments in human embryo growth, we will examine some of the processes in more detail at the cellular level, after which we will examine the event on which the theory of biological evolution is based: reproductive cell mutation. Reproductive cell mutation and biological evolution can be considered one and the same.

The processes we have been looking at are really the story of you. This is how you were made. Evolutionary Theory states that it was by accident, without a plan or reason and for no purpose. But what is the science telling us? Is this the random evolutionary making of your existence or is another scientific process at work? Let's find out.

A Detailed Analysis Of Cell Formation

The most likely scenario in an evolutionary world where cells constantly mutate into other unplanned configurations would be a mass of unstructured body parts and functions scattered randomly in some sort of organic blob. But this is not what happens with any creature, as every cell is operating under instruction from a controlled genetic program. In our examination of humans, the appearance, order, frequency, and timing of trillions of cells in our bodies happens in a perfect linear sequence. Each finishes up as the correct cell type, in the correct place and at the precise time that they were required to be. Evolutionary Theory says this is a process of random chance events without any intended outcome.

Our starting point is the first gamete cell, ova, in the female human that has just been fertilized by a sperm. This diploid cell is generally regarded as non-specific, meaning it is capable of becoming any particular cell for any particular body part – ears, eyes, stomach, blood, brain, muscle, and so forth. These have become more widely known in the general population as generic stem cells.

The first thing this cell does is duplicate itself to form an identical cell. Then both cells duplicate themselves totaling four cells. These duplicate to eight then sixteen then thirty-two. At this point, we assume the original cell has duplicated itself five times and the second cell four times and so on. These duplications are set out in Figure 198: Cell Division and are labelled as 'G' cells representing generations. The first cell is 1G and the second generation is 2G while the third is 3G and so on.

No. of Divisions											Original Cell																			
											1G																			
1 (2 cells)											1G	2G																		
2 (4 cells)										2G	1G	2G	3G																	
3 (8 cells)								3G	2G	2G	1G	2G	3G	3G	4G															
4 (16 cells)					4G	3G	3G	2G	2G	2G	2G	1G	2G	3G	3G	4G	3G	4G	4G	5G										
5 (32 cells)	5G	4G	4G	3G	3G	3G	3G	2G	4G	3G	3G	2G	2G	2G	2G	1G	2G	3G	3G	4G	3G	4G	4G	5G	3G	4G	4G	5G	4G	5G	5G 6G

Figure 198: Cell Division

The thirty two stem cells on the bottom line of the table in Figure 198 are comprised of six generations including the original fertilized cell. They are all stem cells at this stage of development. From this point, new cell duplications will change form. That is, they change into a different type of cell, initially embryo and placenta cells, then every other type of cell such as brain, muscles, skin and so forth. Whilst the exact sequence is unknown, it is the mechanics of the processes that are of interest.

If we assume that all stem cells multiply at the same rate, doubling each generation, after the fifth duplication of the original cell, new outer cells suddenly change themselves into different cells that start to form the placenta/umbilical cord. This means that after fertilization:

- The very first stem cell duplicates five times then one of these, or subsequent duplications, changes its form to something else.

- The second generation stem cell duplicates four times then changes occur.

• The third generation duplicates three times then changes occur.

• The fourth generation duplicates two times then changes occur.

• The fifth group changes occur after one duplication.

The observation is that the number of times these early stem cells duplicate is not the driver causing the changes from stem cells to specific cells, as they happen after different numbers of duplications. As cells change into many different types from this point, they somehow know what type of cell to change into. As each stem cell has exactly the same DNA, it is unknown what is causing these changes. One explanation is that there may be some relationship to the total number of cells and/or the time involved, but that would mean the newly formed stem cells in the later generations (3G-5G) are aware of the historical flow of time that has happened to previous generations before they themselves existed. Science does not specifically know what events or communications are happening to instruct stem cells to change, yet it must be some phenomenon common to all cells as they all are at different stages of duplication.

Research so far has indicated that there are possibly two types of triggers. There may be an internal signal generated by the stem cell's own genes and external signals that come from both physical contact with, and/or chemical secretions from, neighboring cells, as shown in Figure 199: Change Signals Are Sent To Stem Cell.

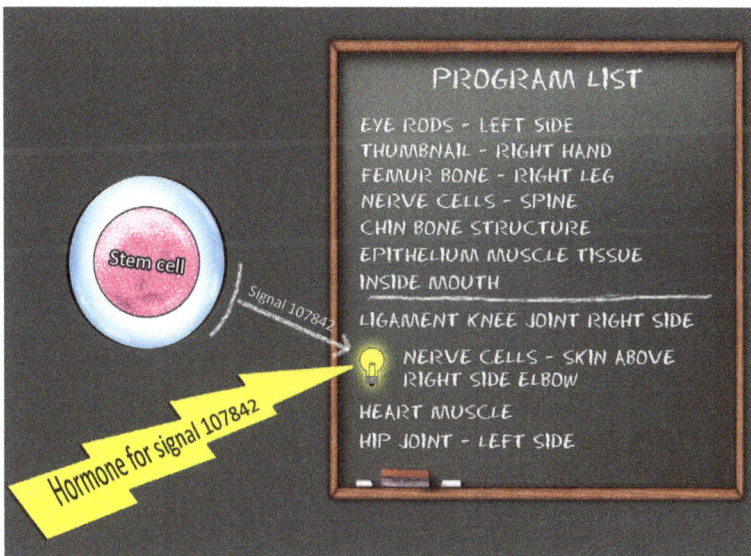

Figure 199: Change Signals Are Sent To Stem Cell

Studies are now concentrating on the types of signals and whether they are specific for different types of cells. This would appear the most logical conclusion because the same signal to any generic stem cell would most likely result in the same change. In this instance, if each cell receives a specific signal, unique to that cell, to induce change into a new and different but specific type of cell, potentially trillions of instructions would be issued throughout every human body, a staggering volume of instructions.

At the point of cell division in which there are thirty-two cells, the cells have arranged themselves into inner and outer layers. The outer cells will duplicate to about thirty cells and then physically change their structure and operation to eventually form the umbilical cord by increasing their types and numbers. The inner cells duplicate until there are about seventy cells, making a total of about one hundred blastocyst cells. The outer cells secrete an enzyme and produce hormones to secure the forming embryo to the endometrial lining of the uterus. This is a very specific and necessary process,

notably:

1. These cells had to be capable of producing these hormones and enzymes at the time of their making.

2. These cells are programmed to deliberately release these hormones and enzymes on day six thereby having time awareness. As yet, scientists do not yet know the mechanisms behind this.

The outer blastocyst cells have, at this point, specifically formed a new link to the uterus while the inner blastocyst cells are duplicating into three layers – ectoderm, mesoderm and endoderm – that will themselves become different parts and functions of the developing embryonic body. The science and biology at this point of embryonic development is somewhat indeterminate and it is difficult to know exactly what happens from here, so we are going to make some assumptions (and, of course, will stand corrected if future research proves a different sequence).

The embryo consists of about one hundred cells in three layers and is implanted into the uterine endometrium by cells that have already started to change. These one hundred embryonic cells now also start to change their structure. To do this, cells duplicate, then specific signals are issued in a specific order so that the new cells change form, ultimately into every type of cell required to make a living human body. To put this in context, the cells of a developing fetus necessarily duplicate themselves at the correct rate, in the correct number, in the correct sequence, and in the correct location with every other one of millions of fetal cells, then stop or slow duplicating when their number reaches a predetermined point. For example, if the cells for the right leg continued to duplicate at a different rate from the left leg, the legs would grow to different lengths, shapes and thicknesses, naturally resulting in mobility difficulties.

To give a better understanding of the complexity involved in organizing trillions of cells to become specific types, in predetermined numbers and perfect locations, we will follow the process in more detail by examining the development of the skin on a microscopic level.

At weeks five and six, a thin epidermis covers the embryo, which is about 25mm long. The epidermis, or skin, is comprised of many components which are shown in Figure 200: Cross Section Of Human Epidermis.

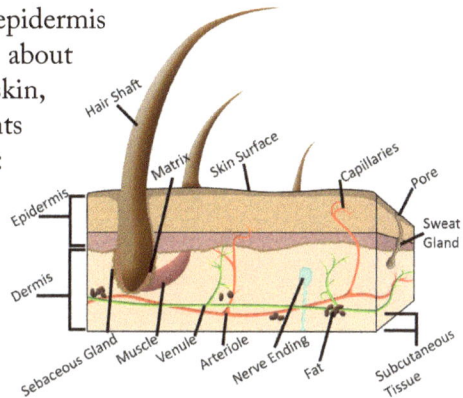

The epidermis is considered an organ[59] and has a high level of intricacy with many inter-acting components. There are different thicknesses, porosities, cell types, strengths, functions,

Figure 200: Cross Section Of Human Epidermis

positions, and interactions. Scientists do not know the order in which these cells form. With so many different parts, they have to develop in harmony with each other, ensuring that each is in the exact location that matches the controlling program. Not only does this program have to lay out the positions of each body component, it must also have intimate detail about each of the millions of cells constituting these parts.

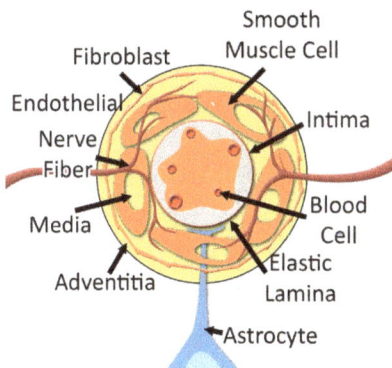

As an example we will select arteries as a starting point. Initially, a newly made generic stem cell, attached to all the other cells in the early embryo, will be instructed to change into a muscle cell, which is the major component of arteries[60] and is displayed in Figure 201: Artery.

Figure 201: Artery

---

59. A part of the body that performs a specific function or set of functions.

60. Arteries carry oxygen rich blood from the heart to the rest of the body, with the

Assuming there are two complementary signals, they must be generated *at exactly the same time* and these must *deliberately instruct this one particular cell* to physically change into a muscle cell. Then, this cell will change its own internal and external constitution to become a muscle cell. Whilst this is an everyday occurrence, it is an astonishing event in itself, repeated trillions of times in sequence.

The significance of this action is that every new stem cell is pre-programmed with knowledge of the structure, design and functioning processes of every different type of cell that makes up the entirety of the human body. This is because any newly formed stem cell can become any type of cell. It cannot change unless it contains the preconditions for the new type of cell to change into and how that type of cell works in conjunction with everything else.

Recent medical experiments with skin cells show that they can be reprogrammed to change into a variety of different cells, which indicates that multiple programs are contained in one cell. Therefore, the signals are some form of unambiguous code directing a deliberate action to occur to one specific cell. In this case, the signal is 'activate artery cell program', directed toward a particular stem cell. Consequently, it appears that every newly made stem cell has complex information about all of the cells' structures in the body, embedded in its DNA from the moment of its making so that it can activate the appropriate program upon instruction. Further, each of these cells has to have all the necessary biological components to enable it to become every other type of cell. There can be significant differences in cell structures and functions, and therefore the genetic programs and cellular components, as seen in Figures 202: Nerve Cell and 203: Fat Cell.

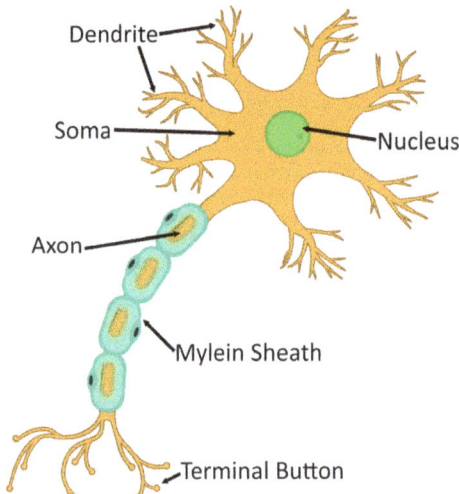

Figure 202: Nerve Cell

exception of the pulmonary artery, which carries deoxygenated blood from the heart to the lungs, and the umbilical artery, which carries deoxygenated blood away from the developing fetus to the mother via the umbilical cord.

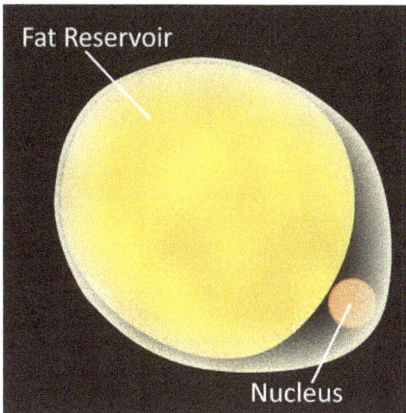

Figure 203: Fat Cell

Once the change to the cell has occurred, it is assumed that the first artery muscle cell then starts to duplicate itself into more muscle cells, supposedly operating under instructions from an artery cell program for the skin in that area of the body. All cells must therefore have some predetermined knowledge about where in the body they are located, as the programs for artery components will differ depending on their locations. As new artery muscle cells are formed, they also continue to duplicate themselves into more and more and more artery muscle cells. These don't, however, swell up as an uncontrolled mass of cells, which would be cancer. Rather, each and every new cell appears in a specific location so that eventually, in coordination with other cells, they will combine to form the structure of a thin tube that connects with the rest of the body's arterial system. Furthermore, in this particular process, the cells are required to make a hollow tube that will extend lengthways in both directions, to link up with other arteries being made in other parts of the embryo.

At some point, this lengthways duplication will stop when the artery meets the artery network. Therefore, not all cells duplicate continuously, or in the same direction or location. Some will cease developing once the correct thickness or length has been achieved for that point in the fetus's development, then once the basic arterial system has been formed they then adjust their rate of replication to match their increasing numbers, shapes and sizes proportionately with the rest of the body as it grows[61]. This requires a highly complex program for the position and duplicating rate for every single cell – not just here but in the entire body. For this to occur, some method of cellular monitoring is required.

The epidermis in Figure 200: Cross Section Of Human Epidermis is comprised of more than twenty different components – hair follicles, capillaries, fat cells, melanin, and so forth – each of which requires different types of cells, structures, lengths, and thicknesses. This requires over twenty different sets of signals to over twenty generic stem cells, each of which

---

61. For example, the arterial system supplying connection to the legs must grow in proportion to the muscles, nerves, bones and other tissues of each leg as they grow until longitudinal growth ceases in early adulthood.

reacts in a completely different way by changing into over twenty different types of cell. Each new cell duplicates identical cells in the correct locations, thicknesses and lengths amid a myriad of other replicating cells to achieve their predetermined structures. Furthermore, these structures will differ depending on the location, because skin is not the same in all areas of the body. For example, skin on faces or hands and bases of feet have no hair and are of different thicknesses compared with skin on the back torso or scalp, each of these having vastly different hair densities.

Ensuring the different components of epidermis are aligned in the correct places at the correct times requires a significant level of coordination and timing. Consequently, instruction programs are theorized to exist for each cell in each of the twenty different components for every micron of epidermis on the body. This program has a predetermined sequence of signals to send out as new cells are developed; it tells them where to appear, how many to make, when to stop, and when to restart again. That is, this program includes signals for both the location and type of cell required.

There is also a program controlling the rate at which new cells are being made. That is, a timing program. If, for example, a new stem cell is made every second, then, when this cell is required to change its makeup, a matching signal (or possibly multiple signals) is released at the same time that the cell appears, that is, every second.

Accordingly, two programs are operating – signal program and timing program – and it is critical to the development of the fetus that these are precisely coordinated in order for the proper structure and formation of the developing cells. These two sequential programs coordinate clear, unambiguous signals to be sent to specific cells, one by one, at the exact time they are formed, wherever they might be in the growing structure of millions, then billions, then trillions of cells.

Because every individual cell knows the correct time to duplicate, there must be some concept of temporal flow and some way of measuring it other than by the brain, which only starts building its basic structure between weeks five and ten. Other cells must know what signals to send and when and where to send them, requiring a knowledge of time and three dimensional layouts (height, width and length).

These programs are embedded in the original fertilized stem cell. Each female gamete cell, ovum, and each of the billions of male sperm cells contain a share of these programs. When they combine, theoretically, the programs blend into a super-program, with trillions of instructions, which is passed on to each subsequent cell as it is made. Of the trillions of mutations that could occur in cell reproduction, these programs rarely falter.

As such, there is a compelling argument that any faults or mutations are caused by external influences such as diet, drugs, genetics, or environment, not the programs themselves.

All areas in the human body have complex layouts that cells must adhere to and accommodate. Cells forming the skull, for example, are required to make complicated shapes that include areas for two eyes, two ears, a mouth, a nose and a brain, some of which only appear in later development stages, from baby through childhood to adulthood, as seen in Figure 204: Human Skull.

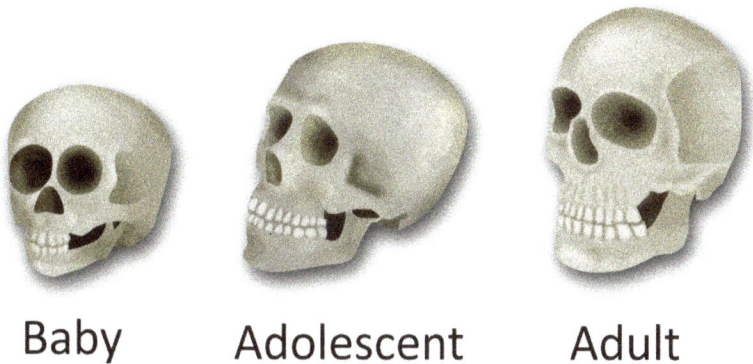

Baby   Adolescent   Adult

Figure 204: Human Skull

Interestingly, apart from signaling and timing, these programs also allow for the mirror image duplicates of many body parts, shapes and sizes for each side of the body. Consider the ears. According to Evolutionary Theory, an ear developed by accident through a random cell signal program, then this exactly shaped and functional design appeared in mirror image on the other side of the head. Other mirror image organs found in the body include hands, fingers, wrists, elbows, shoulders, toes, feet, ankles, ribs, and knees.

The Human Body's Major Systems

While the development of cells into organs and structures is occurring, there is another, higher level of organization taking place. Our bodies have coordinating groups of organs and parts that perform specific functions, such as the muscular system that is programmed to allow us to move in a coordinated and balanced manner. These groups complement all other functions in order for the body to operate efficiently and survive, as well as accommodating each other within the physical body. Each component has a specific shape, size and location so that it, and everything around it, remains in its specified location and can operate effectively.

Following are images of the human body's major systems, their basic components and a brief description of their functions. Each of these is constructed from billions of individual cells that developed into the correct types and appeared at the appropriate time and place to allow the body to mature from an embryo into an adult. These operating systems individually keep the body alive and allow it to function whilst maintaining interdependency with one other. The programs that allowed these systems to develop from one cell into a full body, as a coordinated group, and keep the body alive while doing so, are highly sophisticated.

Integumentary

The integumentary system:

- Is comprised of the epidermis (skin), nails and hair.

- The epidermis has three main layers that cushion and protect deeper tissues. It prevents micro-organic infections from entering, regulates temperature, detects damage and identifies it as pain, and senses things in contact with the body (such as touch, air pressure, heat, and cold).

- Nails made of keratin, a protein, protect the sensitive nerves found at the extremities of fingers and toes. Nails grow in length from the base allowing damaged areas to be recovered. Layers of dead cells become compacted during growth and provide strength and flexibility.

- Hair, such as scalp hair, insulates the head, from which eighty percent of all body heat loss occurs. Scalp hair grows at a different rate from the rest of the body because its growth phase (anagen) is several years compared with a few months for other body hair. All hairs also go through a resting phase (telogen) when the dead materials fall out.

Muscular

- There are three basic types of muscle – skeletal, heart and smooth.

- Muscles control body movement, hold the body posture in place, maintain blood circulation, and generate internal heat[62].

- There are approximately six hundred and fifty muscles in the human body.

---

62. For example, shivering generates heat when we are cold.

- Muscles account for over half of the weight of the average human body.

- Skeletal muscles are attached to the human skeleton via tendons.

- There are two types of muscle fibers: fast-twitch (Type I) and slow-twitch (Type II). Protein strands (actin and myosin) pull the muscle fibers together producing the same force regardless of muscle type. Fast twitch fibers fire more rapidly using anaerobic metabolism to break down glucose (stored energy) without oxygen allowing bursts of speed but produce lactic acid as a by-product, which limits the length of high performance activity. Slow twitch fibers use oxygen to produce energy over longer periods and result in better endurance capabilities.

- Smooth muscles are found in organs such as the intestines, lungs, blood vessels, the bladder, and uterus.

- Smooth muscles are autonomous, in that they operate involuntary and spontaneously, as with cardiac muscles.

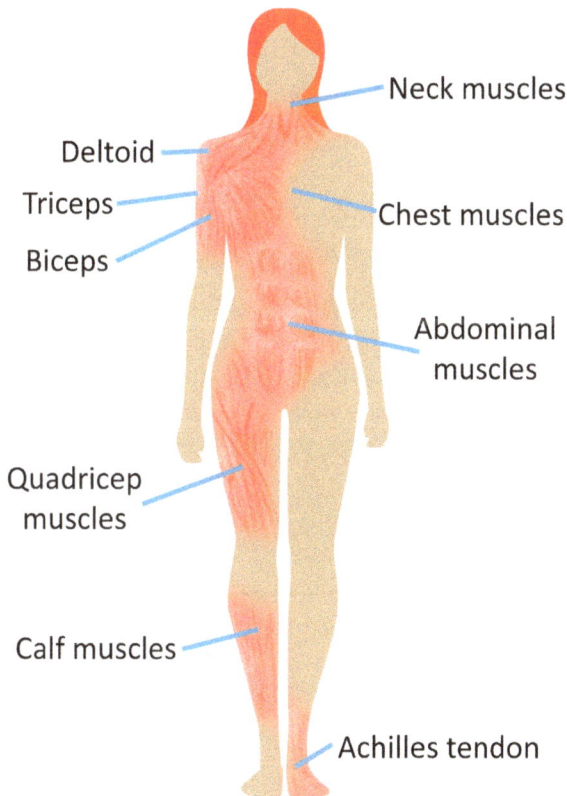

Figure 205: Muscular System

Circulatory (Cardiovascular and Lymphatic systems)

• Comprises arteries, veins and capillaries.

• The circulatory system carries nutrients, hormones and gases around the body, controls both blood and lymph (blood plasma) circulation and balances body temperature and acid levels.

• Veins generally carry deoxygenated blood from the body to the heart[63].

• Arteries generally carry oxygenated blood from the heart to the body (see footnote 63).

• Capillaries are the minute blood vessels that assist the transportation of oxygen and carbon dioxide to and from cells in the body.

• Some of the major arteries of the body include the aorta, the carotid arteries, the pulmonary artery, the renal arteries, and the femoral arteries.

• Some of the major veins of the body include the superior vena cava, the inferior vena cava, the jugular vein, and femoral veins.

• Red blood cells are produced in bone marrow and are released at an astounding rate of one million cells per second, all of which are perfectly formed because if only one cell was imperfect it would be the beginning of a cancer.

• The average lifespan of a red blood cell is one hundred and twenty days, during which it journeys one hundred and seventy thousand times through the heart.

In summary, arteries carry freshly oxygenated blood from the heart and then branch off to smaller arterioles, then to capillaries. Nutrients and oxygen permeate the capillary walls into tissues, and conversely take in waste and carbon dioxide. The waste products are carried by the capillaries into larger venules and then veins, which return blood to the heart. The blood is then pumped to the lungs where carbon dioxide is released and oxygen is taken up to be cycled through the circulatory system again.

---

63. With the exceptions of the pulmonary and umbilical veins, which are rich with oxygen.

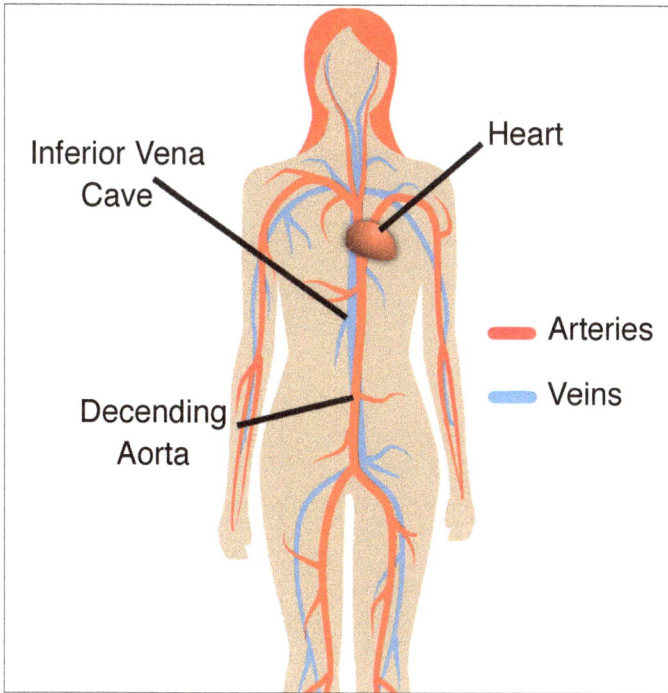

Figure 206: Cardiovascular System

Skeletal

- The skeletal system provides body support and shape.

- An adult human has two hundred and six bones in the body.

- The femur (thigh bone) is the longest bone in the human body.

- The stapes is smallest bone is found in the inner ear at 0.25 – 0.33 centimeters in length.

- The predominant mineral in adult human bone is calcium.

- Women and men have the same number of ribs, twenty-four, contrary to popular myth that females have one more rib than males.

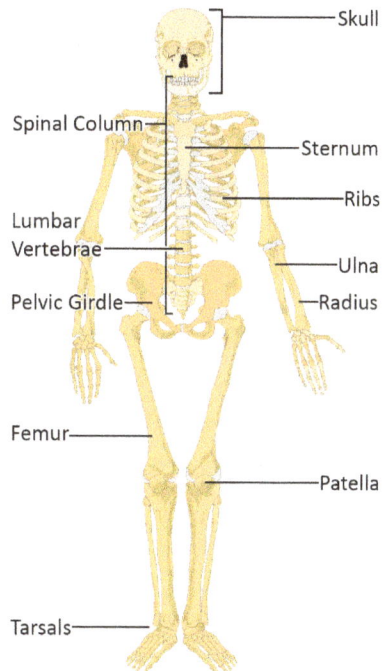

Figure 207: Skeletal System

Respiratory

- The respiratory system, or ventilatory system:

- Consists of the nasal cavities, nasal sinuses, mouth, trachea, larynx, bronchi, and alveoli (duct cells).

- Is responsible for the exchange of oxygen and carbon dioxide into and out of the body.

- Gas exchange, oxygen and carbon dioxide, occurs through the alveoli and capillary membranes.

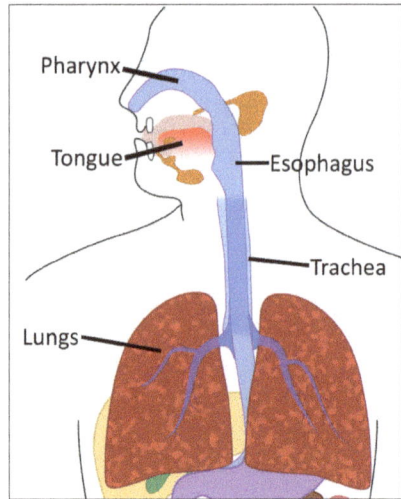

Figure 208: Respiratory System

- Alveoli produce a lipoprotein called surfactant, which lowers the surface tension of the alveoli and prevents collapse during respiration.

- The right lung has three lobes – upper, middle and lower – and the left lung has two lobes – upper and middle – in order to accommodate the position of the heart.

Nervous

The nervous system:

- Collects and distributes information on the body's status, such as temperature, fluid balance, blood pressure and so forth. Information from the nervous system activates immediate responses such as dilating or constricting the eyes' pupil according to light levels, or activating sweat when body temperature rises to cool the body and prevent overheating.

- There are two major parts of the nervous system:

    a) The Central Nervous System (CNS), which is made up of the brain and spinal cord. The CNS connects via twelve cranial nerves to the head and neck, and spinal nerves to the rest of the body.

    b) The Peripheral Nervous System (PNS) is a system of nerves that perform various functions, such as voluntary and involuntary movements. The PNS comprises various sections:

- Sensory Nervous System – transfers information from internal organs and external actions on the body to the CNS.

- Motor Nervous System – transfers information from the CNS to muscles, glands and organs to the CNS.

- Somatic Nervous System – monitors external sensory organs and skeletal muscles.

- Autonomic Nervous System (ANS) – monitors involuntary muscles. As previously noted, the ANS is divided into the Sympathetic and Parasympathetic systems.

Nerve cells comprise specific structures that enable them to transfer electrochemical signals at great speed. The construction and activities of these cells in the brain are detailed in Chapter 25.

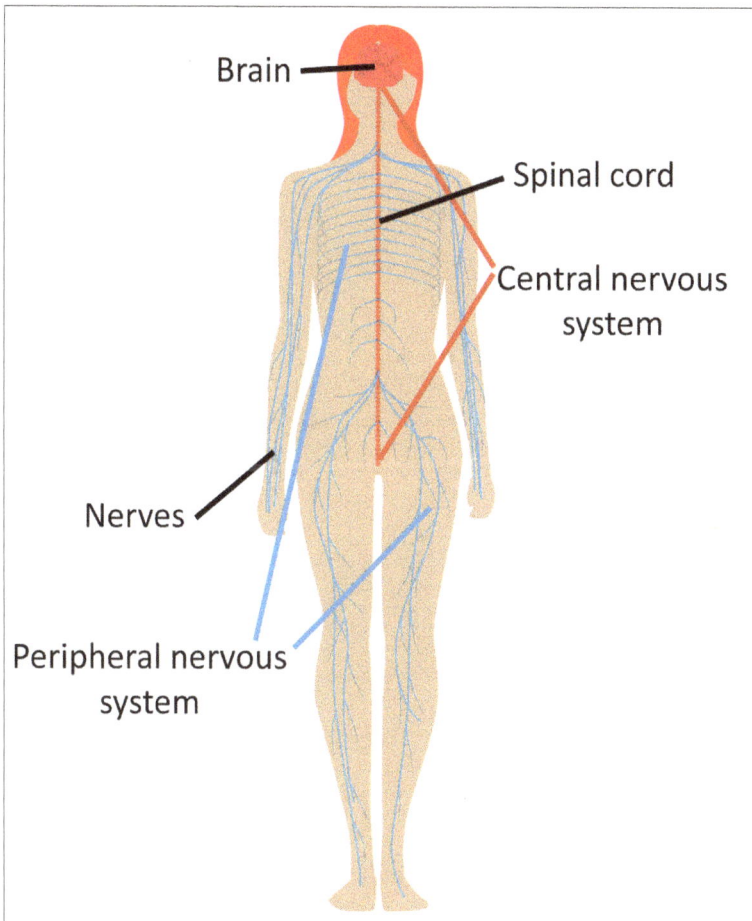

Figure 209: Nervous System

Digestive

The digestive system:

- Is a hollow tube that extends from the mouth through the body via the esophagus, stomach, small intestine and large intestine, and finishes at the anus.

- Digestion is the process by which food particles are broken down into smaller components that are absorbed into the bloodstream, which in turn distributes the nutrients.

- When chewing food into smaller pieces, saliva enters the mouth from salivary glands and starts the process of starch digestion. On arrival in the stomach, food is exposed to hydrochloric acid and other gastric juices that break down proteins. The stomach wall is covered in mucus that protects it against the effects of these acids.

- As the mix proceeds into the first part of the small intestine, the duodenum, enzymes from the pancreas are added, then it proceeds to the rest of the small intestine where most of nutrients are absorbed into the blood, with the remainder being taken up in the large intestine.

- The remaining waste material is expelled as feces from the body through the anus.

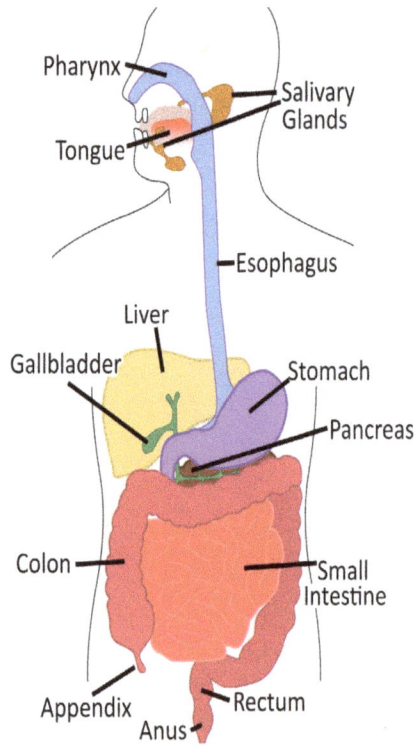

Figure 210: Digestive System

Excretory

The excretory system:

- Is responsible for waste removal of solids, liquids and gases caused by chemical reactions in the body (metabolism). This system allows the internal levels of the body to remain at stable temperature, acidity and alkalinity (pH).

- Specific sensors determine changes in bodily processes and can, if necessary, instigate rebalancing and correction through waste product removal.

- Without this deliberate removal of waste products, the body would have difficulty functioning and eventually break down and fail[64].

- Organs of the excretory system include:

- Kidneys – remove excess water, salts and urea as urine.

- Liver – breaks up toxins and chemicals and makes bile that breaks up fats into energy and inert waste.

- Skin – removes sweat and cools the body by evaporation.

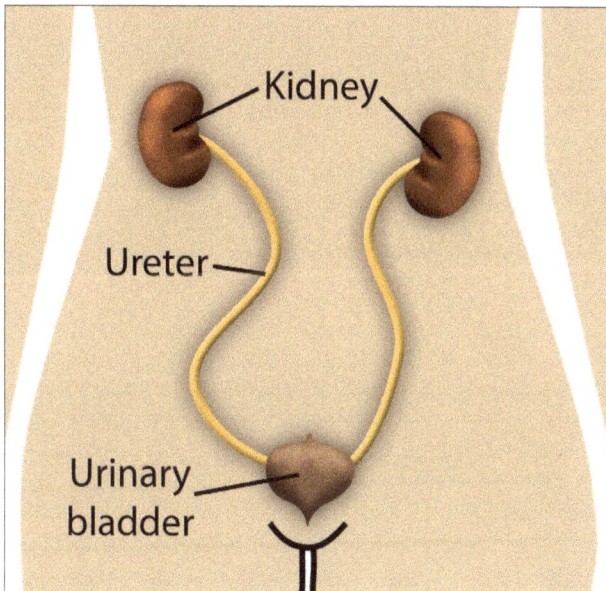

Figure 211: Excretory System

---

64. For example, both renal (kidney) failure and liver failure are ultimately fatal if not medically treated, such as with dialysis and organ transplantation.

Reproductive

- The male and female reproductive organs are both internally and externally located and are complementary in shape, size, location, and function.

- There are limits to the operation of these organs. Males require excitement to allow sperm to leave their bodies while females are optimally fertile for up to four days every twenty-eight day cycle during which only one ovum, in general, is released.

- The union of sex cells produces new and unique genetic blends of human cells, which grow in the female uterus before birth through the vaginal canal after a forty week gestation.

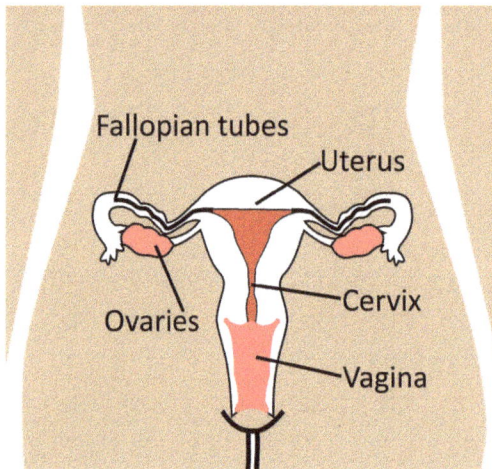

Figure 212: Reproductive System

Endocrine

The endocrine system:

- Comprises glands such as the thymus, adrenal glands, pituitary gland, gonads, kidneys, pancreatic islet cells, which release specific organ-related hormones into the bloodstream.

- Hormones are specific biochemicals made by glands, tissues and cells. They are carried by blood to various areas of the body and affect physiological activities such as digestion, metabolism, growth, and reproduction and they also influence moods. Hormones bind to cell receptor proteins that activate specific cellular responses.

- Examples of hormones are: thyroid hormone, growth hormone, testosterone, estrogen, progesterone, anti-diuretic hormone, and insulin.

• The endocrine system balances the body's temperature, acidity, fluid levels, sex hormones, growth, and metabolism by targeting specific tissue cells that contain specific receptors for those hormones.

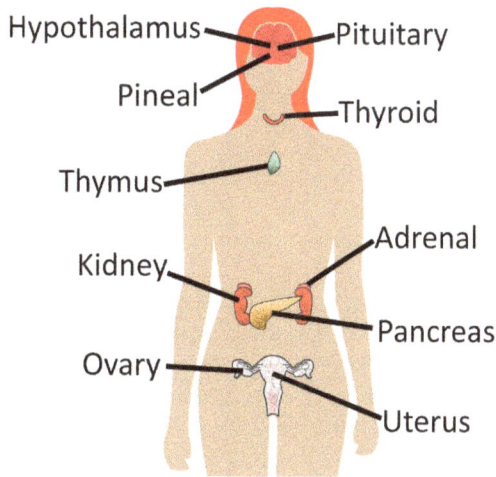

Figure 213: Endocrine System

Immune

In all animals there are automatic functions that help to protect and defend the body on a permanent basis, usually from microbial attack. Collectively, they are known as the immune system. These components remove or destroy infections, bacteria and viruses that attack the body. Some of the major components that operate as part of the human immune system are:

The Skin

The primary role of skin is as a barrier to external intrusions. The skin stops the entry of unwanted bacteria and also secretes antibacterial materials that prohibit the growth of harmful bacterial colonies.

The Thymus

The thymus is an endocrine gland found in the area between the sternum and the heart. The thymus makes a particular type of white blood cell called T-cells. A type of T-cell called a Killer T-cell destroys virus-infected cells when they come into contact. Killer T-cells can identify infected cells by changes in their external walls and they have a system that traps and removes them. This requires sensing and assessment components, knowledge of what determines abnormal status of a cell and components specifically designed to kill infected cells and remove them without harming healthy ones.

The Spleen

Blood is filtered through the spleen, located in the upper left side of the abdomen, which removes old and foreign cells. The spleen can specifically identify, trap and expel the unwanted cells.

Bone Marrow

Bone marrow is responsible for making white and red blood cells. Not all bones have blood marrow; most blood marrow is found in long bones such as the femur and humerus. The marrow monitors how many of each to make and when they are required. Scientists don't yet know how this occurs or where the information comes from or how it is transmitted or assessed.

Interestingly, in bone marrow transplant operations, donor bone marrow is injected not into bones but into the veins of the recipient patient, like a simple blood transfusion. The donor blood marrow then travels via the bloodstream and embeds itself in the appropriate bones. Somehow, the bone marrow knows exactly where it should go, as if guided by a hidden force or map.

Proteins

Proteins exist throughout the body. Some help digest food, such as enzymes. Some help form bone, such as cytokines. Some proteins in the immune system are known as antibodies, or immunoglobulins, which are produced by a type of white blood cell known as a B-cell. Immunoglobulins work by identifying and binding with different antigens, or foreign particles, that enter the body. This effectively tags the foreign particle, enabling other immune cells, such as Killer T-cells, to recognize the invader and neutralize it. Immunoglobulins are present in the bloodstream, as well as secretions such as found in saliva, vagina, and breast milk.

From birth, the immune system has to learn what is foreign to the body and what is not. Autoimmune diseases can occur when the immune system learns the wrong thing, recognizing the normal body's tissues as foreign and attacking them. Diabetes Type I, Rheumatoid arthritis and Systemic Lupus Erythematosus (SLE), are examples of such autoimmune diseases. Allergic diseases, on the other hand, arise when the body effectively recognizes a foreign body or allergen, but overreacts to the threat, sometimes to such an extent that the life of the allergy sufferer is put at risk[65]. Peanut allergies and bee sting allergies are common examples of such over-reactive or hypersensitive immune systems.

---

65. Known as an acute anaphylactic reaction.

The immune system is pre-programmed to allow the body to operate in good health. It is a kind of subliminal intelligence that has the ability to know when unwanted microbial intruders attack the body. If a wound occurs, immune cells rush to the spot and kill any invading bacteria while the wound is healing, or they will fight it in the bloodstream or any organ under threat. Without an immune system, once a body becomes infected, it would not have the ability to amount a defense and recover.

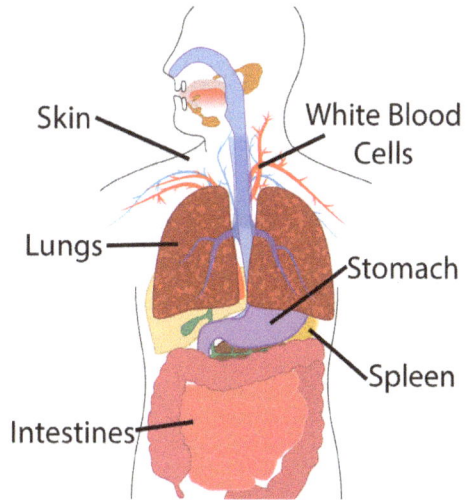

Figure 214: Immune System

Combined Systems

The human body is comprised of a multitude of interconnected systems that allow it to function and attempt to keep it healthy and operating at maximum efficiency; systems like those we have just discussed – integumentary,

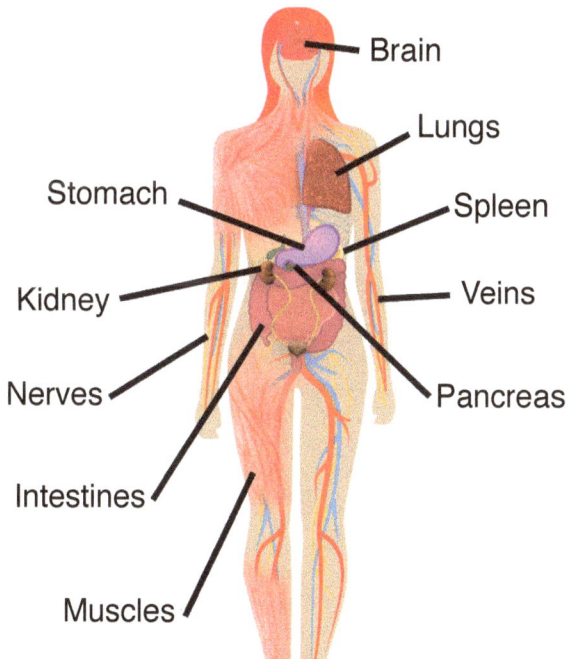

Figure 215: Combined Systems

muscular, circulatory, nervous, digestive, excretory, reproductive, endocrine, and immunological. All systems, and all of their components, operate in tandem cooperating with, and allowing for, each other. The body consists of sensory organs, defensive components and functional systems that operate automatically without any conscious thought, activities with which we have little awareness. They are automatic, pre-programmed and efficient. They recognize different faults such as excess acidity or water levels or bodily intrusions and have defenses in place to counteract them. This indicates a preexisting standard or base of safe operation for every cell, every component and every function in the body against which everything is constantly tested. If there are any variations against the standard, then triggers are activated to specifically combat that intrusion or fault. Specific responses are made to specific faults, which means the body is not only programmed to identify and react but has the appropriate and effective defenses in place.

Every cell is part of a master program that controls every bodily system and is allocated to a specific position and purpose. In other words, there is a specific number of each and every cell type. Each cell has a specific function and a specific location and is programmed to appear in a very specific sequence so that, as the embryo is being made, it has the correct parts to allow other body components to develop without affecting what has already been made or is yet to come. For example, if the muscular system developed before the skeletal system, muscles would develop unsupported. Furthermore, bones may begin to form around the muscles.

As with all other creatures and plants, whilst every embryo is a genetically unique copy of its parents, the newly fertilized egg cell contains genetic programs consistent throughout the species. The program supplied by the parents' combined DNA contains all the necessary data it will ever need inside the uterus, then in the world after birth, then for it to age and eventually die. Individuals can live in excess of a century of years, so these programs will be required to operate over large periods of time.

During the lifespan of a human individual, the signaling programs not only accelerate cellular growth and development but also decelerate and cease cellular development when required. Such a time is birth. The embryo will develop into a fetus, which then grows to a certain point, when it is expelled from the mother. Not only does the embryo have to run a timed sequence of events, but this sequence must synchronize with a separate maternal preset program that is working in conjunction with the fetal program in order to meet the same timetable[66]. The program then enters a second phase following birth, restructuring the baby (bones, respiratory, ingestion and excretion systems, sexual organs, and so forth) in the outside world and continuing to grow in size. A further change is implemented at puberty, where sexual maturity occurs and linear growth is halted. Lastly, in adulthood, the body maintains itself for a period then, excluding external factors or disease, it gradually reduces the renewal rates until eventually it expires naturally.

This long-term program of instructions has a very specific order and timing of events. There are also fail-safe systems to check that proper growth and development are happening appropriately. If we could imagine this program as a set of written instructions, like a computer program, starting in the mother's ovum just prior to fertilization, it might go something like this:

Human Life Program 1

- Human Life (HLF) Program 1 installed in female egg cells – awaiting activation.

- Maturity code activated and ovum released from ovary.

- Ovum travels through fallopian tube.

- Ovum's outer cell wall alert to contact by foreign body (sperm).

- Ovum releases fertilizin.

- Anti-fertilizin detected and checked for human security code.

- Correct code confirmed.

- Ovum opens cell wall to allow access.

- Entry confirmed.

- Ovum releases chemicals to seal outer wall.

- Lockdown of outer surface.

- Change ovum's electrical charge from negative to positive.

- New connection made to hard drive by foreign body – compatible data downloading.

- Data secured and re-sorted for new joint program: Human Life Program 1.1.

- Commence production of trophoblast coating.

66. Sometimes there are faults caused by a variety of factors, environmental and genetic, where the baby is born prematurely and is under-developed, and sometimes they are born occasionally later than the normal forty-week gestation period.

So far,

- An egg has been released and is travelling towards the womb.

- A sensor has recorded the arrival of the male's sperm and confirmed it is human.

- The sperm is allowed entry.

- The outer surface is sealed.

- The male and female DNA programs have blended together and merged into one program.

- An external coating has commenced.

The next step is to follow the new combined program in the fertilized cell:

Human Life Program 1.1

- Copy all internal cellular parts, functions and programs.

- Separate these into two halves of cell area.

- Outer cell wall to partition the middle of the two sets of parts and rejoin at the bottom.

- Maintain outer surface connection with new cell.

- Produce additional trophoblast coating.

There now exists two identical cells. What happens next?

- Repeat HLF Program 1.1.

- Maintain connection with all new cells.

- Repeat HLF program 1.1.

- Maintain connection with all new cells.

- Repeat HLF program 1.1.

- Maintain connection with all new cells.

- Entry into uterus.

- Impact alert between uterine endometrial wall and new life cell group.

- Alert all cells in new life group to dislocate trophoblast coating from cell group.

- Ensure instruction to reorganize next three replications of all cells to form inner and outer layers.

- Repeat HLF program 1.1.

- Repeat HLF program 1.1.

- Repeat HLF program 1.1 until time clock check is 6 days.

- Day 6 alert: send instruction to outer cells only for enzyme release.

- Send instruction to outer cells for hormone release.

- Send instruction to cell numbers seven to fourteen to change to capillary cell program.

- Repeat HLF program 1.1 and include instruction to place itself as inner endodermal cell.

- Repeat HLF program 1.1 and include instruction to place itself as middle mesodermal cell.

- Repeat HLF program 1.1 and include instruction to place itself as outer ectodermal cell.

- Repeat middle cell program three times as often as inner cell program.

- Repeat outer cell program twice as often inner cell program.

Now, for ease of understanding, we will assume at this point that there are thirty outer ectodermal cells, sixty middle mesodermal cells, and ten inner endodermal cells.

This fetal cell group will now send individual instructions to many other cells, along the lines of:

Send signals (possibly the release of individual hormones) to contact

- Ectodermal cell No. 1 to change its makeup to brain cell format.

- Ectodermal cell No. 2 to change its makeup to nervous system format.

- Mesodermal cell No. 33 to change format to cartilage format.

- Ectodermal cell No. 15 to change its format to muscle format.

- And so forth.

From this point, instructions are sent to hundreds then thousands of different fetal cells in the ever expanding group. Scientists assume that hormones are released with a specific instruction to change a generic stem cell to a specific cell format (brain, heart, muscle, and so forth). These hormones are required to locate the correct cell in the correct position. If they stimulate the wrong cell or the wrong location, the development of the fetus will proceed incorrectly and unstructured body parts will grow in unintentional locations.

Mathematically from this point forward, the HLF Program instructions become incomprehensibly complex. We can see that, just to get to the point where there are stem cells changing into new types of cells, the HLF program not only has masses of instructions but they must occur in linear sequence.

Furthermore, not only must the HLF Program contain the necessary data for the size, type, number, location, and sequence for every one of the billions of different cells, it also contains the data to make a shape for every body part, requiring a matching number of every type of cell attached to it in any specific area. As we discovered earlier in this chapter when looking at how cells make up skin, there is an inherent understanding of how many of each type of cell is required to make thickness and then to expand lengthways, then terminate.

Notwithstanding these important processes, there is also a point in these cell duplications around weeks six to seven where certain cells receive signals to change into gamete cells, such as female ova. This is perhaps the most important moment in embryonic development – gamete cell mutation.

The Great Evolutionary Impasse –
Gamete Cell Development In The Embryo

The single most significant claim by biological evolution regarding how millions of complete, fully functioning, different life forms evolved is that changes from one living species to another happened when increasingly more complex and sophisticated mutations occurred in gamete cells while they were forming in an embryo. In layman's terms, the programs that were meant to make a specific type of living thing mutated at the moment of making the gamete cells and created something different, yet better, from the previous generation.

It is this point upon which the theory of biological evolution is hinged. The claim is that, while forming gamete cells, there are mutations that then transfer to the following generation and its successions. Evolutionary Theory proposes that every permanent change in every life form on earth from the very first nucleated cell to fish, reptiles, birds, plants, insects, and human kind is a result of cells mutating to something different from what they were 'meant to be'. Biological evolution happens at that moment when the gamete cells are forming in an embryo during cell division.

In the female, the gamete cells are the ova or eggs, which are formed while the embryo is developing in-utero, then they are released on a monthly cycle after birth, following puberty. The female does not make new eggs after she is born; she is born with the entire collection of ova that she will ever have and then progressively loses them over time. Unlike females, the male fetus does not produce gamete cells, or sperm. Males only begin to produce sperm in their testes around puberty. In the journey of human embryonic development, somewhere around the sixth week, a fetal cell receives a unique and specific signal to change into a gamete cell, such as a female ovum. It is assumed this cell will reproduce itself into all of the female's egg cells for this fetus and then stop after some predetermined number has been reached before she is born. If this egg cell undergoes a mutation then, when it duplicates, all of its copies will also carry that difference. All of the egg cells will be different from previous generations and, when fertilized after birth, the new infant from this particular female (that is a developing embryo now) will be born with that difference. This sequence is displayed in Figure 216: Mutations In The Fertilized Embryo

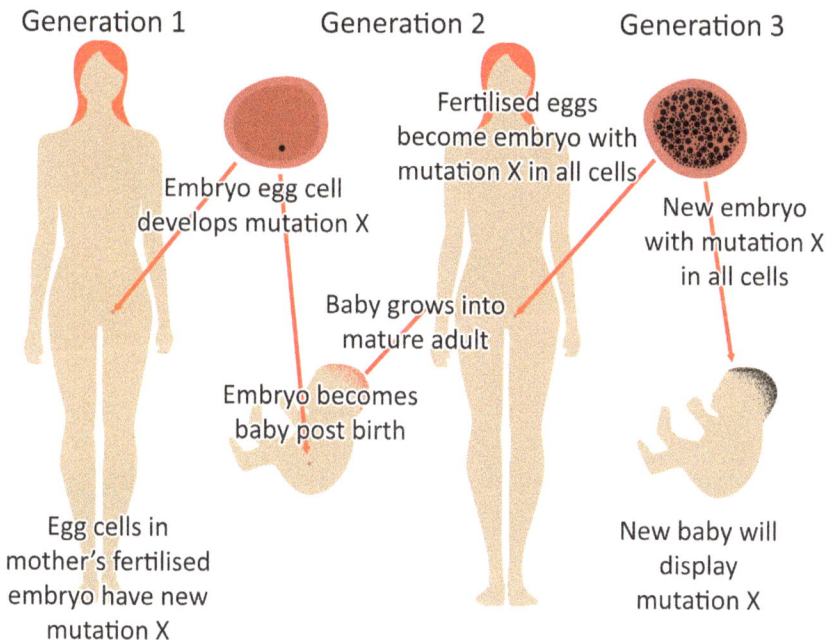

Figure 216: Mutations In The Fertilized Embryo.

We can understand how a mutation might cause a variation in the future generations, but what is it that actually mutates in the newly developed gamete cell? Is it DNA or RNA or something else? To understand how this takes place and gets implemented, we will go forward in time to a point where this female embryo with mutated egg cells has been born (as a female) and is now sexually mature to reproduce.

When her egg cell becomes fertilized by male sperm, the combined cell will start the same replicating process of previous generations, following the sequential program we have described above, to form a human being. The fertilized ovum replicates into sixteen stem cells then, as new cells are made, they sort themselves into inner and outer cells, attach to the uterine wall and replicate until there are about one hundred cells in three layers. Then, as more new cells are made, they receive a coded sequence of internal and external signals to change into different but specific cells to start making all the body parts.

Following the evolutionary random process of mutation, we will nominate, as an example, that the particular mutation in the female's egg cell will create the first step to improved human hearing, the result of increased sensitivity of the skin covering the internal eardrum on the right ear. The skin (which we will refer to as 'S' for superior) will be thinner, more sensitive and tighter than the current skin ('N' for normal). The new embryo will get S-cells rather than N-cells when its program arrives at the instruction for these cells.

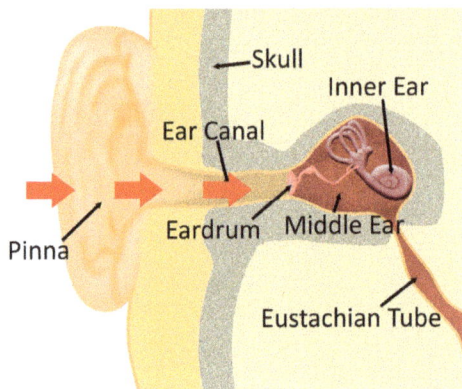

Figure 217: Right Ear With New Supersensitive "S" Skin Covering

THE KEYSTONE FOR BIOLOGICAL EVOLUTION

We are now at the critical moment for biological evolution as this is *the keystone process* on which evolution bases its theory that every life form was the result of mutational changes to some other different life form[67]. Every single step of trillions of steps that are proposed by evolution to have made progressive mutational changes to life forms is wrapped up in this one activity.

---

67. For instance, ape to human.

We will now examine in detail what a mutation really is, how it works and whether or not it is possible. No longer do we need to blindly accept generalized statements that mutations made creatures and plants change[68].

What then is the process whereby a cell, previously programmed to change into an eardrum skin type N-cell, becomes a type S-cell? The N-cell itself does not mutate into an S-cell, as the mutation occurred in the gamete cells (ova) of its mother. Direct N- to S-cell mutations will not be passed on, as it is a somatic cell and not a gamete cell[69]. So, what actually has to happen?

It had to be given new and different signals to change to an S-cell, instead of the signals for an N-cell.

We need to be clear; the mutations claimed to result in evolutionary changes affect the signals that instruct existing cells what to change into. It is the signal that has to change. Otherwise, if the signal program was the same, the embryo would be the same as its parents. For evolutionary change to occur, the mutation in the gamete cells of the developing embryo can only effect change in the signaling program it contains.

In other words, the mutations happening in the gamete cells of a forming embryo are actually changing the signal program that is used by that embryo after it is born and after its own gamete cell joins with a partner's gamete cell that then create a new embryo (as shown in Figure 218: Female With New Mutation Develops Embryo With Coded Mutation). The egg cell in the embryo that gets the mutation doesn't in itself change: it is the program it holds that is changed and it only becomes effective when it is fertilized at a later time.

Several points are raised with this issue. Firstly, is it possible for a random mutation to change a set program without causing problems? Secondly, where is the signal program located and how can it be changed?

Scientists don't know absolutely where the program is. Research indicates that DNA is the most likely location[70]. Let's examine what is

---

68. If this process does not hold up then the self-evolution of life forms is exposed and another non-evolutionary solution must be found.

69. That is, not part of the reproductive system.

70. There is a new field of research called 'epigenetics', which considers factors other than the genes of DNA, such as alternating carbohydrate molecules attached to the chemical base units of DNA – adenine, guanine, cytosine, and thymine – and that these extra-DNA factors can in fact induce physical changes to individual beings, even suggesting that they can influence the biochemical makeup of progeny several generations in the future such as diabetes and obesity.

required to change the program above so we have eardrum S-cells rather than N-cells.

There are specific events that have to occur one after the other in embryonic development. In our sequence of events, the signal program is a linear set of instructions to change specific cells into new types of cells that is timed to match the rate at which new cells are made. If we go back to Human Life Program 1.1 above, there will have been thousands, perhaps millions, of sequentially timed instructions to change new cells into different cells in exactly the same timeframe that new cells appear from duplications. Consequently, a new cell appears and an instruction goes out: "Change to this specific cell". It then follows a program for that type of cell at that location and replicates itself a specific number of times at a predetermined length and thickness. Then, for all the adjoining cells, the instruction includes a process to make sure all of their components are proportional to the size of the embryo. Then they are instructed to grow together at proportional but different rates until the body reaches maturity. Every cell must have these instructions and possibly some way of measuring the number and position of every other cell. It is also assumed they are checking their growth and development against a template to maintain structure and proportionality.

For our theoretical scenario, if we show this long string of instructions at a point where the eardrum skin N-cells are required to be made, say the one hundred thousandth instruction, we will be a long way down the list. For example, let's start where we have a hundred or so cells and they are about to attach to the uterus wall on day six. Here is the string of instructions we looked at previously:

- Day 6 alert: send instruction to outer cells only for enzyme release.

- Send instruction to outer cells for hormone release.

- Send instruction to cell numbers seven to fourteen to change to capillary cell program.

- Repeat HLF program 1.1 and include instruction to place itself as inner endodermal cell.

- Repeat HLF program 1.1 and include instruction to place itself as middle mesodermal cell.

- Repeat HLF program 1.1 and include instruction to place itself as outer ectodermal cell.

- Repeat middle cell program three times as often as inner cell program.

• Repeat outer cell program twice as often inner cell program.

This cell group now has to send individual instructions to multiple other cells along the lines of:

Release individual hormones to contact

• Ectodermal cell No. 1 to change its makeup to brain cell format.

• Ectodermal cell No. 2 to change its makeup to nervous system format.

• Mesodermal cell No. 33 to change format to cartilage format.

• Ectodermal cell No. 15 to change its format to muscle format.

• And so forth.

Let's say we come to the point of the one hundred thousandth instruction when it's time to make these particular eardrum skin cells for the right ear. We will have gone through a list of instructions like those above but, to keep it simple, we will just number them sequentially as shown:

Instruction No.1
Instruction No.2
Instruction No.3
Instruction No.4
Instruction No.5
Instruction No.6
Instruction No.7
Instruction No.8
Instruction No.9
Instruction No.10
Instruction No.11

.

.

.

Instruction No.99,999
Instruction No.100,000 – first cell for normal skin (N) over RHS eardrum
Instruction No.100,001
Instruction No.100,002

The random chance mutation that occurred in the gamete cells of this mother when she was an embryo occurred at this particular instruction No.100,000, which previously started the change to a right side N-cell for the eardrum. So, the mutation would need to alter the signal program to read: right side eardrum skin cell S-program. The change is required to be made at this specifically numbered instruction in order for the newly formed cell to acquire the upgraded S-program. Otherwise, if it went to any other random instruction number then the 'right hand eardrum skin cell S-program' may have been given to a cell previously scheduled for a toe nail program, or any other cell program for that matter.

Consequently, not only must mutations be changes to the program of instructions but they must also be directed at a specific instruction, not a random one. This, however, runs counter to the premise of Evolutionary Theory that mutations are random events.

There is, however, an assumption we make every time we attempt to work out how any changes to living bodies could occur. The assumption is that mutations are actually capable of changing the preset program that already exists in fertilized cells. There isn't any scientific proof identifying where this program exists within a cell, that is, where it is stored. There isn't any evidence telling us how the other surrounding cells know what's going on or how they send and receive signals to make sure the sequence is correct. There is also an assumption that the program can be changed, even though scientists know extremely little about it.

What we do know is that whatever form the program has, it must be sequential, that is, one specific step after another and, when certain steps have been taken, there is a termination instruction to cease growth at a predetermined limit[71]. We know that it must be a string of instructions written in a specific cellular language that allows for a range of genetic variables to ensure the outcome of a complete human being. The program must be very specific, clear and structured.

Any change to the program would need to be written in the same effective language and with allowance for the existing program instructions. The insertion of a different set of instructions would need to be such that there is no disruption to the main function of the program, which is to achieve the predetermined outcome, life. Just as importantly, new instructions would not be followed if inserted in the wrong place; by necessity the new instructions have to be inserted in exactly the correct sequence.

Further complications also arise for which Evolutionary Theory fails to explain.

1. All cells that receive signals to implement cell programs require priming before they receive a signal in order to activate the

---

71. For example, for any of the leg bones there is a restriction on size and shape that must be adhered to in order to ensure the leg proportional to the rest of the body and is situated in the correct location.

program. In our example, a new signal instructs a cell to implement the S-cell ear skin program but such a program does not actually exist. The only program that exists is normal N-cell program, as the mutation has only changed the signal, not the program itself.

Therefore, when a mutation changes the signal program to send out instruction to implement Plan S instead of Plan N, it also has to write the S-program so that the receiving cell can put it in place. At the time of their making, all newly made generic stem cells already have at least some of the programs for every type of future cell in that body. Somehow the mutation has to insert the newly written S-plan into the master plan so that when every new cell appears, it will have the S-plan in place should it receive an instruction to implement it. If it doesn't have this changed program then it can't be passed on to following generations.

Evolutionary Theory needs to explain how such a change could occur by random, chance events, particularly as these new instructions would include millions of downstream cell changes which, by default, require some prior knowledge of the outcome – something that is not considered possible by Evolutionary Theory. In other words, a chance mutation is a complex string of millions of timed and coordinated instructions with an outcome that must be known before implementation.

2. When these new, mutated cell plans are actioned, the entire master program needs to allow for all of the necessary changes and must also be rewritten. In Chapter 21, we learned that one mammal had to evolve/mutate into both a platypus and a giraffe. Consequently, the programs for making every component in these new bodies had to change from their original number, location, sequence, and type of cell for every body part. But particularly, they had to allow for the massive size difference between these two creatures. A giraffe is vastly bigger than a platypus, so the entire master program from the very first mammal had to change the number, location and sequence of every cell to also allow for the size differences. For a giraffe, the original master program would have to change so that it included more of every type of cell in every single location of the body (such as long legs, long neck, thick tongue, and so forth), whereas for the platypus the original program would need to change to reduce, comparatively, the total number of cells every time there was a mutation.

This in essence means constant but specific rewriting of almost every instruction in the master plan every time a mutation occurred without knowing what the final outcome would be, considering each mutation is a random event. Statistically speaking, the concept that every mutation in a giraffe would be an increase in cells to match its final size relies on a 50/50 chance that there could be more cells or less cells. Similar chance applies to the platypus.

Furthermore, all changes to both the specific cell program and the original master program had to be coordinated and perfectly written. The biological arm of Evolutionary Theory relies on random mutations accidentally reprogramming specific and correct instructions for new cells and body parts, without knowing the language in which the program was originally written, and inserting the mutated changes in the exact location of trillions of instructions, all by chance and without meaning to.

The stance of Evolutionary Theory, in terms of biological evolution, therefore goes something like this:

There is a massive and complex genetic program containing trillions of specific, sequential and timed instructions that made itself for no particular reason. It is in an unidentified location and form, yet billions of random and accidental cell mutations are able to find and reprogram it in an exact language and sequence so that trillions of advancements occur over billions of years. The result of which is millions of unique and independent life forms that are unable to be linked to each other by any known biological processes.

As it happens, science has already determined that the above mutational processes proposed by Evolutionary Theory are not possible. The quotation below states that it is not possible for any living form to change into another living form without the deliberate intervention of an exterior intellect. Accordingly, life forms cannot change themselves into anything else because their DNA functionality forbids it.

"'Survival of the fittest' and 'natural selection.' No matter what phraseology one generates, the basic fact remains the same: any physical change of any size, shape, or form is strictly the result of purposeful alignment of billions of nucleotides (in the DNA). Nature or species do not have the capacity for rearranging them, nor adding them. Consequently, no leap (saltation) can occur from one species to another. The only way we know for DNA to be altered is through meaningful intervention from an outside source of intelligence: one who knows what it is doing, such as our genetic engineers are now performing in their laboratories". *I.L. Cohen (Officer of the Archaeological Institute of America, Member New York Academy of Sciences), Darwin Was Wrong: A Study in Probabilities (New York: New Research Publications, Inc., 1984), p. 209.*

Summarizing the above analysis, we have determined that random mutations, the keystone activity on which the evolutionary development of life is based, require prior knowledge of both the existing programs and the end programs, where they are located and the ability to rewrite billions of complex, timed sequential instructions without fault. These random instructions also need to invent new, perfectly timed schedules with different numbers and types of cells never previously encountered.

There are no scientifically validated transitional steps between any life forms in the fossil records, nor are such changes observed in life today. The evolutionary concept of the self-creation of millions of independent and different creatures on our planet is akin to a scientific biological myth.

Chapter 24 Unraveled

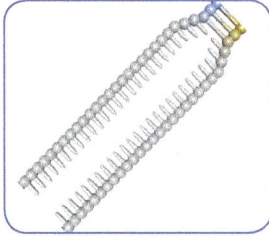

- Science has identified that male and female gamete cells contain chemical codes that must match before fertilization can occur. These systems, along with matching electrical charges, are fail-safe security checks that prohibit embryo contamination from other creatures.

- The formation of an embryo, then baby and adult human is performed by the replication of one original fertilized cell into fifty trillion cells. These cells operate under programs that control their rate of appearance, the type of cell they will become and their specific location.

- There are many different types of cells that make up each and every creature. These require programs for their types, sequence and locations to pre-exist prior to their appearance otherwise each body will not form properly.

- Each creature has a different series of timing, cell type and location programs. This master program ensures that creature is made to a specific design.

- Bodies consist of various systems that operate under their own programs yet cooperate with other programs in the body.

- The programs controlling body functions attempt to maintain a healthy body.

- Cell programs make shapes that grow proportionately with the rest of the body.

- In order for evolutionary development to progress, mutations are required to deliberately change cell timing, location, and signaling programs at specific points of instruction.

- Science does not support the tenets of biological Evolutionary Theory. According to evolution, changes to living creatures is random, accidental, chaotic, and uncontrolled; yet such a process, if true, demands that any change would have to be clear, concise, deliberate, and timed.

- The keystone principle on which biological evolution is based is that uncontrolled mutations occur randomly during the combination of the two parents' gamete cells when making a fertilized cell. For this to result in 'evolutionary' changes, these mutations must occur by precise changes to the instructions in preset cell programs. These changes must be at the exact instruction point in trillions of instructions and in the exact code as the original program. The scientific evidence dismisses this claim and thereby challenges the potential for life forms to change themselves from their intended outcome.

Can different programs from males and females blend together in an embryo?

In the above analysis, there remains a further issue; the female ovum with a program mutation for a type S-cell right-side eardrum skin will be fertilized by a male sperm with a type N-program. When these two programs combine on fertilization they will not match, as had previously occurred, as the male instruction No.100,000 is for a thicker, less tight skin cell. Any mutation must therefore also cater for two different programs to merge, despite different instructions.

Further, if a creature were to change from one type to another, it would mean that the number of cell changes and their sequence would also have to change, along with the timing set out for each cell to appear. For example, the sequence of instructions for a bird with wings and feathers in a hard-shelled egg where there is no connection to the mother's uterus would be completely different from a placental reptile with four legs in a uterus that feeds off the mother (such as *Trachylepis ivensi*, a skink). The total timing programs would also have to be completely different with the shortest gestation period being the Virginia opossum at 13 days and the longest being the frilled shark at 3½ years. Some creatures have millions, billions, or trillions of cells, some take a couple of weeks gestation, others years, so every creature has to have different instructions on all these program levels.

If every living creature came from one original sea creature, as claimed under Evolutionary Theory, then something had to be able to change the sequence program, the cell type program, the cell replication timing program, and the total timetable program – all coordinated together, at the same time, for each creature. Evolutionary Theory claims random mutations made all of these program changes. Science now tells us otherwise.

Yet, there will be many arguments that we do see changes by adaptation in creatures. Yes, there are permanent changes. What are these? For the most part, living entities will produce a range of differing features for that species[72]. One or more of these

---

72. Remembering that 'species' is a human definition.

might be a better fit or adaptation to the environment in which it exists, or the environment may change and it is better suited than its ancestors. But, these changes do not alter the basic type of living entity. There is a range of variables which any living entity can undertake but it cannot go outside preset boundaries.

One of the biggest issues for Evolutionary Theory is that there is no scientific proof of any living entity changing from its basic, preset, structure, design, and boundaries to some other sort of creature. This includes the previously mentioned blendings, which allow specific predetermined mixes (such as lions/tigers). This story of mutation and adaptation isn't supported by scientific evidence, particularly when we examine what is actually required to happen. That is why actual mutations that do occur are almost exclusively failures – because they are a lesser 'quality' than the original – a haywire change that has no supporting rewriting of programs to allow for the change.

Points To Ponder - The Complexity Of Cell Programs That Self-Write

In a newly fertilized human embryo, there is a period of about 4-6 days from when a male sperm and a female egg blend their instruction programs together until newly-made generic (mother) cells start receiving instructions to change into new, specific types of cells. During this period, the new combined database has to work out programs containing many variables, for up to fifty trillion cells over potentially 100 or more years phasing through embryo, then birth, then baby, then adult, then death. 100 years of complex control systems organized in the space of a few days maximum - a truly extraordinary accomplishment.

We know that humans vary in size from under five feet to well over six feet tall – differences that can be over fifty percent in size and capacity. Females are generally smaller than males, also being physically and physiologically different. Such differences include body shape, reproductive organs, variations in hair production (for example, on the face), facial shapes, and so forth. All of these require the cell production programs to self-adjust both the numbers and types of cells to be unique to each person, resulting in potentially trillions of trillions of trillions of permutations of cells.

Comprehending how these programs write themselves with such vast and accurate instructions is challenging enough, particularly as they have to allow for female brains being on average ten percent smaller than males yet still capable of operating with the same (or better?) functionality. But, as well, these programs are required to do something quite extraordinary: They must lay out the connections of female and male brains differently. Females have more connections between left and right hemispheres allowing greater depth of thinking and consideration of more variables than males. Males have other advantages, such as better motor skills. Essentially this would require the combined programs in the newly fertilized embryo to decide almost immediately which gender is to be made via the sorting of X and Y chromosomes and then working out each cell appearance depending on the gender and multiple size variations.

Evolutionary Theory tells us that these programs perform all of these complex calculations and variations without instruction or control, randomly making up programs for no particular reason.

25

BEING HUMAN

Whilst there are many wonderful things that we have observed in nature, and humankind has created many sophisticated and clever inventions, nothing compares to the wonderful complexity that we see in every human being. The human body is probably the most complicated thing on earth. Although we share many physical attributes with other animals, humans have a vast range of emotions, feelings, skills, creativity, ingenuity, productivity, intellectual thinking, and spirituality that makes us unique in the world of life.

In this book we have briefly reviewed some of the operations of the human body and their intricate interconnections with other parts of the body. Now we are going to look in more detail at some specific functions and abilities. Within these, we can see the level of complexity, intricacy, order, and structure required for us to live and enjoy the wide range of abilities that we possess.

The Brain

The brain is the controlling organ of the body, located in a protective skull in which there are eighty-five cavities. The brain is responsible for various organs and functions such as sight, taste, smell, hearing, breathing, ingestion of food and liquids, spinal cord, blood vessels, and nerves. Evolutionary Theory proposes that the brain created itself, its own complex structures, operations and programming, from almost the first moment it appeared as a functioning organ on earth.

Brains are comprised of neurons (nerve cells). These neurons send chemically based electrical signals in one direction only, as shown in Figure 218: Synapses.

According to Brazilian doctor Suzana Herculano-Houzel in 2009, the average human brain contains approximately eighty six billion neurons. While most research has shown that neurons are individually made from special stem cells and do not duplicate, there is evidence that in two areas of the brain, the hippocampus and olfactory bulb (see Figure 223: The Segmented Brain), which are involved in memory and smell respectively, new neurons can appear. However, the identification of their source is yet to be confirmed. Each neuron is estimated to connect with around eighty thousand other neurons, with potentially one hundred trillion connections. Recent research has shown that humans have more of the genes that code for neural connections than other animals, and it is the number of connections, not the number of neurons, that distinguishes the human brain from animal brains[73].

Although compacted into the brain matter, neuronal cell bodies, or soma, do not touch each other. Instead, they are held in place by special brain cells called glial cells, which give structural support to the brain. Glial cells are also found throughout the nervous system outside the brain. Neuronal cell bodies store chemical information and transfer chemical-electrical signals and messages, such as dopamine and adrenaline, to other specifically targeted neurons. Neurons send information through thin extensions or arms called axons and receive information from other neurons through similar extensions or arms called dendrites. Both axons and dendrites create the electrical 'wiring system' of the brain and transfer information through minute gaps at the ends of their membranes called synapses (refer Figure 221: Synapses). Axonal synapses sending chemical information are called pre-synaptic. Dendritic synapses receiving the chemical signals are called post-synaptic. The pre- and post-synaptic areas each contain an estimated one thousand molecular 'switches'. It is not known how these operate but they may work like the binary switches in computers, which are either on or off. Based on these numbers, the total number of switches in the average human brain would be eighty-six trillion (86,000,000,000,000), a truly phenomenal mass of switches and connections.

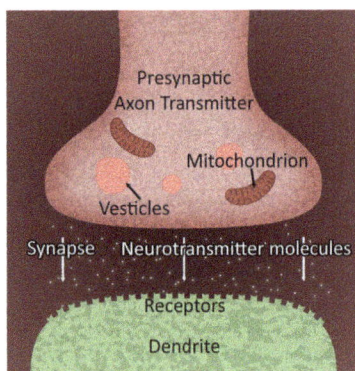

Figure 218: Synapses

73. In one such experiment, scientists re-coded mice DNA to include instructions for greater neural connections. These mice were found to have greater intellect than 'normal' mice, being able to perform functions easier and faster on average. Autopsy on their brains also revealed greater neural connections.

There are two basic types of synapses (gaps) within the nervous system of the brain and body:

1. Electrical

These synapses are extremely small, about three and a half nanometers wide[74] and have special junction gaps that allow transmission of variable chemically charged ions (signals) between neurons. The speed of transfer is relatively high due to the small gap and speed of ionic current.

2. Chemical

These synapses are larger at twenty to forty nanometers. Ionic charges from transmitting neurons trigger the release of chemicals (neurotransmitters) through specific, voltage-gated calcium channels. These channels are highly permeable to calcium ions, which essentially activate enzymes, muscular contractions and other complex cellular activities such as kidney excretion and water absorption.

Neurons can also communicate with each other through indirect electric fields (ephaptic coupling). This appears to be a localized energy field where adjoining neurons can affect the ion charges in neighboring neurons without utilizing synapses. When a wave travels along an axon it can draw ions from a neighboring axon that is resting, effectively causing reverse ionic charges in that axon. Research is currently being undertaken to determine if this is a possible way in which the brain coordinates electrical fields used in the process of thinking where differing areas of the brain can communicate virtually instantaneously.

It is essential that the number of glial cells that sit between each neuron match the width of the synaptic gaps as chemical neurons have six to ten times the gap as do electrical neurons. Considering the mass of eighty-six billion neurons in the human brain, Evolutionary Theory relies on chaotic and random chance events to ensure that both neurons and glial cells are formed in the correct sequence, in the perfect locations and with the correct gaps, during embryonic development. This appears impossible without a complex predetermined layout of each of these billions of cells before the first brain was established.

Figure 219: A Neuron displays the general layout of a neuron. A neuron's operation is based on charged potassium (K^+) and sodium (Na^+) ions. When not in action, more potassium ions seep out through the neural surface membrane than sodium ions seep in, creating a negative charge on the inside of the membrane relative to the outside. This is called electrical

74. A nanometer is one billionth of a meter.

polarization. When the neuronal cell is active, the sodium channels are opened in a sequential wave, where the opening of one channel sets off the opening of the next. The result is that sodium ions enter the neuronal cell, neutralizing the negative cellular charge and thus depolarizing the cellular membrane. The influx of sodium ions has the effect of pushing out potassium ions and thus repolarizes the neuron. The sodium channels have a momentary period post activating where they don't open and this causes an electrical wave to flow along the top of the neuronal membrane. The cycle of change occurs quickly at around five hundred times per second. The brief moments between these ion changes are called 'action potential'. This rapid opening and closing of membrane activity causes neurons to fire electrochemical waves down their length at a speed of about four hundred kilometers per hour. With this immense level of firing, the brain is a constant mix of chemical and electrical (ionic) activity.

In Figure 219: A Neuron, information, in the form of an electrical charge, enters from the top and leaves from the bottom. There are three main segments:

- The input area consists of several dendrites that branch out to connect with other neurons.

- The main neuronal body, soma, handles metabolic functions and includes a nucleus filled with proteins and DNA.

- The axon is the conductive part of the neuronal cell sending information from the soma via the pre-synaptic terminal to another neuronal cell's dendrite(s).

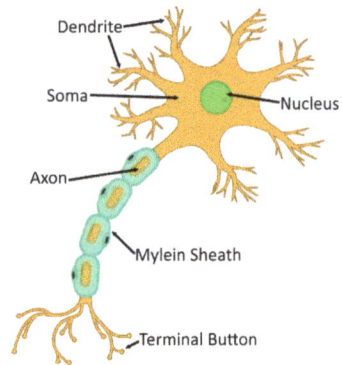

Figure 219: A Neuron

The activities of the brain and its cells involve complex interactions between electrical charges, a range of chemicals, acids, gases, ions, monoamines, biogenic amines, peptides, and other factors. These elements combine and cooperate together to allow complex, yet specific, behaviors of trillions of interrelated components that enable the brain to function. By themselves, the combination of these factors would result in a scrambled chaotic mess in our heads. Yet, for the most part, these connections and components are perfectly aligned in order for the brain to receive information, process that information and instruct the body to react to that information. Whilst scientists have been able to identify many of the activities of the brain on a general level, there is only a very minimal understanding of how a brain really works at a neuronal level.

The brain never stops processing, correcting and instructing, even when we are asleep. It has an immeasurable level of activity, undertaking millions of tasks every second of every day. It is the brain that keeps the body operating. It constantly receives a staggering level of information – measurements, readings and reports – from all parts of the body, works out the best corrective action, usually involving more than one body part, then sends messages to perform those tasks, twenty four hours a day. Such is its constant activity that it uses more than twenty percent of the energy in the human body.

The Brain Is The Monitor

The brain automatically monitors the entire body every second of the day. It does not need anything to monitor its performance; it is the monitor and it is set to operate by itself, knowing in advance how to react to constantly differing levels of change. We are not necessarily aware of a lot of the repair and maintenance work going on in our bodies. It happens at a subconscious level without our knowledge, even though it is our brain. The brain knows which actions are automatic and which require our awareness. If we were consciously required to add bile to our stomachs or to make blood cells from our bones, our time would be consumed with maintenance of our bodies and we would be unable to enjoy life. A brain cannot change conscious tasks like eating or drinking or washing our face into automatic activities that happen without our knowledge as it has pre-determined program where all tasks are already locked in place as being either conscious or subconscious. The question is; where did this program originate?

The brain is not just a mass of interconnected cells, however. It is segmented with various parts controlling specific functions that operate as a coordinated group. These groups are displayed in Figure 220: The Segmented Brain.

Broadly, the main functions controlled by these segments are:

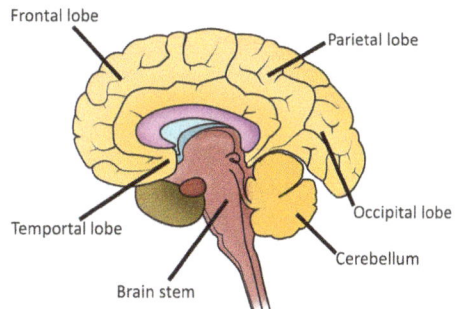

Figure 220: The Segmented Brain

Frontal lobe: thinking, planning, decision making, solving problems, voluntary control of motor functions, such as muscles.

Parietal lobe: constantly reviews the position of the body and its parts in relation to the changing environment and moves it accordingly, such as avoiding an obstacle in our path. It brings to our attention an awareness of surroundings and also notifies pain levels. It is responsible for determining differences in objects we touch, such as weight, dimensions, texture, pressure, and temperature.

Occipital lobe: translates the electrical inputs from the eyes into shapes so we can identify objects, movements, colors, and relative positioning in a three dimensional environment.

Temporal lobes: memory bank building, facial and object recognition, language skills, smells, music, sounds, and emotional reactions.

Cerebellum: automatic voluntary movements, balance, posture, attention, timing and some learning.

Brain stem: maintains body equilibrium, controlling digestion, breathing, blood pressure, temperature, and associated functions of heart rate and perspiration. Associated with sleep, balance and surprise responses. Involved in minute movements of limbs and face.

In the human body, the eighty six billion neurons making up the brain are in charge of managing fifty trillion cells (50,000,000,000,000). Of these, about one billion cells are replaced hourly (over quarter of a million per second). Most of the body's cells live for just a few years, which means that the average human body over a 70-year lifetime is replaced the equivalent of over ten times. If cells replicated at twice this rate (two billion per hour), the body would have too many cells and might keep expanding without any limits (which is what happens with cancerous cells). With such a frantic and specific rate of replacement in operation, the brain must have knowledge of the exact number of cells required at every moment in time, also taking into account that each replenished body is a different size, shape and age during a person's lifetime. Evolutionary Theory says that this knowledge exists only by chance.

To be able to accomplish all of these things, the brain must be pre-programmed during embryonic development. That is, as the embryo grows inside the mother, a range of automatic functions, actions, limits, and standards are programmed into the brain's neurons as they are made, or shortly thereafter. This also requires various sections of the brain, such as those described above, having specific instructions for every neuron in that section so they can perform their predetermined duties effectively.

There is a mass of messages coming into the brain from every part of the body, reporting on cell levels, internal and external temperatures, processing of food, the five senses, damage to organs and cells, levels of chemicals and hormones, emotions, infections, breathing and heart rates, fluid levels, and so forth. The brain has to be capable of knowing what each and every message represents, what area or function it relates to and how far it has deviated from the normal range. It has to sort them into categories and priorities, determine the best responses and send them out in order of importance.

For example, an increase in body temperature could be related to any one of the following, each requiring a slightly different response:

• External temperature (for example, sunbaking)

• Emotion (for example, panic)

• Exercise (for example, running)

• Poor airflow (for example, excess clothing)

• A reaction to foreign chemicals (for example, poisons)

• A biological attack (for example, illness or infection)

• A lack of oxygen (for example, high altitudes)

The brain has to determine which of these is responsible for the increased temperature and, if more than one, the priority for dealing with a solution[75].

The brain is processing information regarding the body's temperature, breathing, chemical composition, perspiration, stress, and energy levels every moment of our lives. For it to recognize a problem, it has to know what standard operating levels are. It also has to be capable of making automatic, predetermined decisions about how to adjust various facets in order to balance the whole body, every minute of every day. It has to be able to give instructions to the relevant areas to balance the body's efficiency – a concept that scientists don't really understand because it involves not only the body's physical structure but emotional and mental components as well.

Evolutionary Theory assumes that these astonishing abilities and programs invented themselves by chance and yet fails to provide any

---

75. For example, in the event of a viral or bacterial infection the brain will instruct the body to increase its temperature and create a fever in order to neutralise the invading germs, which cannot survive in temperatures above forty degrees Celsius. This is why some fevers are a 'good' thing for our health to return to a balanced state.

detailed scientific description about how this could occur. To plug the evolutionary gaps and analyze whether this might be possible, we will return our discussion to the very first living body in history, when the first electrical charge surged through an incomplete nervous system to a proto-brain. We will, through necessity, confront some of the challenges that evolutionary scientists have thus far avoided dealing with.

The First Living Brain

We therefore return to the first living creature in history, a creature that we previously nominated as a proto-fish. Before anything can happen in the brain, a nervous system had to accidentally build two pathways, an afferent pathway (for messages going to the brain), and an efferent pathway (for messages being sent from the brain). These pathways split incoming and outgoing messages to prevent interference with one another.

First, we will need to imagine this partially developed proto-fish body with the first partial nervous system, as displayed in Figure 221: Primal Partial Body. We also need to imagine something that generates the first neuro-electrical signal in history to the partially developed brain, something like an impact with a rock causing a nerve to discharge. This first neuro-electrical current had to travel along the incoming afferent pathway in the correct direction; that is, towards the brain. Most likely there was some interference or distortion, as the nervous system was probably only partially formed and had no intended function as an electrical conduit. This first neuro-electrical charge in history arrives into a mass of accidentally formed, disorganized, unstructured neurons that have no particular role and no programming.

What would have happened next?

Using the tenets of Evolutionary Theory, we can imagine that the electrical charge now coursing through some of the neurons and, by accident, creating some chemical discharges.

Without any rhyme or reason, this disorganized mess of electricity and chemicals in the proto-brain now interprets the incoming charge as having a meaning (for example: pain), even though no recognition program exists. The neurons then send responses back to the body through the efferent pathway (for example: swim away from source

Figure 221: Primal Partial Body

of pain), again without meaning to. In other words, it was not a deliberate action, it was accidental.

Before we can accept that such a thing might happen by accident, we need to understand how the brain operates. Even though it isn't a machine, the concept of the brain's operation can be understood by looking at how a computer works:

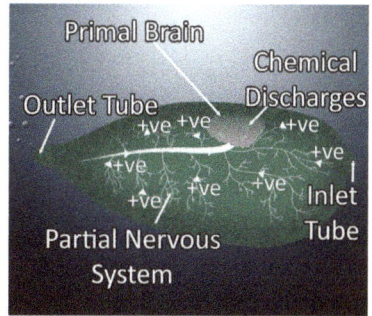

Figure 222: Primal Brain Responses

A computer operates as a series of electrical switches that are either on or off, which are given the binary values of one and zero respectively. Any action or calculation is simply a massive number of switches in a variety of on or off positions. If you are typing an electronic document and you press a key on the keyboard, a specific, identifiable electrical charge is packaged and sent to the operating system. This message is checked against an existing programmed set of on/off instructions called software. When it is matched up, there is a corresponding action that the computer follows and it also sends an image to your screen of the outcome of what it is doing. If you type the letter A, it appears on the screen as a representation of what the program is doing[76].

If, however, you don't load the operating program (software) into the computer's chip, which contains the switches, when an electrical message is sent from the keyboard to the operating system, nothing happens because there is no set of instructions about what to do. Without software instruction

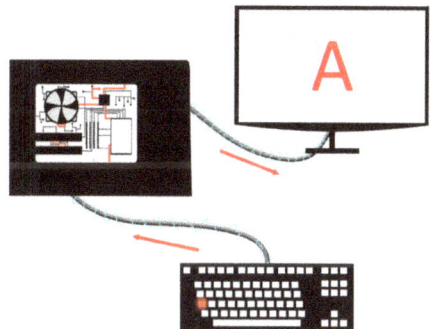

Figure 223: A Functioning Computer

programs, the computer doesn't know where a message comes from or what it means or what it should do. The electrical information simply goes one way. The computer will not make up responses randomly until one is correct, it will do nothing as the message means nothing. It will not and cannot make up its own program.

---

76. In living bodies, however, there is no screen, only electrochemical reactions.

So it is for brains. If there is no program in the neurons against which to follow instructions then nothing will happen. If our proto-fish had rolled onto a sharp rock and an electrical charge went to its brain, the most likely thing to have happened would be to disappear or blend into a mass of charges that meant nothing. The proto-fish's brain had no way of knowing that something was happening. Even if it did accidentally send back random, meaningless messages and one of these ultimately corrected the problem[77], no one is able to explain how this, or any reaction, could be sent back to the brain again as an acknowledgement of the correct survival response, then stored as memory for future similar events. Even if this series of unlikely events were possible, there is no way to pass on these new actions to successive generations other than deliberate teaching as they cannot be back-coded into any creature's existing programs. The only way for a brain (or computer) to work is for it to be programmed from the outset; that is, before it can perform its intended functions.

Displaying The Brain's Inability To Invent Its Own Response Program

To determine if self-programming is the process by which the proto-brain developed, we will examine how the human brain works now.

Consider a new action or event happening to your body, for instance, the first time you hear an alarm clock. Your brain, not recognizing the event, doesn't react by sending out thousands of responses until the right one happens, which is to turn off the alarm clock. If it did react with thousands of responses, your body would be jumping, turning, rubbing, rolling, scratching, kicking, or twitching every time something new happened. But this is not what happens. Upon hearing the new event, your brain will run an existing assessment program and review all of the information it has about the elements in that event (such as the type, direction, intensity, and duration of noise). It will then make a considered response, often looking for additional information. It will not react with multiple random responses until one corrects the situation because it is programmed to act in a specific, controlled manner.

Bodies also display reflex actions that are enacted before being processed by the brain. We are all aware of our bodies instantaneously reacting to sharp pain. In these circumstances, muscles will move reflexively to protect the body before the brain sends a response. This is a primary defense mechanism built into cells that must immediately collaborate together without brain instruction. Reflex actions show that cells are interlinked as a group despite being controlled by the brain.

---

77. For example, it moved a muscle and swam away from the rock.

The scientific evidence reveals that the brain, in all creatures, has always behaved with a set program of operation from its outset. Within that program it can add information, such as how to balance the body to ride a bike or walk on a tightrope, or play a musical instrument, but the program to sort, store and update this data already exists and is designed to allow and perpetuate the growth of knowledge. Whilst scientists don't have any evidence that a brain can write its own programs, we do know anything that happens to the body, and is recorded in the brain after birth, cannot be passed back to its gamete cells. This means personal experience cannot be updated into the brain programs of subsequent offspring. The evidence suggests therefore that preexisting programs have always existed in brains from their beginnings and that they did not, and could not, make themselves through an evolutionary process.

Using another computer example, we can get a better understanding of how our brains update information. For instance, when we prepare a letter in a Microsoft Word document we are adding information into an existing software program called Word. Then we file the document by saving it to a specific location on the computer hard drive. But if we never write the letter, the information will never exist on that computer even though it has the ability to receive and store the new document. If we change an existing letter, we access it from its saved location and refile it, effectively adding, sorting and storing updated information. However, before this can work, someone had to install Word software on the computer's processing chip. The Word program will not and cannot do anything until a user adds information. This is the equivalent of learning how to ride a bike for our brains. Unless there is a program that allows us to attempt the task at a basic level and send corrective messages to our brains, there will not be any method for storing bike riding skills. The program has to exist beforehand and it can't be updated simply because one of our ancestors learned to ride a bike.

This point is highlighted by the ability of humans to read. Some estimates have been made that early human brains had already developed the ability to read over 1.5 million years ago. This implies that the early human brain had a program that enabled it to translate the written word before written words were invented. This isn't just pattern recognition, which part of the skill of literacy requires; this is the full ability to understand the complex aggregation of human-made lettering systems. In regards to reading, the program came first.

Interestingly, however, the early alphabet wasn't invented until about 3,000 BC. This means that the human brain had within it a hardwired system of understanding written sounds over 1 million years before it was needed. The drawback with this theory is that it is anti-evolutionary. By evolutionary standards, the 'useless' ability to read words should have become redundant and obsolete, phasing out of the human gene pool well before the alphabet was introduced to society.

To return to our computer analogy, if someone installs anti-virus software on their computer, it automatically checks every new piece of software (and files) that is installed without the operator knowing. When it detects a virus, it may be programmed to remove it and/or to advise the operator. The computer performs many automatic operations without the operator necessarily knowing, such as updating time, weather, new software, and so forth. This is similar to how our bodies do many things, such as fighting infections, but the brain may not alert us unless the infection reaches a certain level of danger to our overall health.

The point is that these similar processes can only occur in the computer and the brain if programs have already been installed. If only a few parts of the programming software are installed in the brain, however, it cannot perform its intended function. Evolutionary Theory's suggestion that it gradually made up its own operating programs over billions of years means that creatures would not have survived. For instance, only having half of an 'infection response' program developed through trial and error is as useless as having none at all. It is therefore a biological necessity that creatures' brains had their programming in place from the very beginning of life.

The Five Senses

We will now examine the scientific evidence for the five senses – sight, smell, taste, hearing, and touch, including pain – in reference to the proposals put forward by Evolutionary Theory.

Sight

Sight allows us safe mobility, interaction with other people and creatures and to explore the earth and all its wonders. Although sight is not essential for survival, it does enhance our own and other creatures' survivability and lifestyle. Vision allows us to see different shapes, sizes, colors, and movements around us, but some form of light is required in order for vision to occur. Initially, we will review what light is and how it transmits images.

What Is Light?

As briefly discussed in Chapter 4, *Stellar Spectra*, light is energy in a form of electromagnetic (EM) radiation that is transmitted both as waves and as particles[78]. Electromagnetic radiation comprises a spectrum of electric and magnetic fields that oscillate (move up and down). This effect can be observed in waves in water. A release of energy in water, such as dropping a stone in a pond, will cause a wave. From side on, the water can be seen to rise and fall as the wave passes. From our vantage point, the wave may seem to be approaching us or moving away. But this is an illusion. The water is not travelling to us or away from us, only energy. What actually happens is that water molecules are displaced in a vertical motion as the wave energy passes their point. Only the energy of the wave is moving.

Light behaves much the same as continuous ripples in a pond. For example, the energy of light from our sun travels to earth at a speed of about three hundred thousand kilometers per second, taking approximately eight and a half minutes to arrive. It is able to travel as a waveform through the absence of physical matter in the vacuum of space because of its dual nature as energy particles, which act like water molecules conveying the energy of the ripple on a pond. Light travels slower in liquids and solids, which is thought to be caused by absorption and re-emission by charged particles. The phenomenon of light acting as a wave is seen when using a glass prism[79] to diffract a beam of light into its seven constituent colors – red, orange, yellow, green, blue, indigo, and violet. More commonly, as we know, it is seen on a rainy day when sunlight passes through water droplets in the sky, creating a rainbow.

In Figure 224: Components Of A Light Wave, the components of the light wave are represented as:

- **Amplitude:** The vertical distance from B to D which is its height (and depth) from the middle point and represents brightness.

- **Wavelength:** shown as E, being the distance from A to C, which is the length it takes a wave to make one full undulation (up and down motion). The wavelength of light determines its color.[80]

- **Frequency:** The rate at which the full wave takes to pass a given point in a given time, essentially its speed. The

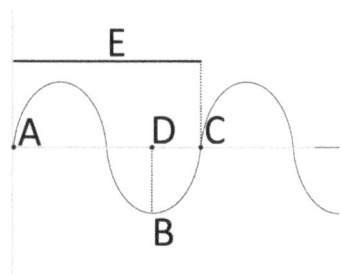

Figure 224: Components Of A Light Wave

---

78. Light particles are known as photons.
79. Refer to Figure 18: Stellar Spectra, Chapter 4, sub-chapter *Stellar Spectra*.
80. For example, the color red has long wavelengths compared to blue, which has shorter wavelengths.

entire spectrum of light travels at the same speed; it is the wavelength that determines the frequency and is perceived as color.

Converting Light Rays Into A Visual Image In The Brain.

Beams of light enter the eye through the pupil and its lens. As light passes through the lens, it is inverted to make the image appear upside down at the back of the eye on an area called the retina. The light then passes through two layers of neurons in the retina, then strikes minute photoreceptors (rod and cone cells) that convert the images into electrochemical signals. There are about one hundred million photoreceptors in the retina.

The rods are extremely sensitive and can react to just one photon of light. They allow us to see in poor lighting, such as night time, and are generally responsible for our black and white vision. Within the rod cells are disk membranes that contain a receptor protein called rhodopsin that is sensitive to light.

The cones complement the rods by working in bright light and provide us with color. The cones allow humans to distinguish over one million different shades and colors. Rods and cones are not spread evenly. At the center of the retina is an area called the fovea that has a high concentration of cones and is less light sensitive than the rest of the retina. Apart from the optic disc where the optic nerve enters, which is devoid of photoreceptors and is known as the 'blind spot', the rest of the retina has about ten times as many rods as cones. Rods and cones also vary during different stages of life. For instance, babies are born with the ability to see only in black and white, as the retina consists only of rods. Cones do not develop until after birth.

Once activated by light, the rod and cone signals are then sent through the optic nerve to an area in the brain called the lateral geniculate nucleus. This is like a distribution center that breaks up the fundamental aspects of the image signals and directs them to other specific parts of the brain that process components such as motion, color, size, and shape. Without this complex separation process, we would probably just see a blurred image.

At the back of the brain is the occipital lobe, which interprets the signals from the optic nerve. The image arriving at the occipital lobe from the optic nerve is upside down, which is then reconfigured and turned the right way up. Because we have two eyes sending two images to the occipital lobe, the end result is an amalgamated three-dimensional image of the scene outside our bodies[81]. People who have lost sight in one eye have difficulty in depth perception because they have lost one of the two

---

81. 3-D films take advantage of this dual-image process. Two cameras using different filters, such as blue and red, record the movie, which is then played on the cinema or

necessary input signals required for three-dimensional sight. Figure 225: Light Rays Converted To Electrochemical Signals By Optical Nerve displays the above process.

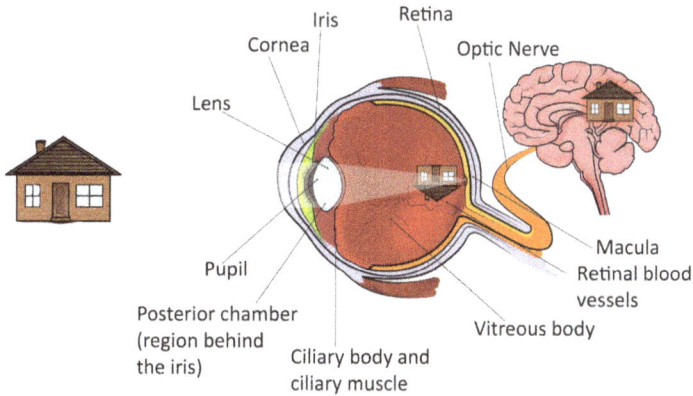

Figure 225: Light Rays Converted To Electrochemical Signals By Optical Nerve

Our eyes automatically focus on objects at differing distances by subtly adjusting the shape or contours of the lens to ensure the incoming picture translates clearly in the brain. This is an instantaneous process where the brain instructs small changes to the eye muscles. For those with imperfect eyesight, remedial options such as glasses, contact lenses or surgery can alter the alignment of incoming light to match the manner in which the brain interprets and sees them. This implies that our brains have a standard for processing electrochemical signals that represents what we should be seeing, but when our sight organs (eyes) vary and send slightly skewed signals, the brain's programming is unable to compensate for it. Any corrections therefore have to be made in the eyes. The program in the brain that converts photoreceptor signals into visual images is set and cannot be altered.

Evolutionary Theory fails to explain how a program such as this, could have changed and self-evolved when scientific observation shows that it cannot be altered. In this case, the intended outcome (how to interpret perfect sight) had to be known prior to establishing the sight program, as it is unchangeable.

First Eyes

There is no detailed evolutionary explanation about how the first eyes in history could have built themselves over millions of generations,

---

TV screen. To view the movie in 3-D, we must wear glasses with the same filters (for example, red over one eye and blue over the other) in order for our brain to recognize the two different filtered images and process them into a single 3-D image.

originally starting with sea creatures. It must be remembered that, under Evolutionary Theory, there was no plan for an eye, or how it might be constructed or connected to the brain, or any previous knowledge that there were electromagnetic rays that could be used to visualize the outside environment. The fundamental concept of self-development requires that small changes occurred across millions or billions of generations by mutations in embryonic gamete cells. Considering that eyes contain millions of specific parts located in specific areas, the following general events must have occurred in an evolutionary sequence:

- Mutations in the gamete cells of an early sightless creature resulted in the first cells of an eye appearing in the following generation. This also required the skull to make a protective orbital depression for the eye (or two eyes) to sit in without knowing an eye was to appear millions of years later.

- Subsequent mutations added newly invented ocular components without knowing they would have any use or where they should be located. Millions of cells for rods and cones, the lens, retina, neurons, and muscles appeared by pure luck, without meaning to, in the correct sequences, numbers and locations for later sight to occur.

- An accidental neurological connection to the brain appeared at some point, also capable of carrying signals to the correct areas of the brain that would interpret sight.

- Rods and cones were made in a manner that was capable of translating incoming light waves into matching electrochemical signals that the brain could recognize as sight, even though they had no knowledge there were light waves, what they were comprised of, that color existed, or how to convert one form of energy into another suitable for a brain they didn't know existed.

- The brain had to know that eye muscles existed and then develop a system of instructions to perfectly match muscle movements to change the shape of the eye in order for it to correct incoming signal changes.

- The brain had to interpret the incoming signals from the optic nerve as sight and also recognize that the image was upside down, three-dimensional, moving or stationary, and in color. Then it had to invent its own method for turning millions of electrochemical inputs through one hundred and eighty degrees the right way up.

Further complications arise when attempting to consider how this random process created either two identical eyes at the same general time or that a second eye also invented itself in precisely the same manner sometime after the first eye evolved. Either way, Evolutionary Theory predicts that the same outcome is not possible from two random series of events over millions of generations with millions of different components. Furthermore, the Evolutionary Theory of eyes contradicts the scientific evidence found in the fossil records. There is no evidence of one-eyed development of life into two eyes, as almost all fossils show creatures with an even numbers of eyes. On average, Evolutionary Theory predicts that fifty percent of creatures on earth should have an odd number of eyes, but there are only a couple known to exist in this category. For instance, starfish have an eye (more of a light sensor) on the end of each leg and many of these have five legs, or odd numbers of legs, while centipedes have odd numbers of eyes. Tadpole shrimp have three eyes, evident without change in the fossil record for over 70 million years. There are many different types of eyes in creatures. Some are so complex they have thousands of lenses per eye (for example, trilobites).

Eyes also have another feature: eyelids. Along with tear ducts and eyelashes, eyelids protect our eyes from minor damage and dust, and clean the surface of the eye. Eyelids also block out the light if it is too bright or to facilitate sleep. According to Evolutionary Theory, eyelids were created from millions of lucky accidental cell mutations that also coincidentally happened to make an identical eyelid for a second eye.

Smell And Taste

Smell

Olfaction is the term applied to the sense of smell. Both animate and inanimate objects release chemical compounds called odorants into the air (or water) that are able to attach themselves to receptors onto sensory cells in animals' noses called olfactory receptor neurons, which are then recognized as smells. In humans, there are five to ten million olfactory receptor cells located in a mucus membrane in the roof of the nose called the olfactory epithelium. The axons from these olfactory receptor neurons penetrate through perforations in the covering bone and terminate in two olfactory bulbs, which are situated under the brain's frontal lobe. Specific odorants appear to be aligned with specific olfactory receptors, some of which can react with multiple odorants. Effectively, each of these millions of receptors is tuned only to specific chemical compounds. The matching of chemical molecules with their olfactory receptor triggers information patterns or formulas in the brain, each of which represents different smells.

Humans' main sense is sight but we do have sensitive levels of smell, being able to recognize around ten thousand different types of odors. Dogs have very high sensitivity to smells with beagles, for example, having around two hundred and twenty-five million smell sensors. About a third of dogs' brains are allocated to smelling activities.

The internal process for determining smells is the same as sight, hearing, taste, and touch. Sensors convert smells into matching electro-chemical impulses that are sent to the brain that records, categorizes and files them as smells. As we smell things all the time during breathing, our brains constantly compare these results with those previously encountered, which are stored as memories. The brain cross-references new incoming data against its memory records and will bring to our consciousness the memories of those events and the actions that occurred at that time. This is the process of nostalgia, for instance, of reliving memories from many years, if not decades, ago.

Animals react to others' smells, mostly without realizing it. Mothers and babies identify each other by smells. Smells allow animals to make determinations about their surroundings both close up and from a distance, such as other animals, smoke, rotting food or bacteria.

Taste

Humans have around five thousand to ten thousand taste buds. These each have about fifty to one hundred sensory receptor cells that can identify chemicals in foods as acids, salts or sugars, recognizing four basic types of taste: sweet, salty, sour, and bitter. Taste buds are located in papillae, which are very small bulges on the tongue, and in the back of the mouth and palate. Taste receptor cells transfer taste information as electrochemical impulses via axons to the thalamic center in the brain. From the thalamus, the information is then sent to the olfactory cortex, which is located under the frontal lobe. Information about smells and taste are then combined in this area to make the sensation we call flavor.

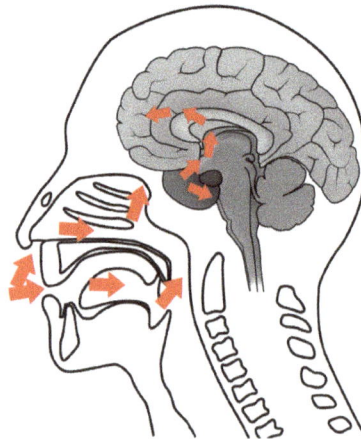

Figure 226: Smelling Process

When chewing food, chemicals are released into the mouth with some finding their way through the nasopharyngeal passage at the rear of the throat to the nose. When we have a blocked nose, we often think we cannot taste the flavors. However, it is the smells that cannot reach the nose sensors so our overall recording of flavor is disturbed because the combination of smells and tastes is incomplete, as only the tastes are being detected.

Healthy foods release chemicals that our brains categorize as pleasant whereas rotting food aromas are registered as bad or foul.

Teeth

Teeth are completely different from any other body part. Human teeth provide both cutting and crushing actions and require the structure, functioning and muscles in the mouth to be complementary.

In Figure 227: Structure Of A Human Tooth, the basic structure of a human tooth is shown.

Teeth are secured by roots made of dentine that attach to the jaw (alveolar) bone. This bone has W-shape that allows for the root on either side of the tooth to embed into it. Nerve cells run between the bone and roots and also inside the roots connecting with the main nervous system sending information to the brain.

Between the roots and bone is a layer of cementum, which is excreted by the teeth roots and fully covers them. It is similar to bone and comprises around forty-five percent

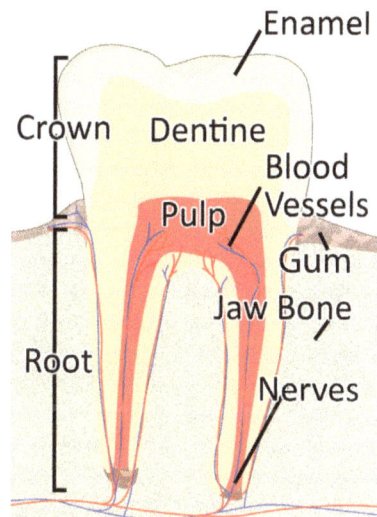

Figure 227: Structure Of A Human Tooth

hydroxypatite, thirty-three percent collagen, with the balance being water. Its thickness varies depending on its location such as at the root apex where it is thickest. It is held in place by periodontal ligaments that attach to the bone. These are complex constructions with three different types of epithelium that perform various tasks such as support, bone formation and reabsorption of bone calcium into the blood system.

The central section of the tooth is comprised of connective tissue called pulp, which encompasses blood vessels and nerves. Dentine is the next layer containing mostly collagen and two protein classes not found elsewhere in the body (amelogenins and enamelins). It is a mineralized connective tissue containing microscopic channels (tubules) that run from the internal to external borders. It is made by the dental pulp and acts and as a protective layer while supporting the crown of the tooth. It is softer than enamel and more prone to decay.

The gums cover the jaws and are made from mucosal tissue with three different types of epithelium. The part of the tooth that is visible is the enamel, which covers the internal components down to the gum line. It is over ninety-five percent mineral crystalline calcium phosphate which is the hardest substance in the body, although it is relatively brittle. Enamel density varies over the tooth surface being thicker at the ends where cutting occurs. Mammals develop two sets of teeth with baby or milk teeth replaced during childhood.

Although it is obvious that teeth have to be located in the mouth, random evolutionary mutations give no guarantee of such things.

Hearing

The ability to hear is a complicated process utilizing a range of differing components and mediums that carry, translate, transfer, and interpret sound waves into hearing in the brain. The process in humans is displayed in Figure 228: Hearing Process.

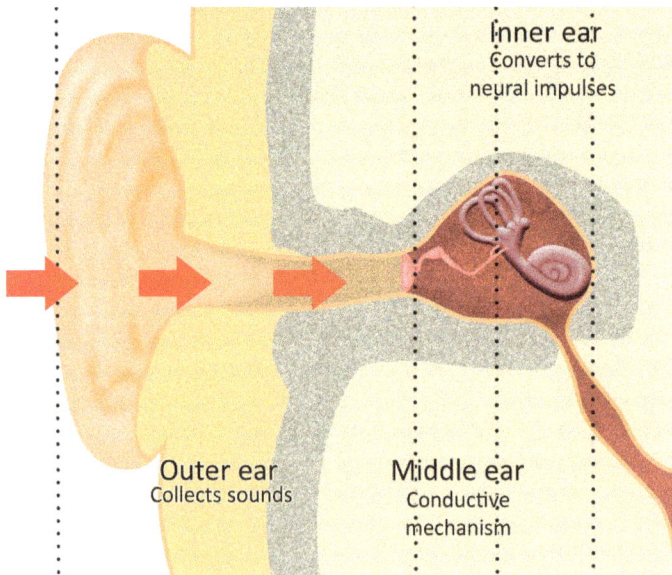

Figure 228: Hearing Process

The outer ear receives sound waves transmitted through air (or water), which then travel through the auditory canal and strike the surface of the eardrum, the tympanic membrane, which is comprised of tight skin over a circular opening in the skull. Sound causes the tympanic membrane to vibrate, which transfers the movement to three tiny bones called the malleus, incus and stapes (hammer, anvil and stirrup respectively). These three small bones of the middle ear in turn vibrate and amplify the vibrations onto the inner ear. This contains liquid and a sensory epithelium (thin tissue) in the cochlea that is studded with microscopic protein hair cells that poke out into the fluid. There are about sixteen thousand hairs, being only a few thousandths of a millimeter long, which are arranged in rows of V-shaped patterns. These wobble and convert the vibrations into nerve impulses that release chemical neurotransmitters that translate the level, frequency and length of sound through special auditory nerve fibers running from the bottom of the hairs to the temporal lobe of the brain. The sensitivity of these hairs is so great that even an infinitesimal movement will register as a sound. The reason that we can hear high-pitched sounds is because these hairs can switch the currents on and off at an astonishing rate of twenty thousand times per second. Bats' hearing is even more extraordinary as their hairs can operate at ten times this speed.

The ability to hear comes from a series of complicated processes:

- From an original vibration in air to
- A vibration in skin (on the eardrum) to
- A vibration in three bones to
- Waves on a membrane surface (inner ear) to
- Vibration in a fluid to
- High-level vibrations in hairs to
- Impulses in fiber roots to
- Chemical transmissions to the brain to
- Electrochemical reactions in the brain that then sorts and interprets the differing sounds against a preset list of noises and records and stores new noises.

Points To Note About Hearing Components And Processes

- After the eardrum vibrates, it goes back to its original static position almost instantaneously. This stops sounds from echoing in our ears.

- The middle ear has two small muscles that automatically control the volume of sound from the hammer, anvil and stirrup bones.

By involuntary motion, they limit exceedingly high levels of sound from reaching the inner ear where the hairs can be damaged and hearing can be permanently affected. This happens by muscles contracting and reducing the force of vibration.

- The components in the ear are so finely tuned they can react to one five-thousandth of a millionth of barometric pressure.

- For the eardrum to vibrate properly so that sounds are interpreted correctly it needs to have air pressure equal on both sides, which is usually at atmospheric pressure. The middle ear space behind the eardrum contains air that is constantly absorbed by the cells on its lining, requiring occasional renewal. There is a hollow tube[82] running from the inner ear to the back of the throat and nose that maintains the middle ear pressure the same as atmospheric air pressure. This tube is normally closed but will automatically open at those times of chewing, swallowing or yawning when there are pressure differences. This allows air to flow into the middle ear and any mucus to flow out. For this to occur, there needs to be a control mechanism to determine differing air pressures and an automatic process to correct them.

- The three bones at the back of the eardrum are perfectly engineered for the purpose they perform. Ear bones stop growing after birth, being the only bones to do so.

- When the head is covered by a liquid, the shape of the ear canal generally tends to hold it out, not allowing it to reach the eardrum, as liquids can cause damage.

- There are around ten types of problems with ears that can cause the human body to lose balance or experience various forms of dizziness, such as vertigo (the sense that the room or outside is spinning). For example, the utricle contains calcium crystals that send information about head movements to the brain via the inner ear. If these move from their normal positions they can send incorrect data which the brain will misread, causing a spinning sensation. The utricle is an oblong shaped organ located in the inner ear that has hair cells covered by otolith membranes that pull on the hairs when motion occurs and effectively determines variations in tilt of the head. While the utricle measures horizontal movements there is a matching organ, the saccule, that measures

---

82. The Eustachian tube.

vertical movements. The brain compares information from both of these organs, the eyes and the receptors that indicate stretching in the neck to determine whether it is only the head or the whole body that is tilted.

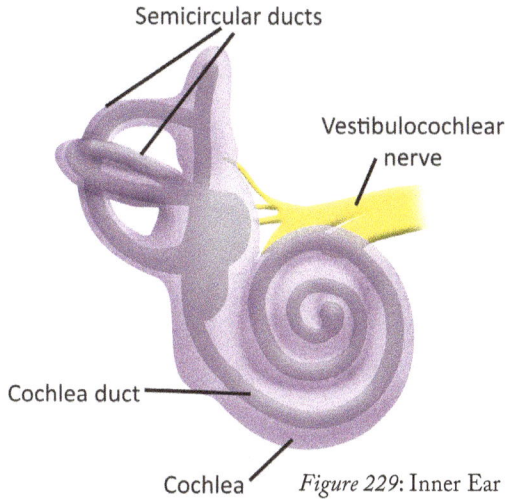

Figure 229: Inner Ear

- The inner ear is vital for humans to balance and walk upright. One of the most significant differences between humans and other creatures is the ability to walk upright on two legs. This means that the inner ear (on both sides) must have developed before humans walked upright, otherwise upright bipedal walking would not have been possible.

There are at least ten different components used for hearing located in the outer ear, middle ear, inner ear, and the brain:

- Cartilaginous ear.
- Canal.
- Eardrum.
- Skull bones.
- Tympanic fluid.
- Cochlea.
- Hairs.
- Nerves (and electrochemical impulses).
- Muscle fibers.
- Receptors.
- The brain.

Although Evolutionary Theory states that the appearance of hearing components is accidental, there is no evidence to confirm which appeared first or in what order they originally appeared. There is also no description about how the process of converting sound waves into chemical neurotransmissions occurred, or how these neurochemical transmissions were first recognized by the brain as hearing. Nor is there any explanation about how the components of the ear formed into the perfect shapes, sizes, locations, sequences, connections, and processes with each other.

If we consider just these ten components in hearing, Evolutionary Theory would suggest that billions of different and unworkable mutations with no results would occur until one worked and suddenly sound was recognized for the first time. Not only was a vibration recognized as a sound but it also meant that random mutations in fertilized embryos had managed to accidentally over-write instructions so these ten components would develop themselves as a group into the perfect shapes and sizes, and into the correct sequence that allowed the vibration in air to ultimately be converted into a chemical neurotransmission. If the order in which they are laid out had been different, or the timing of their development had not matched, or the brain had failed to recognize the signal, then hearing would not have occurred.

If we look at this self-development of hearing in a little more detail, the two most likely ways that the components necessary to hear could have formed by chance are:

Option 1 – *it all happened in one embryo.*

The perfect auditory components occurred at the same time in the same embryo. Then, that became the last time any of these cells mutated. As soon as sound was recognized and processed, the mutation process stopped. Even Evolutionary Theory does not propose this one-off self-invention as it defies the very principle of gradual mutational development and requires knowledge of how a hearing system would work beforehand.

Option 2 – *it happened in multiple embryos, which ultimately joined in the same family line.*

Each part needed to form into its perfect structure over a series of billions of embryos. The auditory components then stopped mutating while other anatomical parts were still mutating into other shapes and other types of creatures.

This, however, contradicts the process of random, chance evolution. For example, how would a cochlea know that it had attained the perfect design for the inner ear and stop mutating when other components such as the hairs or middle ear bones had not yet formed or had continued to mutate? Even if there was a process by which the cochlea did know it was fully formed, how could a cochlea consciously stop mutations in following generations when the process is random and assumedly unstoppable?

Also, like most creatures, humans have two identical, mirror image ears in the same area on opposite sides of the head. Evolutionary Theory, however, claims that these were made by separate chance mutations. Evolutionary Theory states this even though the process of hearing is so complicated that, if it did happen once, it could not have happened identically twice in mirror image.

When we consider the complicated series of processes required to actually get sound transmitted to the brain, if it did happen by millions of chance mutations, there would still be billions of ongoing mutations changing the design and shape of the components, hopefully so that we don't ever lose our hearing. Or perhaps there could be a development that would allow us to hear in the same way as bats. But there are no identifiable mutations to hearing processes anywhere.

The Mathematics Of Evolutionary Theory

We are going to look at the mathematical probability that hearing self-formed by chance. There is no absolutely correct manner for doing this, as there is always a 'guesstimate' of the likelihood of any particular event. To simplify the exercise, we will initially apply some odds that are much better than reality would suggest.

We can say with confidence that the ten reasonably complex components (or structures) that allow us to hear do successfully work together and perform their task perfectly well. That is, we can clearly hear and differentiate sounds. Therefore, we can say that as far as hearing is concerned, the components used are essentially as perfect as is necessary for the outcome. Let us then assume the likelihood of any one of the ten hearing components (inner ear, canal, bones, middle ear, and so forth) forming perfectly by itself, by chance, is one in ten thousand (1/10,000). These are extremely generous odds, just as likely to be one in a million (1/1,000,000) or even one in a billion (1/1,000,000,000). In this example, the odds are only representative of the chance happening of an event; it is not known that such changes can actually occur biologically.

The probability that all ten would be perfect is one chance in ten thousand multiplied together ten times. That is, the number one followed by forty zeroes, numerically displayed as:

1/10,000,000,000,000,000,000,000,000,000,000,000,000,000.

That, in itself, is a staggering number (one in ten thousand trillion, trillion, trillion) that we will call 'mega-chance'. There is, however, a further consideration. The chance that all ten were accidentally in the correct order is 1/10 multiplied by 1/9, then 1/8 and so forth resulting in a chance of about one in about three and a half million (1/3,500,000). But of course, each part did not move about from first to third place and so on until it was in the right sequence. There was only one chance to be in the correct order as they had to develop into the perfect shape and connection with the next part at their very first point of development. This makes the chance of all components forming themselves perfectly and in the correct order as one in three and a half million (1/3,500,000) times mega-chance, shown numerically below:

1/35,000,000,000,000,000,000,000,000,000,000,000,000,000,000

In layman's terms, this is one in thirty-five thousand million, trillion, trillion, trillion.

Furthermore, a probability such as this does not actually mean that hearing did occur in this manner just because there were trillions of cell mutations over billions of years (which is an evolutionary forecast, not a fact). Probability is only a numeric manner of expressing a likelihood. For all practical purposes, the likelihood in this case is zero.

Of course, if we were to assume that hearing could self-invent once, despite such an unlikely mega-chance probability, then we can also expect to find some of the other trillions of trillions of trillions of variants that do not have perfect components or sequences for hearing. Under evolution, for every being that can hear, there should be trillions that have incorrectly formed components. None of these trillions exists, either alive today or in the fossil records.

Two final points:

1. If we accept the mathematical chance that the process of hearing inventing itself was one in three and a half million times mega chance, then the chance of two identical mirror-image systems, perfectly aligned on either side of the head, is this number multiplied by itself (at a minimum). That is, the number twelve with eighty zeroes behind it. Numerically:

1/1,200,000,000,000,000,000,000,000,000,000,000,000,000, 000,000,000,000,000,000,000,000,000,000,000,000,000

2. As stated, these odds are very generous. In reality, even one in a million (1/1,000,000) for any particular hearing component is

still generous, meaning there would be one hundred and twenty zeroes in our number:

1/12,000,000,000,000,000,000,000,000,000,000,000,000,000,000, 000,000,000,000,000,000,000,000,000,000,000,000,000,000,000, 000,000,000,000,000,000,000,000,000,000,000.

The conclusion is really very obvious. There should be trillions of trillions of trillions of mutants (plus countless descendants) with different or partially formed hearing systems; but there aren't any. Therefore:

There is no real mathematical possibility of hearing forming itself by accident.

Nor is there any evidence of any transitional steps in the self-development of hearing.

If we were to extrapolate this type of probability for all of the components that were necessary to be accidentally formed from the first single cell through all life forms to humans, we would finish up with such a large number that the mathematical possibility of biological evolution explaining the existence of life on earth is effectively and emphatically zero.

Touch

The skin is the largest organ in the human body. This extraordinary covering is waterproof, it stretches and contracts, repairs itself, controls thermoregulation (heating and cooling), comes in a range of different colors, and fits the body it surrounds. The skin differs in various areas: for example, the skin on fingers and palms have higher sensitivity and unique patterns, called prints, giving better grip, particularly in the wet. Scientists would expect there to be a limited number of fingerprint designs that had carried down the generations as there is no evolutionary need to have different prints, and yet everyone has a different design. Evolutionary Theory also does not explain why, if prints are accidental and random, they are not on many parts of the skin or indeed the whole skin.

The skin is covered in millions of minute duct openings for sweat glands. They are evenly spaced across the entire surface. When the brain records an increase in body heat, it reacts by instructing sweat glands to excrete water onto the skin surface to cool the body, the process known as perspiration.

Further, when the temperature drops, goose pimples[83] rise on the skin to trap warm air and keep heat in. These bumps result from tiny muscles

---

83. Medical term: cutis anserine.

at the base of each hair contracting and pulling the hair upwards. The erect hair traps air and creates a layer of insulation. This is believed to be a reflex action started by the sympathetic nervous system.

Evolutionary Theory does not explain the accidental existence of one-way skin apertures (holes) in a waterproof covering. It is left to us to investigate the process. Did perspiration occur accidentally in one area alone or did it happen simultaneously over the entire body? If only one area occurred, this would suggest the process was copied over the rest of the skin surface; but this couldn't happen as there is no known evolutionary process to copy a previous mutation[84]. This then suggests that the remaining millions of sweat glands and apertures formed by accident, independently of each other.

This also doesn't take into account that humans don't just have one type of sweat gland, but two: eccrine glands, which occur over most of the body, and apocrine glands, which occur in specific locations such as armpits, nipples, ear canals, eyelids, and genitalia. There are even modified apocrine glands that don't produce sweat but other excretions, such as milk (mammary glands in breasts), wax (ceruminous glands in the ear), and lipids to prevent evaporation of tears (ciliary glands in the eyelids at the base of eyelashes).

Again, Evolutionary Theory fails to explain which accident came first:

• The skin apertures?

• The availability of perspiration?

• The sweat glands?

• The heat sensors?

• The neural connections to the brain?

• The brain's understanding of thermoregulation?

• The brain's ability to respond by perspiring?

Most mammals have sweat glands in their skin. Whilst humans are not unique in using skin sweat glands to cool there are only a few animals that do so (for example, horses and cats). Dogs and cats have apocrine style sweat glands on their footpads, which probably contribute to keeping them supple, and in their mouths that cool them when panting. Cattle and rabbits and a few others sweat through their noses.

---

84. Because it would be deliberate and nothing under Evolutionary Theory is deliberate.

Other Brain Functions

Memory

Our brains are fantastic resources for coping with the enormous complexity in the world. They have to recognize, analyze and respond to:

- The physical demands of our bodies, including an almost overwhelming array of sounds, smells, sights, tastes, and touches. Just the incoming pictures from the eyes involve millions of colors, shapes, sizes, and movements per second.

- A world of diverse creatures, plants, surfaces, temperatures, dangers, and opportunities.

- Our own body's emotions, feelings, thoughts, calculations, and memories.

- Contact with other humans: physical, verbal and emotional.

To cope, the brain is programmed to allocate everything into categories with several memory classifications that we know about.

Working Memory

The working memory assesses recent events and chooses where information will be stored. This memory must have some predetermined criteria against which it assesses and evaluates information. It must be capable of identifying which signals are sounds, sights, tastes, and so forth, and be able to categorize and cross-reference them against each other, including the times that they occurred. It must also allocate signals of lesser importance to deeper memory files.

This memory only caters for recent events of a few minutes and then passes the memory (signals) on to other sections of the brain. This process requires a mass of electrical impulses and chemical interactions to be held, identified, separated into categories, priorities and levels of importance, following which they are then released, separated and redirected to a variety of specific areas of the brain, such as the hippocampus. Then the signals have to be retained in specific neurons[85]. This means each one of millions of signals, travelling at four hundred kilometers per hour, has to be instructed to stop and be retained in a specific neuron in a mass of eighty-six billion neurons, every living moment. To do so, there must be a highly sophisticated program, far beyond our current comprehension or

---

85. Scientists do not know how memory is stored, or in fact where precisely. The assumption is that memory is stored in neurons, probably in and/or around the hippocampus, but this has not been proven.

our ability to construct, that knows exactly what to do and how and when to do it. A program that Evolutionary Theory claims invented itself, its own systems and functions by accident.

Procedural Memory

Procedural memory is a long-term memory that stores the repetitious processes and actions for doing things such as riding a bike or playing a musical instrument. This memory can be changed by repetition. The memory can be accessed, updated and refiled. For this to happen, the brain must compare the new information with all of the existing information and overwrite only those instructions that are different. This process reveals that the brain's programming has a significant level of both knowledge and understanding.

Declarative Memory

Declarative memory is the super long-term memory that allows you to recall everything you have experienced or learned from childhood. As new inputs arrive every moment, this memory must be capable of reprioritizing the importance and timelines of all data, effectively reordering older memories towards the bottom of the list while new memories are allocated to a higher level. This requires a constant process of refiling, categorizing and prioritizing. When the brain recalls a memory, it can take any part of that memory and then draw the rest of the memory stored in other parts of the brain. It can firstly find the memory relating to the sounds of an event and then draw up other aspects such as sights, smells, feelings, and so forth. Somehow the brain separates these parts of the memory and then cross-references them with each other so that they can be combined to present the total memory. This means that it is only necessary to recall one part of a memory and the brain will find the rest automatically. We are sometimes aware of this happening in our own brains when it takes a little time to find the full recollection of a previous event. Then suddenly, once found, it is brought up to our consciousness. This process is endless as everything that comes in through our senses is immediately assessed and compared with previous similar events. For example, when you see faces in a crowd, your brain is comparing every one of them against its 'register of faces' to see if any match. If so, it will bring to your consciousness the name and relationship with that person and, usually, the most recent interactions.

It has not yet been determined whereabouts in the brain this massive database is located but obviously it needs a highly sophisticated referencing system to readily find where things are stored. Considering incoming information is data in electrochemical form, it must be consistent in the way it is received otherwise the brain will allocate it to incorrect areas. When finding memories, the brain is looking for matching data. An example of how we might apply this type of referencing is with fingerprints. These days, key points (or markers) in fingerprints are coded and held as data in a computer. If trying to match a fingerprint, the computer will look for a match to the data, which is effectively a number sequence. The brain is theorized to employ a similar type of process, matching its data by some form of sequence. Then it must invent its own competent cross-referencing system so that it can access relevant data by complementary data recognition.

We also know that the way humans remember is different from other animals. Research shows that animals tend to remember by repetitive actions. Humans remember in the sequence that events take place and in visual and word forms. The ability to remember in word forms gives us enormous potential to think and develop responses and understandings that other animals cannot do. We can also add information to existing databanks so that our understanding improves. Because of this greater understanding, we can adjust our behavior to allow for the increased knowledge.

According to Evolutionary Theory, human brains developed from reptilian then mammalian brains. What we can observe about the main operations of the brains of these three land-based creatures is that:

1. Reptilian behavior is controlled by instincts and they communicate through images and actions.

2. Mammals have advanced brain capacity that allows them to adapt their instinctive behavior based on events and they show feelings of a personal nature.

3. Human behavior is affected by knowledge, the ability to reason, and insight (understanding).

    • We communicate using words, signals and images.

    • We display intuition, logic, evaluation and imagination.

    • We show a level of concern and feelings for our families, other humans and creatures, for truth, fairness, the environment, and so forth.

    • Our ability to develop and understand language as well as writing is a significant advancement.

Carl Sagan (1934 – 1996), former professor of astronomy at Cornell University, originally proposed the concept of stepped development in intellectual levels. A Pulitzer Prize winner for the book *The Dragons of Eden: Speculations of the Evolution of Human Intelligence* that "the mind... [is] a consequence of its anatomy and physiology and nothing more".

There is much debate among evolutionists about how the vast differences in these abilities came about, as they represent major jumps in levels of both knowledge and understanding that do not comply with the evolutionary expectation of gradual increases. The major and instant increases in brain abilities align with the arrival of new types of creatures without any evidence of intermediate developments.

Although the following quote from Sagan was not directed at the theme of this book, it fits in with the concept highlighted at its beginning; that our society has been so highly exposed to evolutionary theories that we find it difficult to dismiss them despite the weight of contrary evidence.

"One of the saddest lessons of history is this: If we've been bamboozled long enough, we tend to reject any evidence of the bamboozle. We're no longer interested in finding out the truth. The bamboozle has captured us. It is simply too painful to acknowledge – even to ourselves – that we've been so credulous. (So the old bamboozles tend to persist as the new bamboozles rise.)"

(Carl Sagan, from his essay "The Fine Art of Baloney Detection", *The Demon-Haunted World: Science as a Candle in the Dark*, Random House / Ballantine Books 1995/1997).

Speech

Once we have learned the skill of speaking, it becomes something we do automatically. We don't consider it to be special despite the process being fairly complex, requiring many things to function so that we can make sounds, words and comprehensible communication. The most important organ for speech is the brain which has to send complicated and coordinated instructions to muscles once we have decided what to say. It has to translate thoughts into specific timed actions, instantaneously. The brain can alter these instructions to allow for a person's level of excitement, such as cheering at a sporting event, by increasing the volume level or making a sound linger for effect. This also allows the body to release built-up emotions on a physical level. The voice, therefore, is not limited to sounds, it is also a factor in releasing tension in the body generally.

How does speech work?

As displayed in Figure 230: Organs of Speech, air has to be forced from the lungs up through the trachea[86] to the vocal cords. In between the vocal cords is a space called the glottis. During normal breathing, the vocal cords are held open by arytenoid cartilages that keep the tissues taut. When speaking, tension on the vocal cords is loosened allowing them to vibrate, using a principle called the venturi effect where, as air (or liquid) passes through a constricted passage, it increases in velocity resulting in a decrease in pressure. Because the pressure is lower, the vocal chords come together which causes the pressure to build up again and the cycle can repeat itself. The opening and closing speed of this process can be between eighty to two hundred times per second in humans and it is these variances that cause the different pitch in our voices.

To make specific sounds, a range of different parts and movements can be used. Constant changes in the position, shape or angle of the tongue, lips, mouth, teeth, jaw, nasal cavity, and hard palate are among the tools we use. As an example, say the letter S then, while still breathing out, change it to TH. You can feel your tongue push forward in your mouth forcing the air onto your teeth. This is an automatic movement instructed by your brain to your tongue muscles after you had simply thought of the different sound. To produce a broad range of sounds, the brain instructs muscles to contract or release themselves by small, coordinated changes.

Evolutionary Theory assumes the existence of the components for speech and the brain's ability to instruct muscles are the result of billions of accidental mutations as speech was never meant to happen. Naturally, for speech to be effective, we have to be able to hear and then comprehend the words and respond. For this, we need components for receiving sound waves and translating them into a form that the brain

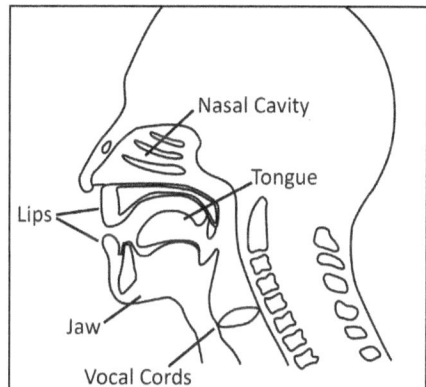

Figure 230: Organs Of Speech

can interpret and for the brain to be programmed to understand and have the ability to respond. Further, we have developed writing, art and numerical skills that enable us to record events so others can be aware

---

86. Commonly called the windpipe.

of them without speech, requiring the ability to define words, thoughts, images, and measurements into other more permanent forms.

Pain And Pleasure

We will now examine the process involved in 'feeling pain' as an evolutionary concept.

Physical Pain

As much as we don't like pain, it is a very effective way of getting our attention. Pain is an indication that something differs from the standard operation or function of a body part. We can feel physical pain when there is some sort of damage to almost any body part or when it is not operating properly[87]. A message is sent from the source of the problem to the brain that then assesses the message against a set of normal criteria and it is the brain that sets off the pain alarm. For example, when you cut your finger, you actually feel the pain in a part of your brain called the 'pain center', which identifies where the damage is occurring. This is seen in the phenomenon of phantom limb pain, whereby pain can still be 'felt' in a foot that has been amputated.

Medical research has identified two areas where people fail to register pain:

1. Congenital analgesia: which exists from birth and is a condition where the pain sensing nervous system does not work.

2. Neuropathies: conditions where the pain-sensing nervous system does not develop in the body or are made redundant through disease processes or surgical procedures.

When people with these conditions damage their bodies in collisions or cuts, they do not feel any pain because their pain systems are not operational. Whilst this presents personal problems with not knowing what is happening to their bodies, it is again evidence that the brain is the instigator of pain. It can identify the location and intensity of pain by raising an alarm without that component actually feeling pain. Therefore, each and every signal continuously sent from every cell of every component of the body is recognized by the brain and any variance from its normal predetermined status is immediately registered and responded to.

Further, if the incident is assessed by the brain as warranting attention it will set off the pain alarm. The brain can also implement other safety responses. For example, a broken bone will cause the brain to instruct

---

87. For example, an infection or a cut in the skin.

muscles to move in a manner that protects the position of the body to minimize the pain level or further damage, seemingly knowing in advance what to do. A minor cut will result in a message for the individual to look at the extent of the problem and to determine a remedy for oneself. A major cut will generally result in one's hand covering it immediately without necessarily seeing the extent of the problem first. Both reactions being evidence that the brain is able to identify differing levels of damage to an area and raise differing responses appropriate to the situation.

Pain is a very specific way of alerting a living being about a problem that otherwise may not be recognized or attended to, as we would probably not know about it until it had resulted in some sort of significant malfunction. Having pain as a warning of a fault means that initially there must be a preset, ideal condition that each and every cell of every body is working to.

This suggests that there are clear and specific operating limits against which the brain is comparing incoming messages for every area of every body part. As such, there are millions of messages from every component of every body part that are constantly being sent to the brain, which is comparing them against a standard that has been DNA coded for that part.

However, under evolution, body parts do not have a final 'end point' or function; they are merely in the process of infinite adaptability and will change again at some future stage. Adaptability is meant to weed out life forms with non-performing or susceptible parts in favor of those with better parts. The lesser life forms are therefore meant to become extinct. This means that, according to Evolutionary Theory, a brain or body part cannot determine its ideal operating platform, as there is no end point by which it can be ascertained. Yet fixed structure and processing limits are perhaps one of the fundamental bases on which the brain operates. It can only do this if body parts and processes do not change.

What we observe in modern creatures is that they have anatomical components that operate in a specific manner and in conjunction with the rest of their body. Each part has a defined role and location complete with all the necessary connections to perform its function. This is observable, verifiable science. These are the facts that we are presented with.

Pleasure

Pleasure is observed in humans and its existence can be communicated to others. Pleasure in individuals varies and can be psychological or emotional, such as satisfaction of achievements, falling in love, and laughter, or it can be physical (for example, eating chocolate, immersion in a hot bath, a massage). Like pain, pleasure occurs by chemical stimulation in the brain in areas thought to include the nucleus accumbens, the septum pellucidium and the hypothalamus and involves a multiplicity of processes and various intersecting neurological networks.

The levels of pleasure experienced by other animals are unknown with little real data available. Many animals display pleasure in physical contact with their young, and also in sexual rituals and intercourse. It appears that pleasurable human activities far exceed those of any other creature, even apes, significant differences that remain unexplained by Evolutionary Theory.

Feelings

Feelings and emotions are different from the process of thinking. We think, measure, calculate, plan, imagine, dream, and learn in our brains, yet we seem to feel things in our entire body, like happiness and sadness, anger and bitterness. The center of our feelings is romantically said to be our heart. Although we display our emotions in our bodies, we have the ability to override these by conscious thought. Psychiatrists will argue that all feelings and emotions are brain functions that can be realigned using chemicals and drugs and/or electrical charges. Research has shown that different feelings display different electrical and chemical reactions in the brain.

The range of feelings is very broad, yet scientists don't know how feelings work. We know that the body releases certain chemicals for different feelings but we don't know what necessarily generates these or how it was originally known that specific chemicals result in specific feelings.

There is no medical consensus about how to define sexual feelings or exactly what creates them. Humans and most creatures show signs of these feelings on a regular basis. Ultimately sexual arousal is the driving force behind sexual intercourse without which the continuity of life would be difficult to imagine.

Consciousness

As individuals, we perceive consciousness as an awareness of our surroundings, thoughts, emotions, memories, activities, and the things that happen to our bodies, yet science does not have an official definition of consciousness. Research by neuroscientists[88] indicates that information coming into the body (sounds, sights, touch and so forth) is dealt with by

---

88. Those who study the body's nervous system, including the brain.

specifically allocated areas at the same time. This combined information will only enter our 'consciousness' if firstly transferred to sensory areas, such as auditory or visual brain cortexes, then into the prefrontal cortex, then held and magnified in these processing areas.

Sigmund Freud, the Austrian neurologist (1856 – 1939) referred to as the founding father of psychoanalysis, believed there to be three levels of human consciousness. Briefly these are:

The Conscious: an individual's awareness of his current thoughts.

The Preconscious: various pieces of information that can be immediately brought into the conscious such as birth dates, names, localities, weather events and so forth. Generally information that one would refer to as an easy recollection that is not currently in one's thoughts.

The Unconscious: a collection of memories, thoughts, events and desires of which we are not necessarily aware because our brains have allocated them to rarely accessed areas. The types of things buried deep in memories include events where we may have failed or felt inadequate or were subject to embarrassment and adverse feelings towards others. Essentially the deeper things tend to be those with a negative connotation. These can affect an individual's behaviour without his awareness.

It is believed that humans are alone in having consciousness – a substantial attribute unable to be defined by science or explained by Evolutionary Theory. The French philosopher, Descartes, probably summed up consciousness the best when he said, "I think therefore I am" (Latin: Cogito ergo sum), or put another way, "I am thinking, therefore I exist".

The meaning at hand is that humans are aware of being aware. We are aware of being, and most evolutionists concur that human consciousness is a direct side-effect of our brains. That is, our biology determines our level of consciousness[89]. We have a higher 'evolved' brain, therefore we have a higher 'evolved' consciousness. An ant with its limited neural network is basically an automaton, a biological robot, with minimal consciousness[90]. A reptile with its larger brain has a higher awareness of its surroundings, but not as much as a dog or an ape, which have proportionally larger brains. Cats too have larger brains but they are not aware of themselves as individual creatures[91]. Humans, being the end of the 'evolutionary line'

---

89. Although this may not necessarily be the case.
90. Although, as previously mentioned, ants are aware of the role they play within the nest.
91. For instance, a cat thinks its image in the mirror is another cat.

thus far, have the biggest brains and vastly superior neural networks[92] of any species, therefore we have the highest consciousness.

However, consciousness cannot be purely due to the size of our brains because there are species with larger brains than us, such as whales (six times that of humans) and dolphins (one and a half times that of humans). Even a walrus has a brain almost the size of ours.

Aging

Over time, bodies deteriorate and become gradually less functional. Why do we age?

Biologically, there is no specific law requiring the aging process to occur. Under stable circumstances, most creatures, including humans, should be able to live for extended periods, if not indefinitely. However, this is generally not observed. Most creatures on earth have limited life spans.

The oldest recordings in common animals such as dogs, cats, horses, bears, and elephants range from 29 to 86 years. Other less common species such as macaws, tuataras, eels, and whales can vary from 100 to 200 years. The oldest known life form is a sixteen foot tall Spruce tree in Sweden claimed to be 9,550 years old.

There are many evolutionary proposals concerning the possible reasons for limited life spans. Suggestions include:

- Reproductive cells consume proportionately more repair resources from the body whereas somatic (all other) cells expire earlier due to less maintenance over a lifetime.

- Natural selection – being influential in youth and less effective in older age.

- Family based circumstances where older relatives support the youngest, thereby reducing overall stress in the familial group. For instance, a grandmother who can no longer bear children has more time to help and assist her fertile offspring and reduce some of the burden of rearing their own children.

- The mitochondria of female reproductive cells passing to offspring and further generations. Those with higher energy levels naturally select themselves and thereby pass on longer lifespans.

---

92. As discussed earlier in this chapter, *The Brain*, humans have a vastly increased number of neural connections compared with every other species, which is postulated as the reason for our higher intelligence and maybe accounts for our higher consciousness.

- Genes that are beneficial in early life but are detrimental in older age.

- Environmental influences – where unfavourable conditions require higher birth rates, thus allowing for shorter lifespans; and where plentiful resources require lower birth rates and longer lifespans.

- Apoptosis: a process in multicellular organisms where cells appear to have a programmed death. These are mostly caused by bio-chemical activities such as bulging in the plasma membrane (blebbing), gradual shrinkage, breaking up of the nucleus or fragmenting of chromosomes. The effects of cell deaths can be seen when human embryos lose the 'skin' between their fingers and toes (around week eight) allowing them to separate.

Despite all of these ideas, it appears that cells themselves have an inbuilt limit to the number of replications that they can undergo[93] causing breakdown of individual bodily functions. Interestingly, a repair enzyme called telomerase adds DNA sequence repeats to the ends of chromosomes when cells replicate. If this didn't happen, cell replication levels would be markedly reduced due to the loss of DNA information, thus leading to shorter lifespans.

The fundamental evolutionary argument for biological self-development is that the best adaptation will survive and the weaker will die off; essentially 'survival of the fittest'. If this were the case, it is reasonable to expect that in any type of life form, a group of individuals with longer lifespans will have a significant advantage over all others. They would have more generations alive at any point in time and be able to adapt their lives with more cumulative information and security than others, perhaps establishing dynasties for eons. However, as detailed in many instances in this book, life forms work together collaboratively, not singularly, and 'survival of the fittest' is not a process that is observed to work exclusively.

There is no agreed biological explanation for what appears to be pre-determined limits to life times.

Summary

We have only taken a brief look at the human body. We could write many books describing the complex operations and interactions within our bodies but we would still be far from a complete understanding. There are many attributes that we have not covered in this book because we

---

93. The Hayflick limit being about fifty to seventy times in humans.

encounter the same barriers of biological impossibility, where evolution cannot explain self-development. Things that require another explanation such as blushing, lateral thinking, or circadian rhythms (internal body clocks also evident in plants, animals, fungi, and cyanobacteria).

What then can we conclude about the human body?

It has many components, functions and processes in common with other creatures. Like all other creatures, the size, shape, location, operation, and connections of all body components suit the body they are in. We do not find a bird-sized heart or a horse-sized tongue in humans. Everything, in every class and species, is in fair proportion to its body. Evolutionary Theory, being a random series of events with no intended outcomes, offers no processes for allocating any individual body part to be in any specific body because there is no plan or design for anything. An evolutionary world will display uncontrolled variations in creatures where the components will not match the location, functions or sizes required for that body. But the world does not display such irregularities.

Whilst humans display similar body functions, we are significantly different from all other creatures. We have significantly advanced brain activities, feelings, emotions, thinking and calculating abilities, morals, appreciation, ingenuity, imagination, inventiveness, and so forth. Humans are unique, complex, sophisticated beings who have inner needs and desires that extend beyond the physical to the spiritual.

Chapter 25 Unraveled

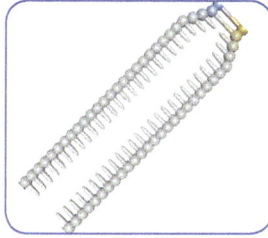

- The brain is comprised of billions of neurons that do not physically touch each other, yet they are able to transfer masses of data between themselves and to identify, separate, store, and cross-reference information from all over the body into specific locations. Even though the brain is a complicated mass of neurons, switches and neurochemical activity, it performs its operations automatically, sorting and prioritizing information with astounding speed. It is controlled by complex programs that cooperate together allowing it to perform millions of tasks simultaneously.

- To operate effectively, the brain requires pre-knowledge of masses of activities, such as evaporative cooling (releasing perspiration onto the skin surface to cool the body). It also requires monitors to send information, prioritized programs to implement corrective activities and functioning systems to perform the necessary tasks.

- The brain is enclosed in a protective skull and its location and surrounding structure differs from the rest of the body.

- The ability to see is extraordinarily complicated. It requires millions of light rays made up as different lengths, frequencies and amplitudes to be translated into electrochemical transmissions that are then received upside at the back of the eye. Millions of body cells have to be in specific locations and numbers and to perform unique functions to achieve sight. The accidental existence of identical multiple eyes in any creature is an impossibility according to evolutionary principles. There is no explanation how the complex, intricate and purposefully laid out structure of the eyes could have made themselves without pre-knowledge of their design.

- There are many different types of eyes in creatures.

- *Smell and taste are performed in the mouth, nasal and throat cavities and combine to provide the sensation of flavors.*

- *Teeth are made from several unique substances. They are ideally located and shaped in the mouth which can move sideways to perform a cutting action. Their construction provides strength on the outer surface which allows hard objects to be broken down to acceptable sizes for digestion. The process of breaking food into small particles releases odors and enhances the sensations of smell and flavors.*

- *The process of hearing uses ten different components to transfer sound waves through eight different processes into electrochemical transmissions that are interpreted by the brain as sound. Within these are mechanisms that protect the most delicate components. The combined structure, operation and order in which these components are laid out are mathematically impossible to occur by chance.*

- *Pain is a deliberate function that alerts creatures to damage or a malfunction to their bodies. For the brain to send a pain signal it has to know the perfect layout of the body and the location and operation of each one of trillions of cells. There are no known processes allowing such sophisticated and complex levels of knowledge to be accidentally known by the brain.*

- *Humans experience pleasure across significantly more activities than any other animal. A different level not explainable by Evolutionary Theory.*

- *Mathematically, the existence of body functions and self-invented life is virtually zero.*

- *Age limits to all life forms exist but are unidentified by science.*

Each of the five senses (touch, taste, sight, sound, and smell) has different receptors that convert external inputs into electrochemical signals that the brain can identify, interpret and categorize. Each type of signal is coded in order for the brain to read and identify its type and source. Each signal is then split into different groups and sent to designated areas in the brain for distribution and memory. Like a biological computer, the sensory information can be retrieved by cross-referencing with each other.

For example, if you enter a bakery and smell the odor of freshly baked bread, eat a sample, meet a friend, shake their hand, speak with them and then buy a loaf of wholemeal bread, all of these actions are recorded and stored in your memory. But they all enter the brain from different types of receptors resulting in different types of signals that are required to be identifiable when they reach the brain. As a consequence, each signal will have a special coding to inform the brain the precise point of origin:

- The eyes take in the details of the shop, the bread you bought, the friend, their appearance and so forth, and send these images through photoreceptors.
- The ears channel the incoming sounds through the middle and inner ear.
- Each nose sensor only picks up certain odors and converts them through olfactory receptors.
- The tongue uses sensors on taste buds to send signals via the thalamus.
- The skin has four different receptors that cumulatively translate touch, heat and pain sensations:

  - Mechanoreceptors – are the sensors that detect light pressure and texture (Merkel), light touch (Meissner), stretching (Rullini) and deeper pressure (Pacinian).

  - As well, the skin has free nerve endings:

    Thermoreceptors – there are two types that deal separately with skin temperatures: hot (30-45C) and cold (5-35C).

    Nocireceptors – are pain receptors that identify penetrating and burn injuries.

At some future point in time, you can remember any part of this particular event at the bakery, or even any associated happening such as meeting that friend in a different setting, and your brain will bring to your consciousness all of the memories of that event and any other associated events. You can then choose which of these you want to have in your consciousness at that moment.

Even though each type of sensor has sent signals to your brain with different codings, and these are broken up and filed in individual locations, the brain has some method of aligning each type of memory code against new events in that category. Then it has some way of cross referencing them against previous events which it can reassemble as one combined event and bring to the conscious part of your brain.

As all signals are new when an event occurs, the brain cannot have already prepared a coding for that event. So the brain must overlay a special code onto, or into, each one of the signals prior to filing them. Whatever type of code this is, it works remarkably effectively. Perhaps this is a time code, but scientists do not know for sure.

The point here is that the brain has completely different types and levels of signals coming from each sensory group. Yet it identifies these signals individually and can divide them into many different areas in the brain and reassemble them as a group later because it is controlling, coding, filing, and cross-referencing all of the information it is receiving – and it knows how to retrieve it.

At the very heart of Evolution Theory is that the brain has learned these things by itself over time. However, the brain can only possess these miraculous abilities by one of two methods:

- Trial and error – the brain learns by trialing actions until they work, then recording that in its programming file.

- Pre-programming – When the brain is developing during embryonic growth, all the programs that make it work are embedded into the neurons at the time of their appearance. The segmentation of the brain into different areas that control different tasks also happens at this time.

The problem for Evolutionary Theory is that if the developed brain learns a new activity there is no method for transferring this knowledge to its offspring. Consequently, the brain would through necessity be pre-programmed to perform all of its tasks, and the only way for this to happen is at the time of its making.

There is much conjecture among evolutionary proponents about brain development across classes and species. Brains appear to be appropriate for the types of creatures in which they exist, their functionality showing no signs of self-development in that creature. There is no scientific evidence that identifies how or when any proposed changes in brain abilities did or could have occurred that allow for such major differences as one might find between a mollusk (for example, a slug) or a fish (for example, a sea perch) and a human.

26

A SUMMARY OF REPRODUCTION

The evolutionary story of living bodies commences with the first ocean creatures at 500 – 570 million years ago and concludes at humans (*Homo sapiens sapiens*) who, from an anatomical perspective, are presumed to have arisen in Africa around 200,000 years ago. There is no record of any new later life forms on earth since humans' appearance despite, statistically on average, at least three hundred and fifty new forms expected over the last 200,000 years. This number could reasonably be in the thousands given the vast variety of different life forms that are now on earth and are said to be evolving through the process of mutation. However, to date none have been identified. The question therefore is: has evolution into new life forms ceased since the arrival of humans?

During the past 500 million year period, there have been only three methods for creating new life in creatures, all based upon the same cell duplication process found in microbes:

- Soft eggs released into water by a female which are fertilized by a male then the embryo then develops into a baby.

- Internal fertilization: whereby male sperm fertilizes a soft egg internally in the female reproductive tract, where it develops into a baby and is released by the mother as a live birth.

- Internal fertilization: whereby male sperm fertilizes a soft egg internally in the female reproductive tract, which then forms into a hard-shelled egg and is released externally. The baby develops in that egg and releases itself upon maturity.

There are rare minor variations from these. For instance, some amphibians release soft eggs that have been fertilized internally. Some hermaphroditic species exist whereby an individual creature has both male and female sex organs[94].

These birth types in this 500 million year timeline are shown in Figure 234: Reproductive Egg Types Over History. Of note is that live births, where the baby matures internally within the maternal reproductive tract, did, and still do, occur in both water and land creatures. However, this process doesn't occur in the creature that has been proposed as the link between the two, *Ichthyostega* (refer Chapter 16), the first claimed amphibian. There are two types of amphibians, African toads called *Nectophrynoides* and *Nimbaphrynoides*, that do have live births but these are not linked to the first land animals, being reptiles. Consequently, Evolutionary Theory proposes that the live birth process invented itself in water, then reinvented itself on land in reptiles without any links between the two lines of creatures. As we have learned many times, evolution as a concept proposes that the same outcome, in this case live births, cannot occur in two unrelated lines because they are both subject to large scale, improbable chance happenings in different environments.

Evolutionary Theory has many contradictory proposals in relation to the birthing process links between creatures. Because of this, there is no clear evolutionary path to follow, as can be seen by the assumed links in Figure 231: Reproductive Egg Types Over History [95].

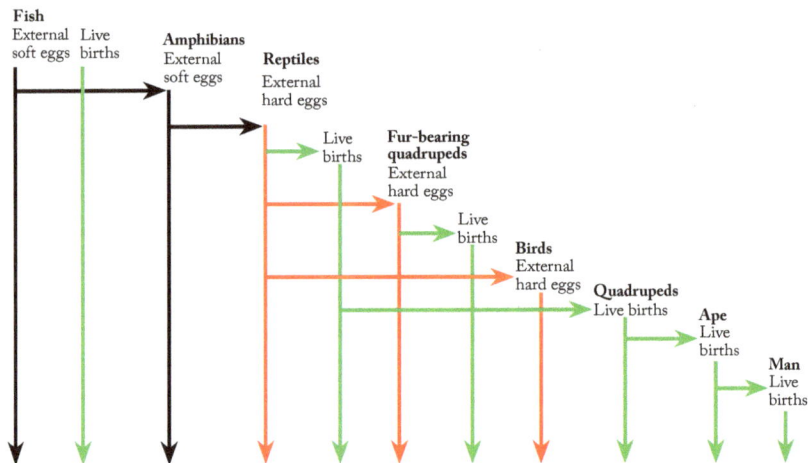

Figure 231: Reproductive Egg Types Over History

---

94. For example, barnacles.
95. In fact, evolutionists are now openly claiming that flight also evolved multiple times throughout history – birds, dinosaurs, insects – in order to match the observable evidence to their theories, despite the anti-evolutionary nature of their claims and the mathematical improbability of it.

Invertebrate animals (those having external skeletons, such as insects, crabs, and snails) have been excluded from the above list as they have no known evolutionary links to any of the creatures featuring in the list. They mostly lay eggs, with a few producing live young after the eggs have hatched inside the mother.

All reproductive processes have significant levels of complexity, relying on series of millions, billions or trillions of specifically timed and coordinated cell duplications. Once any reproductive method had invented itself and was producing healthy offspring, a change to the total process would require possibly trillions of changed instructions, a new timing program and different body parts, all occurring over perhaps millions of generations. Every different step in these changes would require the males, females and embryos to coordinate together to yield a complete, healthy baby creature even though the changes are the result of random, out of control mutations.

The reliance here, as with all Evolutionary Theory, is pure luck, as there is no external influence controlling the changes. For example, the shift from an externally delivered hard-shelled egg to a baby maturing inside a mother would need millions of coordinated and timed changes to cell programs that failures seem impossible to avoid. The sequence of changes would need to be in a specific order and timeline to ensure success otherwise the loss of every new line of creature would regularly be at risk.

Chronologically, this chart tells us that under evolution:

• Fish with external soft eggs mutated to amphibians with external soft eggs, which then mutated to reptiles with external hard-shelled eggs.

• Fish with live births mutated to reptiles with live births by jumping the intermediate step of first life on land.

At this point, where life appears on land, we encounter conflicting evolutionary theories that propose two completely different creatures – fish with live births and amphibians with soft eggs that mature externally – both mutated into reptiles, each with different reproductive processes. As previously noted, one of the principles of biological evolution is that a specific type of creature cannot evolve from two independent series of mutations. Reptiles contradict this principle, appearing from two differing sources, clearly evident by their reproductive processes. The reason for these contradictory theories is because there are no actual known scientific links between any of the creatures listed, leaving gaps to be filled by theory.

The next steps along the pathway are that:

- Reptiles with hard-shelled eggs mutated to birds and fur-bearing quadrupeds.
- Reptiles with live births mutated to quadrupeds and fur-bearing quadrupeds.
- Here again Evolutionary Theory proposes that one group, reptiles, mutated into fur-bearing quadrupeds by two different lines – one with hard-shelled eggs, the other with live births, resulting in two different evolutionary proposals.

Further along the pathway we find:

- There is no common proposal about apes' ancestry, although some research suggests a 'prosimian to monkey to ape' theory, so we will assume they came from some type of quadrupeds due to similar reproductive processes.
- Similarly, the pathway shows that apes 'evolving' into human was seemingly possible as they had the same birth processes.

Various evolutionary theories claim there are constant changes to reproduction suggesting that the three known processes continually re-invent themselves, even though they are chance events. However, what is involved is simply a regular repetition of processes that already exist.

Putting aside any theories, what science actually knows is that live births and (soft) egg releases both operate in ocean life today, and live births and (hard-shelled) egg releases operated among ancient reptiles, the first land-based creatures. The reason that hard shells are needed on land is two-fold: soft eggs are more porous and dehydrate quickly; secondly, additional protection from the air and heat of the sun is needed over and above the protection required for a submerged egg in water. Whilst amphibians live in both environments, they constantly require some form of water to survive and yet reproduce like fish with soft eggs released into water.

Essentially, the three reproductive processes we are discussing can be pared down to two reproductive processes – live births and egg releases. Both operate in water and on land and are suited to certain creatures, depending on their roles and environments. There is no actual scientific evidence either in the past or in the present of any reproductive processes changing into another process. There is only evidence of two basic processes being independent of each other.

Examining reproduction at the cell duplication level, all of the following activities followed each other in the timeline proposed by evolution. These start at the point of first duplication in microbes:

The first 'living' cell (a microbe), which invented itself from non-life, reproduced by division (duplicating itself).

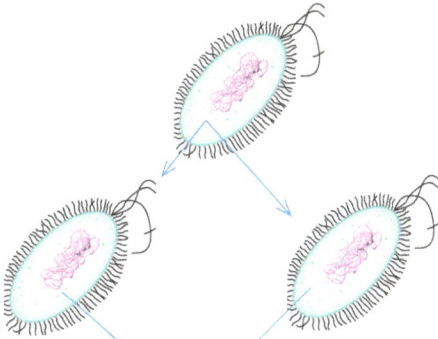

At least two of these cells joined together to make an entirely different and very complex nucleated cell.

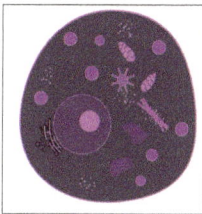

The first nucleated cell then reverted to reproduction by **division**.

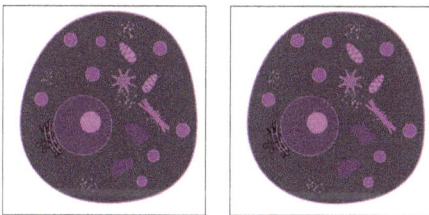

The first living creature in history was made from nucleated cells joining together, then mutating and duplicating by **division**.

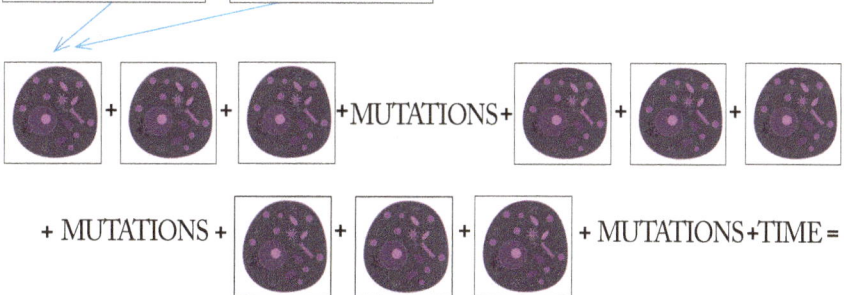

+MUTATIONS+

+ MUTATIONS +

+ MUTATIONS +TIME =

This creature released an egg cell which then made a copy of itself by **duplication** and then **multiplication** into a copy of the original creature.

First-ever egg cell

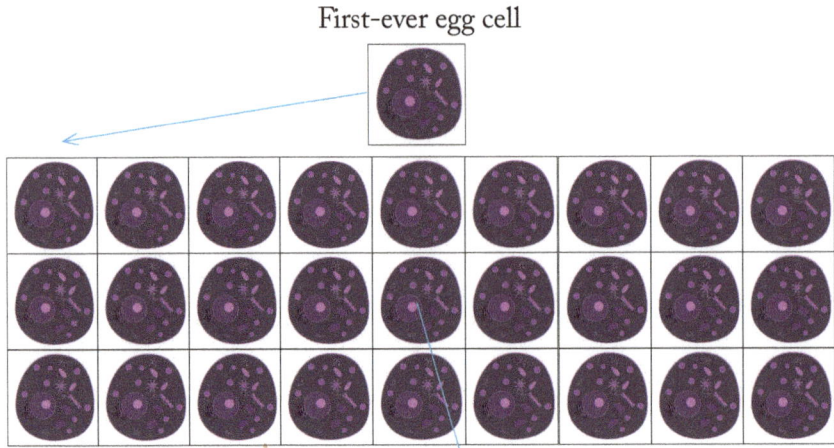

At some point, a creature divided itself into two genders.

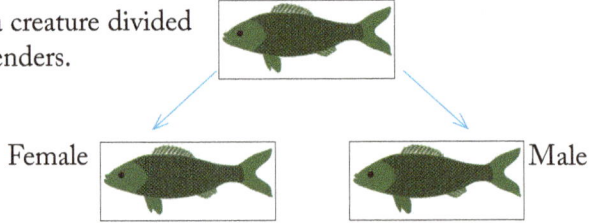

Female   Male

These two genders joined together to continue reproduction, developing soft egg reproduction and fertilisation by a male.

A fish dropping eggs into water.   A fish with an internal egg.

Amphibians maintained **soft egg** reproduction (externally).   Continues to today.

Reptiles changed external soft egg reproduction that they 'inherited' from amphibians to **external hard** eggs and reinvented **internal soft eggs**, without any connection to fish with the same process, while also re-inventing sexual intercourse, this time on land – a **joining** together.

Reptile With External Hard Eggs

Mammals maintained **hard egg** reproduction (externally).

Reptile changed to soft egg reproduction (internally).

Birds maintained **hard egg** reproduction (externally).

Mammals either reinvented or maintained **soft egg** reproduction (internally)

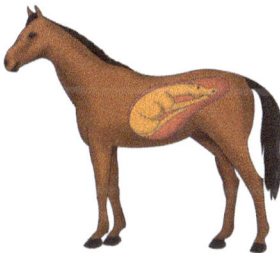

Humans maintained **soft egg reproduction** (internally).

Apes maintained **soft egg** reproduction (internally).

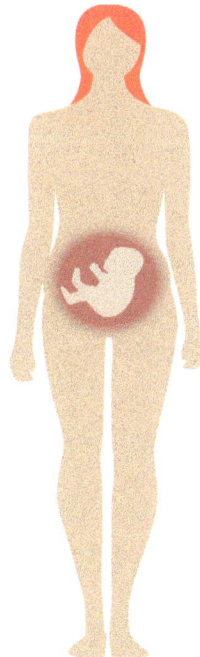

Invertebrates, which aren't shown above because evolution claims they formed separately, mostly reproduce by the female releasing fertilized eggs externally. Those on land tend to be **hard-shelled** while those in water tend to be **soft-shelled.**

Summarizing these steps in chronological order, Evolutionary Theory proposes that millions of chance mutations altered the highly complex, specific and important activity of reproduction from:

• Cell division to

• Cell joining to

• Cell division to

• Cell joining, then division to

• Cell release, then duplication, then

• Division into two genders to

• Soft egg and sperm joining, then duplication, then

• Shelled egg to

• Soft egg again whilst still maintaining cell division as part of the process.

If we look at these differences, each type of creature has a reliable reproductive system that compliments its own anatomical design and place among living things. If we strip all of these systems listed above back to their basics, a picture begins to emerge:

• Single cells reproduce by cell division to make a 'copy of the parent'.

The full entity of a microbe is just one single cell and it reproduces from this cell.

• Bodies, made of nucleated cells, predominantly reproduce by the joining of male and female sex cells to create one fertilized cell[96] that goes through a process of cell division to make a 'small copy of the parents'.

The full entity of a creature combines with a partner to make one single cell and it reproduces from this cell.

Both systems start with one cell making a new entity that suits its role and environment. Evolutionary Theory complicates the process by trying to link each chronological arrival on earth to the next, whereas there are

---

96. There are rare exceptions where just a female can do it by herself.

no proven scientific links and no known biological processes that actually allow these changes to occur. As previously discussed, it was Louis Pasteur who proved that cells can only be made by other cells and are incapable of creating themselves, leaving the first single cell and the first nucleated cell ever to appear as scientifically unexplainable events.

Chapter 26 Unraveled

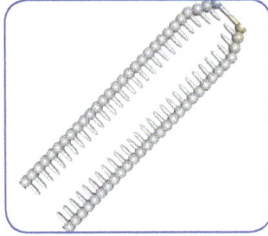

- *There are three basic methods of reproduction used by life forms on earth, processes that have appeared on multiple occasions over time without known connections to each other –*

  - *Soft eggs released into water by a female then fertilized by a male.*

  - *Soft eggs in a female, internally fertilized by a male, then producing live births.*

  - *Soft eggs in a female, internally fertilized by a male, then producing hard shelled eggs released externally.*

- *There are no known scientific links between the various creatures that share the same reproduction processes.*

- *Complete changes to reproduction processes require many, many changes to produce healthy living offspring, and failure to do so results in the end of that line of creatures.*

- *At its most basic level, reproduction results in a new life form from just one cell, regardless of whether it is a single or nucleated cell.*

27

SUMMARY OF THE THEORY OF EVOLUTION

We have learned that there is no official 'Theory of Evolution' as a distinct and singular theory of science. It is only a general concept grouped as a collection of ideas, thoughts and hypotheses from different sources which are strung together into a chronological sequence. Furthermore, none of the individual theories that make up the Theory of Evolution are facts; they are only theories.

As new information becomes available, existing theories can and should change, but this does not always occur. New information will not necessarily change a theory, while accepted changes can get lost in the mass of other information. As there is no official Theory of Evolution to change, there is often no need to notify the public of new information, as nothing has technically changed other than an idea has been proven unlikely if not impossible. Many scientific theories on evolution have already been changed but we, as the general population, are usually unaware of the changes or their significance. A good example of this is the expansion of the universe. Since 1999 the universe has been confirmed as expanding from within, not contracting as previously taught, yet many people have not been made aware of this change and it is has not been amended in some books on the topic.

Similarly, in 2014, *Scientific American* released a Special Collector's Edition book titled *Secrets of the Universe, Past Present, Future* in which the title article acknowledges the inability of science to explain how the "physics destroying insanity" of a big bang does not match our stable and predictable universe. So there was no big bang. However, not to be deterred, scientists from the Perimeter Institute for Theoretic Physics in Canada have proposed a different scenario suggesting that our universe was, more or less, ejected from the three-dimensional edge of a collapsing four-dimensional star via a black hole from a different universe. The basis for such an idea is that it would fit the mathematics, something a big bang doesn't do. Of course, who is there that can substantiate they have knowledge of mathematics from a star with an extra and unknown dimension, or a different theoretical universe about which we have no information at all, or that matter can escape a black hole where we had previously been told nothing could get out?

The reality is that none of the theories about the universe forming itself are known to be possible. But there is no official statement that the big bang is considered redundant because, as previously noted, there was never an official statement that it was an accepted position of science.

The essence of theories regarding evolution is that they are alternate proposals to explain the existence of everything, assuming the absence of a Creator. These scientific proposals are meant to be based on the operations of the natural world but are mostly speculation that cannot be proven, such as the current wave of ideas that we live in only one of many universes – a multiverse. In this case, some astrophysicists now theorize about the operations of an unknown number of universes including where they are, where they came from, what they are made from, how they are laid out, whether they contain life, and specifically why they exist and how they are linked with ours. This, and science's inability to explain how our own universe operates in defiance of natural laws, merges into the realms of unprovable belief, a scientific religion of sorts.

Our expectation, however, should be that science will provide us not with another belief system but with commonsense explanations about what is actually known, that is, the observable universe. Science, though, is handcuffed by its own demands that any explanations fit in with the natural operations that have been measured and observed and on which scientific laws have been documented. And that is where there is a blurring of information. Science generally doesn't give us natural explanations about evolutionary theories because they mostly defy natural laws.

This therefore, right at this moment, is a great challenge for all of us:

If science, because of its own self-determined rules, can't clearly explain the existence and functioning of the universe, based on operations of the natural world from which it was meant to be formed, is science the right forum, and are scientists the right people, to be seeking answers from?

By deliberately ignoring other possibilities, science as a whole has limited itself. If science can't prove that matter, the universe and life made themselves from nothing and fails to give explanations based on observable reality, do we now need someone else to provide answers that align with the facts we know?

As we have become so reliant on science, it is difficult to determine the right group of people to give us answers. There are many groups and individuals who have other explanations, often not publicly broadcast, but we are faced with the same issues we have with science, in that everyone has their own perspective and agenda that may distort the simplest explanation. That is why this book has been written. It presents the known facts and compares them logically with Evolutionary Theory so the reader can draw their own conclusions. You are the best person to decide, for yourself, based on the facts and commonsense.

The following summary is our best attempt to look at what has been presented in this book as the most reasonable and up-to-date information that evolution (self-creation) is responsible for everything in existence. The major steps set out below are followed by a comment on the observations that have been made about them.

According to the basic premise that everything happened by chance, of its own accord and without external influence, the following major events happened as a result of random, unrelated accidents. This is the Theory of Evolution in summary, starting from the beginning:

There is nothing but blackness. There is nothing physical. In fact, there is nothing. Huge amounts of matter and energy spring into existence by themselves, out of nothing, in a relatively short period of time and then the event stops.

This is considered impossible according to the First Law of Thermodynamics as neither matter nor energy can make themselves from nothing; they have to be made, which requires someone to make them. Because of this, science does not, and cannot, explain the original existence of matter or energy as it requires a non-scientific explanation.

At some later time, all of the newly created primal matter attracted itself back together through unknown (presumably gravitational) forces into one small dense point that became so hot even atoms could not exist.

This suggests that the original primal matter was comprised of some form other than atoms. It also had gravitational attraction, and resistance to massive heat, but there is no scientific explanation as to what this might be as the matter in our present universe is only constructed from atoms. Furthermore, if the primal matter was able to resist extreme heat there would be evidence of it still in existence.

Yet there is no scientific explanation about what it is and particularly where it is now.

About 15-20 billion years ago, this dense mass exploded at a speed greater than light and then slowed down quickly. This theorized event is now referred to as the big bang.

The assumed big bang speed of expansion contradicts known limits in the natural universe because it is considered impossible for anything to travel faster than light. This theory also does not explain the sudden slowdown of the expansion for this simple scientific reason: gravitational forces weaken as material spreads out and distances increase. Recent measurements show that the 'outward' speed of the universe is actually accelerating, not slowing down.

Surprisingly, science does not have an official position on the big bang other than it is only a concept, in the absence of any natural explanation, to try and explain why matter, energy and light appear all over the vastness of the universe. Many cosmologists now consider the big bang theory inadequate, if not defunct.

Most of this intensely hot expanding matter was again attracted together into smaller groups. These groups became so dense and hot that they exploded under thermonuclear fusion reactions (supernovas).

It is considered doubtful that atoms had been formed at this stage. It is therefore difficult for science to explain how the gravitational force was generated or what the actual material was comprised of that was 'fusing together'. This point is highlighted by the observation that today's supernovas are made from atoms.

An incomprehensibly large number of protons, neutrons and electrons, sufficient to make up all of the matter in the universe, self-created from quarks and leptons from an unknown origin by an unknown process. Then protons and electrons drew themselves together to make hydrogen atoms. At some point in this process, massive levels of energy from somewhere unknown were embedded and held in the cores of atoms by another unknown process. These components balanced themselves by inventing positive and negative charges matched to specific locations and actions within the atom. During the making of the components of atoms, the Higgs Boson[97] self-invented and attached itself to protons in the nucleus of the atom. A medium, or a field, which has been dubbed the 'Higgs field', self-created and filled up the entire universe. The Higgs Boson pushes against this

---

97. A particle known as the 'God particle' and tentatively confirmed to exist in March 2013.

perfectly matched field creating 'friction' and mass (substance), without which things could not hold together and exist.

There is no scientific explanation about how the Higgs Boson or its matching invisible Higgs field appeared from nowhere by accident. This is vital because without the two, nothing would have mass. There is also no known natural process or scientific explanation as to how such infinitesimally small components of atoms restructured the layout, energy and design of the entire universe into an ordered pattern, by accident, whilst also establishing gravity as a new force at the same time.

Hydrogen atoms clumped themselves into billions of billions of separate groups and gravitated together, resulting in controlled nuclear fusion reactions that eventually formed as stars.

This required countless trillions of hydrogen atoms to encounter each other and then group together in separate lots. However, the basic laws of physics predict that the primal hydrogen atoms should have spread out and dispersed rather than congregate together, particularly as they would be propelled away from each other after a large big bang explosion.

The massive variation in star sizes (some being two thousand times larger than others) seemingly defies the known laws of physics because of this scientific reason: a constant equilibrium point is required at which the pressure from compaction of hydrogen atoms causes a star to burst into life, which would result in stars being of equal or similar size.

Furthermore, neutrons were inexplicably drawn in from an unknown location to join fusing hydrogen atoms in stars to create helium atoms.

The concept that stars could self-form in this manner is hypothesis, it is not proven to be possible.

Stars grouped themselves together under the forces of gravity from black holes to form galaxies.

Physics tells us that, following an uncontrolled explosion, the regular occurrence of galaxies is unnatural and the layout of stars should be scattered, not structured into swirling groups. Further, those stars near the centers of galaxies should travel significantly faster than those further out as rotational gravity of the core is causing their speed. However, in defiance of the natural laws of physics, the opposite occurs, with speeds increasing with distance, allowing the galaxies to hold shape while rotating. Black holes are theorized as being collapsed stars and possibly portals to other unknown and unobservable potential universes or dimensions.

The big bang theory predicts that even if stars formed into galaxies, those galaxies should be scattered randomly throughout the universe and moving away from the central point of the big bang explosion.

Some force, described as dark energy, is causing galaxies to group together as if on the outer surface of huge hollow balls. The identifiable universe is essentially cubic in shape, its structure constructed from layers of these ball shapes. It does not show any resemblance to the pattern that would be made by a large explosion. Rather, it has a specific, complex but unnatural layout. The size of these hollow balls, and therefore the universe, is increasing, with science believing that more dark energy (or dark matter) is constantly being inserted into them. In other words, creation of new material is a regular activity.

About 4.5 billion years ago, our solar system formed from clouds of dust and gases. The planets were made from a variety of new elements, which are theorized to have formed by processes in stars because they cannot form naturally. These elements combined into new compounds such as gases and solid materials.

The appearance of new elements (that is, forms of matter other than hydrogen and helium) runs counter to conventional laws of physics. There are barriers (mass gap 5 and mass gap 8) that prevent new types of atoms forming naturally from a base such as hydrogen. Because of these fundamental barriers, there is therefore no scientific explanation for the existence of most elements in the known universe[98].

The concept of planets forming themselves by gravity from matter drifting in space is contradicted by the existence of their moons and rings of close-by materials, which should have already been drawn into the planets.

To explain how the solar system operates in defiance of the known natural laws, Einstein proposed the existence of curved space, which is now a commonly accepted scientific proposal. There is no natural explanation how curved space could come into existence of its own accord and begin to operate, nor why it should exist in the exact place where our solar system just happened to form.

---

98. Refer to Chapter 4, *Stellar Spectra*

The shape of our solar system is unlikely to have developed under the current scientific theories. Even allowing for the sun's rotation, all the planets in the solar system would have to have been at, or close to, the same plane when forming. Many other planets are claimed to have formed at this time but have since disappeared.

Almost all of our solar system's debris that did not self-form into planets or moons was sucked into the sun by gravity or blown away during its formation.

A number of observations contradict this possibility, particularly as the solar system has been in place for billions of years and should, by now, be structured and functioning by the influence of gravity alone:

• The asteroid belt between Mars and Jupiter still exists and is not being pulled in by the sun's gravity or that of other planets.

• The massive gravity of each planet should readily pull most moons and any nearby material into themselves, but this doesn't happen (for example, Saturn and its rings). The speed of the moons is claimed to balance with their planet's pull. For instance, our moon is supposedly drifting away from the earth at about four centimeters per annum, placing it at about twenty two thousand, five hundred kilometers away at the time of its assumed evolutionary making, compared with about four hundred thousand kilometers now. The gravitational attraction at the time of its making would have easily pulled it back to earth, particularly as its speed would have been minimal. With many different speeds, masses, directions and proximities, almost every extra-terrestrial body in our solar system defies the laws of physics and the mathematical equations that determine levels of attraction, and yet they all remain in their normal locations and orbits.

• The components of our solar system do not display the characteristics of being controlled by gravity as the primary source of influence on their movements and locations. There is some other unidentified force operating.

As a result of being formed by natural gravitational forces, all planets should revolve anticlockwise around the sun from their northern axes.

Venus contradicts this theory by rotating clockwise while Neptune rotates almost on its back, indicating their rotations were not started by the effects of the sun and its gravity. Even so, if they began with these odd rotations, they should have changed to an anticlockwise spin after 4.5 billion years of influence from the sun's gravity.

An invisible barrier, called the heliopause, developed at the point where the sun's outgoing rays meet incoming rays from other stars. It creates an area called the heliosphere, which encompasses the solar system up to Pluto. This barrier resists the furious level of radiation bombarding earth and becomes part of a series of barriers that will allow future life to exist.

The sun provides a controlled barrier against the radiation from billions of billions of stars that would otherwise destroy (future) life on earth.

Earth, the third planet closest to the sun, is in the perfect place and contains a perfect range of components in the perfect ratios for life to exist.

To allow life to exist, earth has the ideal location, crust thickness and composition, land surface, temperature, protective atmosphere, level of oxygen and mix of gases, as well as plentiful liquid water and renewal systems, such as the hydrologic and gaseous cycles. The range of elements on earth is not known to exist elsewhere. Earth is made up of very specific components blended together in the perfect ratios matched with the perfect location to support life.

A massive volume of water developed on earth and then stopped. No more water has formed since.

There are no feasible explanations regarding the appearance of so much water on earth, nor why it would stop. There was not enough oxygen or hydrogen on earth to make this amount of water, despite the fact that water can't just make itself out of its base atoms in any case. Water has to be made by an external source[99]. The earth has a specific amount of water and a proficient recycling system (evaporation and precipitation) to ensure it remains constantly clean. Along with the light of the sun, water is one of the most crucial essentials for life to exist on earth.

Data from fossil records show that the earth's temperature has remained constant around twenty-two degrees Celsius for most of its life.

According to Evolutionary Theory, the sun would have increased temperature by fifty percent if it had formed over billions of years, affecting the record of the earth's temperature. The evidence suggests that either (or both) the earth is a lot younger than is theorized, or the sun did not gradually form via the process proposed by Evolutionary Theory; it was made in one relatively short 'moment' in time at, or near, its full output.

---

99. Perhaps in stars, for example, despite there being no explanation how it could escape, maybe somewhere else not specified as yet by science.

A series of gaseous layers formed into an atmosphere around earth. These protect the planet from solar radiation[100] while also burning up most incoming physical matter through frictional heat. They also hold important gases close to earth so that life can exist.

The atmosphere is a complex structure that allows life to exist. It is assumed to have developed over the last 4 billion years. Oxygen is a key component in the current atmosphere but its arrival was only 600-800 million years ago. The atmospheric layers form perfect barriers in perfect sequence and thicknesses to protect earth both from incoming intrusions and from unwanted outgoing departures, such as water, gases and heat.

The sun protects the earth from the massive radiation from other stars and the atmosphere protects earth from the sun's radiation. These invisible structures still allow humans to see the sun and the stars clearly.

About 3.5 billion years ago, the beginnings of life started suddenly on earth following an unknown incident, theorized, as one possibility, a bolt of lightning randomly hitting a precise mix of chemicals in water. These chemicals eventually formed into DNA and RNA structures and then into a living cell after millions or billions of years of accidental coincidences.

Despite the most sophisticated equipment, knowledge and experimentation, scientists over the past 50 years have been unable to repeat such a process and cannot even theorize how it might actually happen. Repeated attempts at infusing electrical energy into life-elements (carbon, oxygen, nitrogen, hydrogen) does nothing more than create sludge. Making life from inanimate materials is known by biologists and chemists to be impossible without external influence. Furthermore, crucial to the development of life is oxidized molybdenum, which did not exist on earth at the time that life started.

Single cell microbes appear in the fossil record at 2 billion years ago and spread by dividing into two separate cells, growing back to their original size, then dividing again.

Louis Pasteur proved that cells cannot be made by spontaneous generation from non-living matter. This suggests that this primal life form arrived on earth by some method other than self-formation.

---

100. For example, gamma rays.

Single celled organisms mutated to perform a vast range of activities, creating products and processes that are fundamental to life today – nitrogen, sulfate and oxygen absorption, fermentation, photosynthesis, oxygen generation, and so forth. They also developed porphyrins that allow us to see color.

Many types of single cells exist, yet this alone is not scientific evidence that one formed from another. Rather it is evidence that, without their processes, later nucleated cellular life (creatures and plants) would not have been possible. Evolutionary Theory stakes its reputation on luck and chance. In this instance, it relies on good luck that microbial development happened in the correct manner to perfectly support other life forms over a billion years before the unplanned appearance of these other life forms.

At 1.45 billion years, the development of single cells stopped suddenly.

Evolutionary Theory does not explain how or why new microbe types stopped appearing after such massive and diverse development for 550 million years. The actual evidence is that the 'arrival' or appearance of new types of cells ceased. There is no scientific evidence past or present that all the different types of microbes evolved one from the other. It is only an assumption from an evolutionary standpoint that single cell development 'stopped' because there is no evidence that they were or could have evolved in the first place.

About 800-900 million years ago, a release of enormous volumes of oxygen occurred in the atmosphere multiplying the level by two hundred thousand times. It then stopped about 500-600 million years later and has maintained itself at that level of twenty-one percent ever since. Evolutionary Theory claims microorganisms such as cyanobacteria appeared at that time in vast quantities, causing water molecules to split into their original components, hydrogen and oxygen.

Evolutionary Theory cannot satisfactorily explain this major event. If oxygen was released from water molecules by algae, splitting it into its base components of hydrogen and oxygen, then both of these components would necessarily have had to exist on earth beforehand. Crucially, there is no record of:

• The components of water existing before water appeared.

• A decline in water levels once oxygen appeared.

• Where the hydrogen (twice as many molecules as oxygen) went once it was separated from water, other than use in photosynthesis.

The only scientific evidence is a large increase in oxygen levels. Other than luck, Evolutionary Theory also does not explain how the oxygen level happened to be perfect for life at twenty-one percent, or how it has remained at a basic steady state at that level for many hundreds of millions of years.

At 750 million years ago, nucleated cell life appeared in the form of sponges. Evolutionary Theory assumes that single cells combined together to form a nucleated cell.

After no action for 700 million years, single cells are theorized to have suddenly united to form extremely complex and sophisticated nucleated cells. Despite knowing in great detail how both are constructed, science has been unable to prove, or duplicate the process, how single cells could form into multiple cells.

As suddenly as it started, the theorized unification of single cells stopped, indicating that it was potentially a once only, or very restricted, event that instantaneously produced the perfect nucleated cell that would support future highly complex life forms and creatures that were not planned.

This process is not observed to happen naturally today, despite there being no scientific reason why it should have stopped. The actual evidence is that organisms consisting of multiple cells exist in the fossil record. There is no actual scientific evidence that they were made from single cells, only that they appeared at a certain point in the earth's history. To further note, single cells did not recommence self-evolution from this point.

What the fossil records actually show is:

- Single cells in various forms appeared from around 3.5 billion years ago with no new forms since 1.45 billion years ago.
- Nucleated cell life forms first appeared about 750 million years ago.

There is no evidence other than this. The evolutionary theories surrounding life appearing on earth are only theories. What's more, they are theories with no supporting evidence or known biological processes.

By 570 million years, multiple cells had formed themselves into living aquatic beings (possibly arthropods). The first original life form then invented a way to reproduce itself in full from a single cell. Then, relatively quickly, it invented male and female genders and the process of sexual reproduction. The oceans then start teeming with multiple life forms.

The timeline of Evolutionary Theory suggests that for up to 700 million years, single cells, which in relative terms are very simplistic, had not been able to evolve further than duplicating themselves. Then, in just 180 million years, single celled organisms united to form highly complex nucleated cells. Nucleated cells then cleaved together and created a living creature of immense intricacy with two genders and complex reproductive processes.

The nucleated cells we see today are already part of living entities and they have specific roles.

There is no biological method or model that permits free nucleated cells floating in water to randomly group together to form anything new, let alone a complete, fully functioning creature with the capability of extracting oxygen from water, a digestive system, sight, movement, sexual reproduction, a brain, a cardiovascular system with a chambered heart and blood vessels, and other complex internal and external organs. This is particularly so as these nucleated cell systems are interdependent and mostly cannot exist without the co-existence of all the other parts.

At 390 million years, millipedes appear on land without any ancestral trace.

The first recognized finding of mobile land life is millipedes, which appeared on earth as if having been dropped from the sky, fully formed. There is no indication they came from ocean life, which was the only other life form in existence. Since first appearing, millipedes show no signs of changing into anything else over the enormous period of 390 million years. This is despite one of the tenets of Evolutionary Theory that living forms are in a state of constant mutational change.

Around 370 million years, a line of lobe-finned fish came ashore and mutated into amphibians.

Life moving from a water environment to land is probably one of the most significant changes put forward by Evolutionary Theory. It would have required a quick and complete change in breathing and eating processes, vertebral and muscle structure, skin type, eyesight, and many other functions, including the reconstruction of swimming aids into walking aids. All of these evolutionary changes would need to have occurred in a sequence that allowed the creatures to remain mobile, alive, nourished, and able to reproduce. The appearance of amphibians' food base, predominantly insects, was not for another 70-170 million years, raising the issue of what they were using for nourishment.

Scientists estimate that the time required for the physical changes to occur from fish to amphibians would have taken hundreds of millions of years. Yet the fossil record shows just 100 million years between the lobe-finned fish and the next stage after amphibians, being fully developed reptiles. That is, two complete changes of creature.

The appearance of land life is considered by evolution to be a result of ocean life moving ashore only because of the timetable where one preceded the other. There are no known scientific links or biological processes, only theories based on measurements of time.

At 350 million years, insects appear in small numbers, then records of them disappear for 50 million years. At 300 million years, insects appear again in large numbers and varieties.

Insects appeared instantaneously with similar functions to other creatures but Evolutionary Theory does not link these creatures with any other life forms. There is no scientific explanation how the sudden arrival of an entire class of life forms with multiple species, made from the same types of nucleated cells, occurred. There is also no scientifically valid explanation as to how so many different insect species may have self-developed with such historical speed, almost instantaneously.

At 270 million years, numerous early reptiles appear from nowhere in a wide range of species.

As with the appearance of insects, the sudden arrival of a range of reptilian creatures across the world cannot be scientifically explained, particularly as, in this case, there is an expectation of enormous number of examples of the evolutionary changes from amphibians to reptiles, which do not exist. Evolutionary Theory cannot explain the variety of reptiles in such a short period.

At 260 million years, the first milk producing animals appear. These are theorized to be repto-mammalian creatures called therapsids.

Therapsids appeared from nowhere, without trace in the fossil record and without explanation from evolution. The timing of their arrival is also disputed within the scientific community.

At 225 million years, eozostrodon mammals appear.

These are considered to be mouse-like mammals appearing from nowhere, without trace in the fossil record and without explanation. As with therapsids, the timing of their arrival is also disputed.

459

At 205 million years, dinosaurs arrive, fully developed in a large range of species.

The arrival of many huge and small dinosaurs is similar to other creatures' appearances in that, for all intents and purposes, they appear to have been dropped on earth from the sky, ready to go with complex physical abilities, including a new type of anatomical joint, the ball and socket joint[101]. The name dinosaur means 'terrible lizard', but there is actually no fossil record of any evolutionary change from amphibians or reptiles into dinosaurs. Despite their heavy structures, they are also theorized to have developed the capability of flight, even though some dinosaur species had wings without feathers. Evolutionary Theory cannot explain the massive range of dinosaur species appearing almost instantaneously.

At 140 million years, the first birds appear with highly complex design features enabling them to fly.

This is a disputed timing and is more likely to be around 65 million years. There is no evidence of birds' ancestors in the fossil records despite the expectation of millions of examples of proto-birds that had fallen from trees in the attempt to take to the air. Birds appeared on earth fully developed with vastly different features from reptiles[102] that allowed them to resist gravity and maneuver in the air.

The only explanation from Evolutionary Theory is that birds are lucky, which is not scientific. Birds are lucky to have compact bodies, lightweight porous bone structures, wings and feathers, which are comprised of hundreds of millions of components, all of which are perfectly suited to flight.

At 135 million years, flora and fauna spread quickly around the world.

There is no previous evidence of any evolutionary development to validate the arrival of plants. They had poisons and toxins to protect themselves from unwanted predators and scents and flavors to attract others. These systems are targeted at animals and insects that are unknown to the plant and come with an inbuilt DNA design to achieve long term growth and colonization.

At 65 million years, the dinosaurs disappear. Mammals appear in vast numbers.

Like fish, insects, reptiles, dinosaurs and birds, mammals are another life form with an unexplained sudden arrival. There are some in the

---

101. Such as found in hips and shoulders of humans.
102. Such as warm blood, porous bones, feathers, and wings.

scientific community who also consider that birds first appeared at this time.

At 20 million years, the great apes arrived fully developed followed by human bipedal predecessors at 2.5 million years. After this, anatomically modern humans appeared around 200,000 years ago.

The apes appeared from nowhere fully developed without any ancestral trace. There is a general consensus that humans, as we know today (homo sapiens sapiens), have only been identifiable for 50,000 years. All the fossil finds claimed to show steps from ape to human have been proven to be other unrelated creatures or falsely manipulated[103]. There are no scientifically proven links between the two lines of creatures. Humans appeared on earth in several different locations with some evidence that they were independent of each other. Humans are the last new life form known to appear on earth.

Today, based on the history of the regular, unexplained arrival of new classes and species of creatures, we could reasonably expect to be part way towards seeing the next stage of development after man (perhaps a superman). We could also expect to find billions of living creatures, plants and microbes changing into different life forms and that the fossil record will be awash with examples of the billions of failed and successful changes in life forms.

We should expect evolution to still be relentless in mutating endless new species, creatures and body parts in all creatures around the world – but none of this is observed. There is no evolutionary development into new forms evident anywhere. Every evolutionary change of every creature on earth into something other than its own kind has apparently ceased and disappeared. There is no record or scientific proof anywhere of any of the endless steps required for biological evolution.

In summary, none of the basic steps of evolution has any evidence to support them. They are just ideas or concepts that do not align with the scientific data that exists. In many instances, the steps are known to defy known scientific and biological laws.

Another Perspective

In reviewing the actual events that have occurred, rather than the theories surrounding them, we have to ask whether they could be explained without considering evolution. Let us attempt to do so by looking at the two major aspects of physical and living matter and consider what science really has to say.

---

103. Refer Page 328 - Figure 189: Table Of Ape To Human Disproven Finds.

Physical Matter

Physical matter exists and we have shown through scientific laws that it cannot self-create; it has to be made. Our vast universe is moving outwards in all directions at an increasing speed. It contains a range of different forms of matter such as physical matter, anti-matter and dark matter, as well as kinetic energy, potential energy and dark energy. There are billions of galaxies and stars that are arranged in a specific, yet unnatural, cubic pattern (cosmic web). Scientists do not know the reason, function or purpose of this pattern but we do know it is not the result of an explosion (big bang) or any known natural event. The structure and behavior of the universe do not represent the chaotic actions expected from violent, uncontrolled self-creation.

The levels of most components and structures in the universe are perfectly suited for life to exist on earth. Worthy of note is that if any of the levels of gravitational force, dark energy, the speed of light, the Higgs Boson particles, or any one of hundreds of other values were only fractionally different from what they actually are, then life in our universe would not be possible. For example, if the ratio between the nuclear strong force and the electromagnetic force was fractionally (almost infinitesimally) different it would have been impossible for stars to form.

Many of the behaviors of the contents of the universe also defy normal gravitational effects. Our solar system is laid out with a central star that has planets revolving around it in one, nearly level plane. The solar system cannot have self-created in its current form through the known laws of physics[104], and the movements of the planets and moons defy normal gravitational forces.

The earth is in the perfect position for life to exist. It has all of the ingredients to support life, including masses of water and the perfect ratios of oxygen, nitrogen and other critical gases. It has a multi-layered atmosphere that protects its fragile life forms from deadly intrusions from space. It has complex and effective recycling systems that ensure constant renewal of life-giving resources such as air, food and water. It has a moon that complements many natural actions including tidal movements that assist with the food chain in the oceans.

The earth is a unique and fully functioning place for life to exist and is the only place in the universe known to support life. The physical matter in the universe exists in such a manner that its locations, quantities,

---

104. For instance, calculations based on angular momentum, a measure of the amount of rotation a particular object has, such as stellar debris and planets, indicate that our solar system components should have been expulsed in many different directions.

components, and functions are perfectly arranged to allow life to function on earth. There are powerful forces holding the structure of the universe together, overriding the natural physical interactions of gravity.

Living Matter

The fossil record evidence shows that every type of living entity suddenly appeared on earth fully developed with a specific design and, with rare exceptions, matching male and female genders. There are no scientifically proven records of any intermediate or transitional steps from one living species to another. Each type of life form has a separate and individual role that fits in with all others and reproductive security systems that stop contamination from other species. All life forms have remained basically unchanged for periods up to 500 million years for creatures and plants, and 3 billion years for bacteria. All classes of creatures have a variety of different species. These variants mostly appeared within the same timeline allowing little time to develop one from another through an evolutionary process.

Creatures and plants, which are unrelated, have highly complicated processes for reproduction that rely on different genders combining to make miniature offspring that grow into adults over time. These processes, particularly in creatures, rely on billions and trillions of cell replications in strict matching sequence and timing programs. These programs are so complex that scientists cannot replicate them and do not know where they are or how they work.

The functions of living cells in a body or plant prohibit any changes outside of a limited range of adaptability, maintaining the entity within its original basic type. Biologically, there are no known methods by which any living entity can change into something else.

The different kinds of creatures on earth are interdependent and rely on the existence of other species. However, the evidence clearly shows that (with rare exceptions) species are independent of each other, do not interbreed and do not show any signs of reproducing anything but their own kind. Furthermore, the scientific evidence shows that they do not mutate into any other species.

Conclusions

As individuals, and collectively, we have to be capable of looking at the facts rather than theories that contradict the facts. If we summarize the scientific observations and research regarding the universe we can say that we know the following:

- Matter did not, and cannot, self-create from nothing.

- The layout, structure and components of the universe reveal that they are arranged in a cubic-based pattern that did not self-create from a big bang explosion.

- The solar system cannot have self-created from clouds of matter, nor can other stars, planets or galaxies.

- The earth is in a very specific place so that life can exist.

- Life did not, and cannot, self-create from inert chemicals.

- One component crucial to the self-development of life (oxidized molybdenum) did not exist on earth at the time that life started.

- All of the different life forms of earth appeared fully developed in an environment perfect for their continuation.

- All life forms operate within preset limits and do not, and cannot, change into other forms.

- Despite the theories of accidental development and chaotic environments, the universe appears to be controlled and balanced and it behaves within limited boundaries.

The scientific facts of life as we know them lead to a conclusion that evolution (self-creation without purpose) is most certainly not responsible for the existence of matter, energy, the universe, or life.

Commentary On The Biological Proposals By Evolution

As we have seen throughout this book, it is often difficult to align evolutionary conclusions with the information used to make those conclusions, often contradicting scientific facts and lacking in reasonable logic. Adding to the confusion is that the manner in which information is presented can also misrepresent what is actually known.

As an example, we are going to now examine how the Tree of Life diagram is presented to us. The Tree of Life is a concept suggesting a flow of evolutionary change where the first appearance of one creature secures its status as the ancestor to a subsequent creature. This concept is based purely on the chronological appearance of each life form. There is no absolute agreed layout of these steps endorsed by science, as many dates and the identification of certain creatures are in dispute.

The following six pages each display an image across a double page spread:

Figure 232: Finds In The Fossil Record is an illustration of some of the more recognizable fossil species that have been found, identified, named and catalogued by science. They are displayed from the bottom up in the general sequence of their first appearance on earth. This is the only valid diagram that anyone should use as it represents the actual finds from the fossil record.

Figure 233: The Tree Of Life is typical of how science visually presents the concept of biological evolution. It is the same as Figure 232 but shows links between species as if they changed into other completely different species. These links are usually worked out on the basis of the most common characteristics between any two species taking into consideration the time they first appeared.

It does not represent evolutionary theory which claims there are millions of minor intermediate steps between any two species. There is no evidence to support the links shown so they should not be displayed. This type of diagram is a scientific falsehood, however it is how most people believe that evolution occurred.

Figure 234: Evolution's Theories Of Life Development shows a representation of how evolution claims that life developed. There is a recent proposal that chickens may have been ancestors to dinosaurs. If the steps involved were set out there would be millions of small changes resulting in millions of slightly different creatures over hundreds of millions of years. These are represented by the small dots in the expanded image on page 471. If we could select five of these at random, we should find the types of creatures displayed. For example, small chicken bodies with large dinosaur legs (#1) or both long and short legs on one body (#4). There should be billions of creatures showing transitional changes with partial blends of new and old body parts but no fossils have been found.

If we were to display evolutionary theory properly in this visual form, it would be a mass of partially changing life forms, all being the same size image as all are in the same process of transitioning by mutation and nothing would be an identifiable final version of a species.

The reality of what exists in the fossil record is Figure 232. Anything else is fantasy not fact.

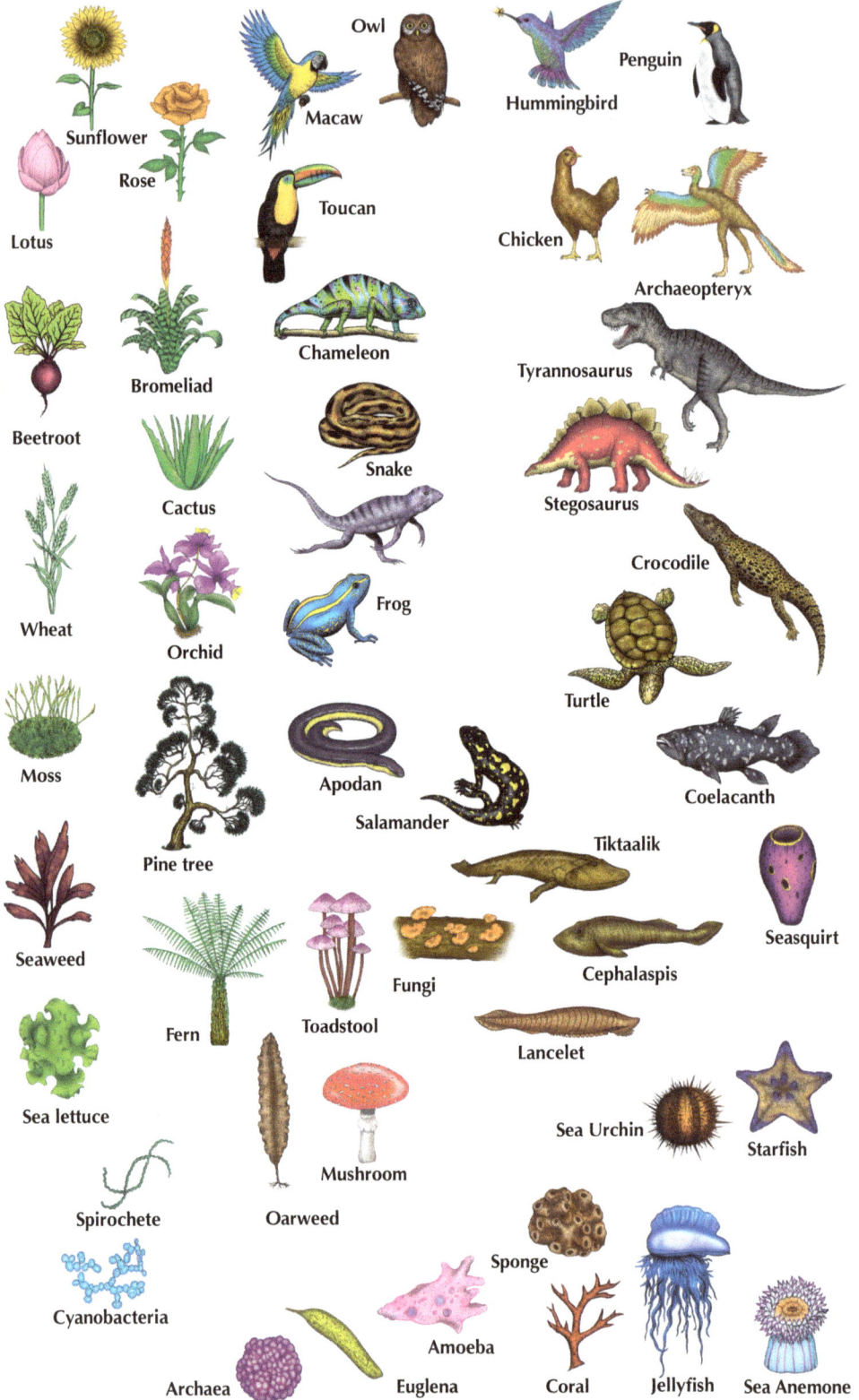

Sunflower

Rose

Lotus

Beetroot

Bromeliad

Wheat

Cactus

Orchid

Moss

Pine tree

Seaweed

Fern

Sea lettuce

Toadstool

Spirochete

Oarweed

Mushroom

Cyanobacteria

Archaea

Euglena

Amoeba

Coral

Macaw

Owl

Hummingbird

Penguin

Toucan

Chicken

Archaeopteryx

Chameleon

Tyrannosaurus

Snake

Stegosaurus

Crocodile

Frog

Turtle

Apodan

Salamander

Coelacanth

Tiktaalik

Seasquirt

Fungi

Cephalaspis

Lancelet

Sea Urchin

Starfish

Sponge

Jellyfish

Sea Anemone

466 *Figure 232:* Finds In The Fossil Record

Cheetah

Panda

Lemur

Chimpanzee

Human

Seal

Wolf

Whale

Rhino

Elephant

Rabbit

Rat

Zebra

Anteater

Butterfly

Grasshopper

Antelope

Dugong

Ant

Bat

Squirrel

Wasp

Dragonfly

Beetle

Bee

Kangaroo

Platypus

Hedgehog

Crayfish

Scorpion

Marine fish

Marlin

Crab

Spider

Salmon

Ray

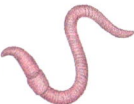
Shark

Nematode

Centipede

Mite

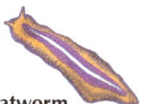
Earthworm

Giantproductus

Nudibranch

Octopus

Flatworm

Mussel

Scallop

Snail

Ammonite

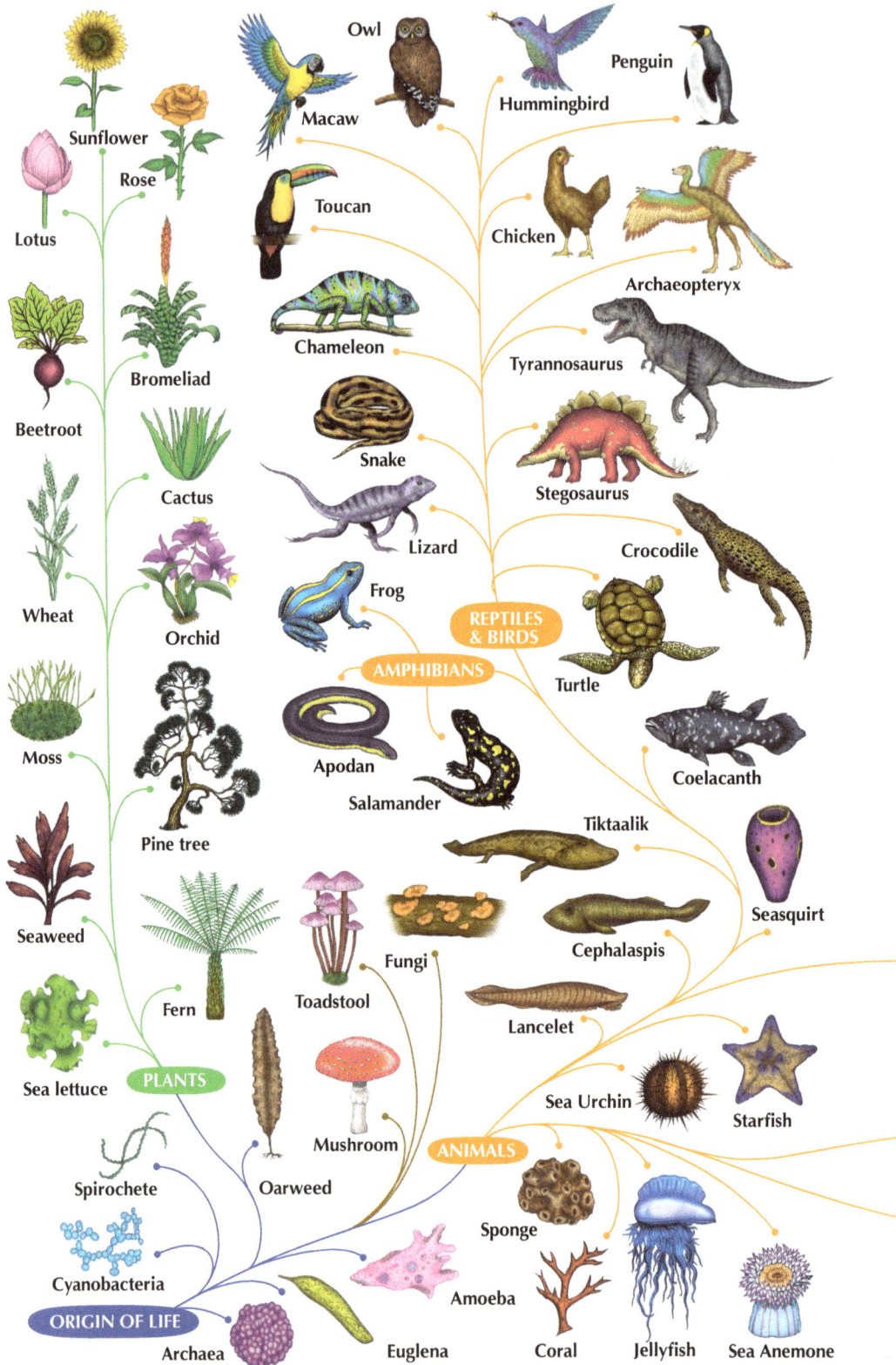

Sunflower

Rose

Lotus

Beetroot

Bromeliad

Cactus

Wheat

Orchid

Moss

Pine tree

Seaweed

Fern

Sea lettuce

PLANTS

Spirochete

Oarweed

Cyanobacteria

ORIGIN OF LIFE

Archaea

Euglena

Toadstool

Fungi

Mushroom

Owl

Macaw

Toucan

Chameleon

Snake

Lizard

Frog

Apodan

Salamander

AMPHIBIANS

Hummingbird

Penguin

Chicken

Archaeopteryx

Tyrannosaurus

Stegosaurus

Crocodile

REPTILES & BIRDS

Turtle

Coelacanth

Tiktaalik

Seasquirt

Cephalaspis

Lancelet

ANIMALS

Sea Urchin

Starfish

Sponge

Coral

Jellyfish

Sea Anemone

Amoeba

468   *Figure 233:* The Tree Of Life

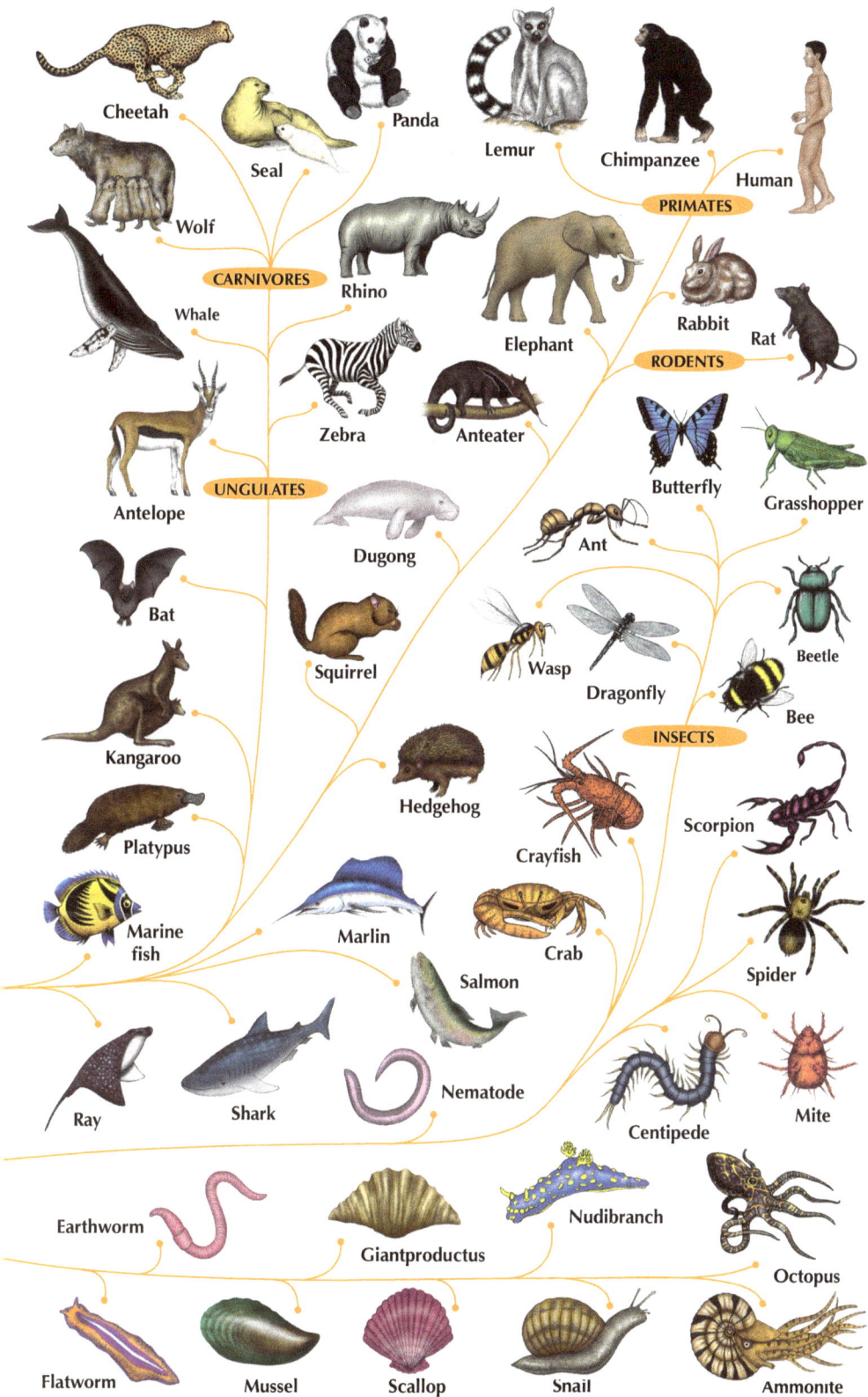

Cheetah

Seal

Panda

Lemur

Chimpanzee

Human

PRIMATES

Wolf

CARNIVORES

Rhino

Elephant

Rabbit

Rat

Whale

RODENTS

Zebra

Anteater

Antelope

UNGULATES

Dugong

Butterfly

Grasshopper

Ant

Bat

Beetle

Squirrel

Wasp

Dragonfly

Bee

Kangaroo

INSECTS

Hedgehog

Crayfish

Scorpion

Platypus

Marine
fish

Marlin

Crab

Spider

Salmon

Ray

Shark

Nematode

Centipede

Mite

Earthworm

Nudibranch

Octopus

Giantproductus

Flatworm

Mussel

Scallop

Snail

Ammonite

469

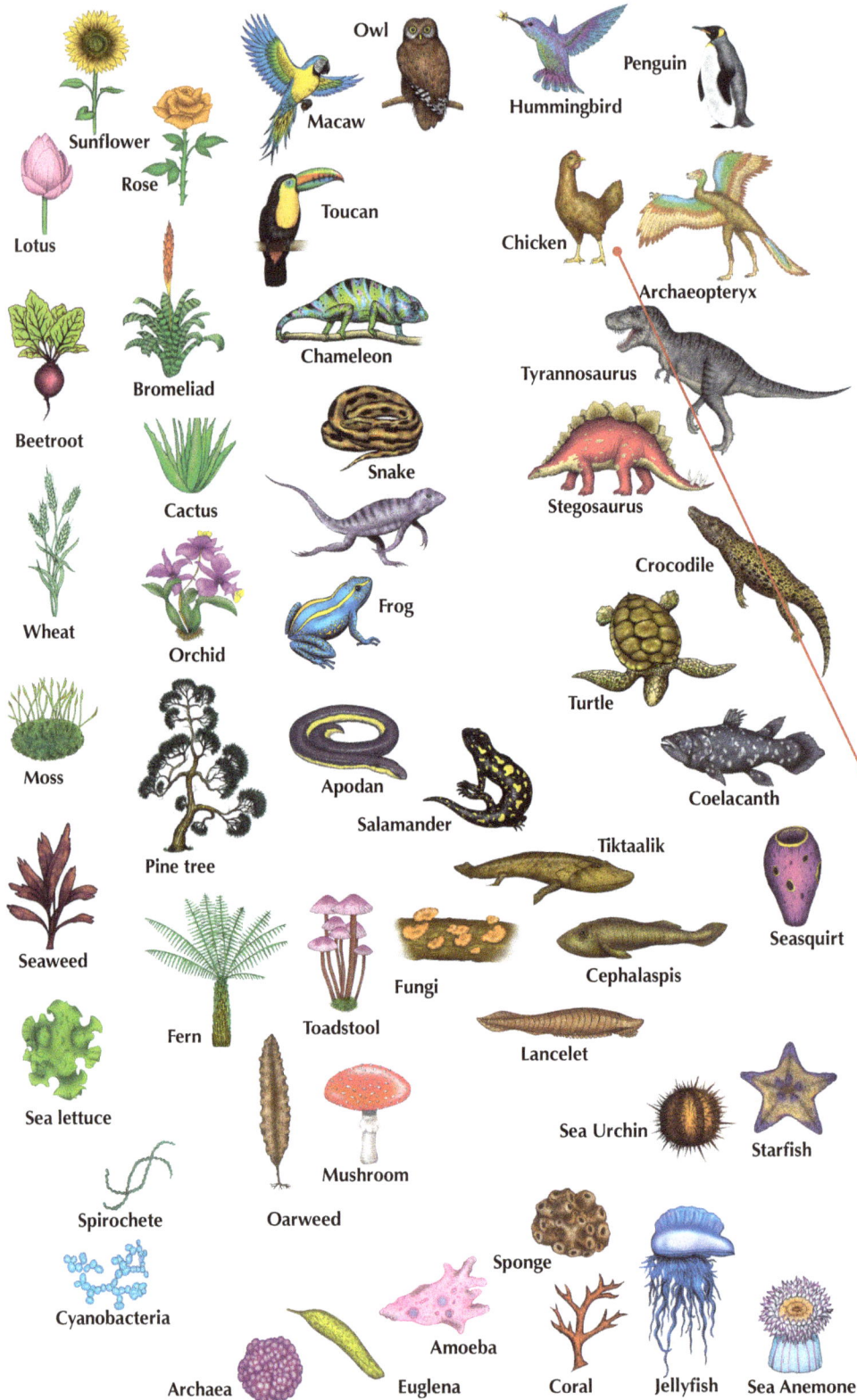

Figure 234: Evolution's Theories Of Life Development

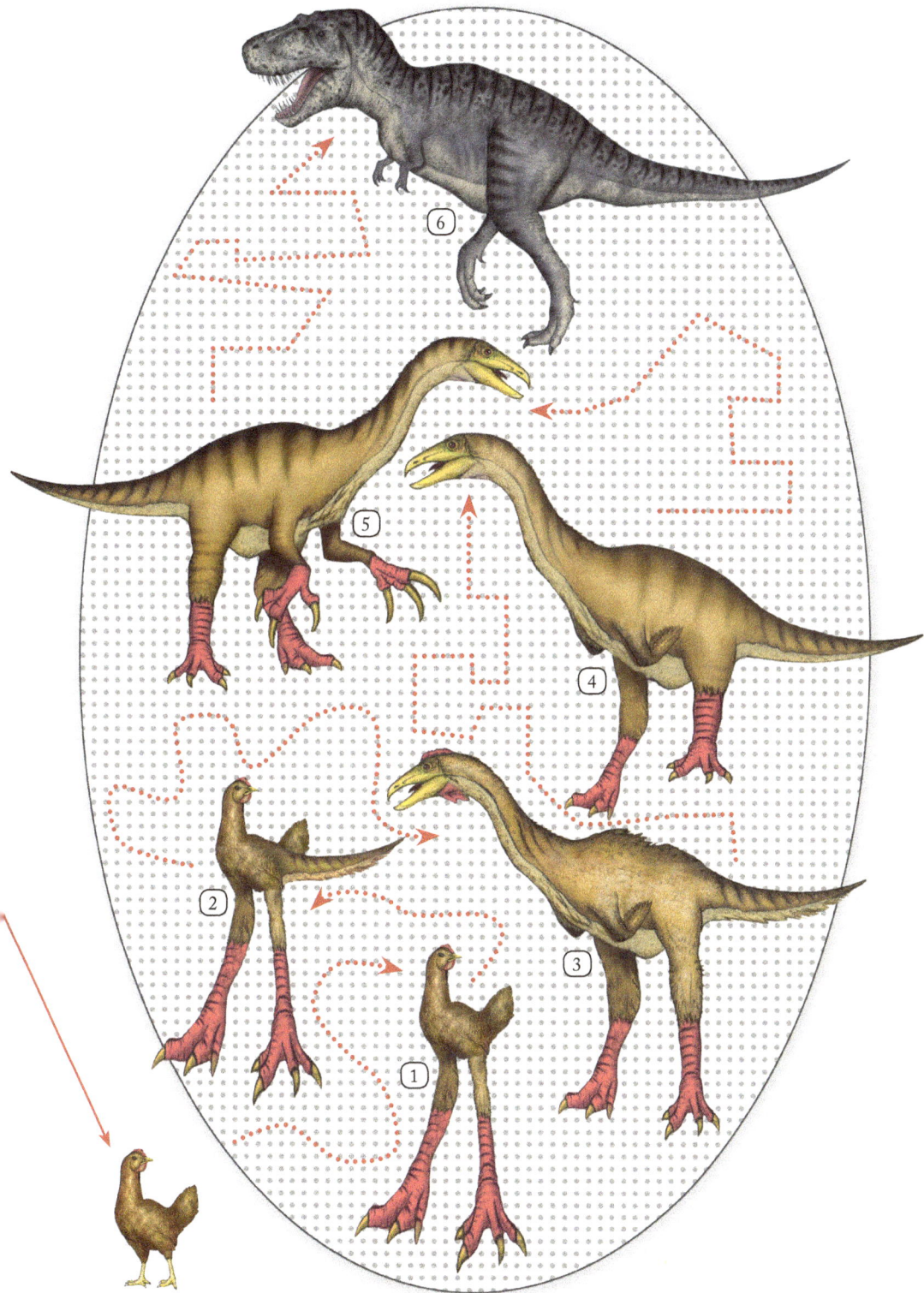

471

It is curious that we are consistently presented with the type of diagram in Figure 233: The Tree of Life that does not represent any theoretical proposal and why science does not strongly object to such false and misleading images. This type of diagrammatic representation, drawing lines between various creatures, represents nothing more than an idealistic attempt to promote a theory that has no factual basis.

Evolution Is A Faith Belief

Despite the evidence that evolution could not have occurred as it is currently proposed, many evolutionists won't change their views because they claim that the alternative, a Creator, is an act of faith. Consider the following, which is paraphrased from *Challenge Newspaper* (Australia) May 2001:

"There are two premises of faith for evolutionists.

1. Spontaneous generation where life forms itself from non-living matter cannot be proven by science, in fact the scientific evidence is that it can't happen. The Australian Academy of Science says that 'There is no reason to believe that spontaneous generation can occur today, yet evolution scientists maintain that it did happen in the past because of different circumstances.' This of course is an act of faith, that is, a belief in something that can't be proven or repeated. It is not a scientific fact; it is a faith belief.

2. All species developed from a common ancestor. This cannot be verified by science yet one would think that it would be easy to do so as we should be surrounded by this as everything should show a common link. But science can't even come up with an explanation as to how it might have happened. So evolution is a faith doctrine and those who accept it are acting on faith whereas those who reject it aren't rejecting science. Quite the contrary, they accept the science that life can't make itself or develop from a common ancestor".

These two issues highlight the basic failure of Evolutionary Theory. One of the main arguments that many evolutionists have put forward is that anything other than evolution is a faith issue, not a scientific one. In reality, the opposite is true. The evidence dismisses evolutionary proposals and shows that it is a belief unsubstantiated by scientific evidence and known scientific laws.

What Do The Experts Say?

Science is the major driving force behind many of the different theories and concepts that bombard our lives. This is because it has a self-determined obligation to present theories and explanations that reflect the natural world and deliberately excludes other options. The reason that virtually every step in the Theory of Evolution is unprovable is because, in the natural world, things do not make themselves and cannot change from one thing to another by themselves. Yet, for evolution to be possible, these natural, universal rules have to be broken, even though there is no known scientific way for this to happen. When science or evolutionists try to explain that these impossibilities must have happened in the past, when it is known that they couldn't have happened, then their theories are exposed, as we have seen throughout this book.

To any outsider, this situation would seem absurd. Here we have an educated and qualified group of people who have spent many years, decades and centuries diligently studying our world in detail, yet when they analyze the information and evidence available they are required to ignore the only and obvious realistic conclusion. There is some good news. Not all scientists are restricted by their disciplines or decline to present the obvious. Following are various quotes from some of the foremost scientists in the world.

Paul Davies, once the professor of mathematical physics at the University of Adelaide and subsequently a professor at the Arizona State University, Director of BEYOND: Center for Fundamental Concepts in Science, summarizes the growing findings of scientists from many fields in his book *The Mind of God: The Scientific Basis for a Rational World* (Simon & Schuster UK, 1992):

"A long list of additional 'lucky accidents' and 'coincidences' has been compiled... Taken together, they provide impressive evidence that life as we know it depends very sensitively on the form of the laws of physics, and on some seemingly fortuitous accidents in the actual values that nature has chosen for various particle masses, force strengths, and so on..."

"Suffice it to say that, if we could play God, and select values for these quantities at whim by twiddling a set of knobs, we would find that almost all knob settings would render the universe uninhabitable. In some cases it seems as if the different knobs have to be fine-tuned to enormous precision if the universe is to be such that life will flourish".

"Through my scientific work I have come to believe more and more strongly that the physical universe is put together with an ingenuity so astonishing that I cannot accept it merely as brute fact. There must, it seems to me, be a deeper level of explanation. Whether one wishes to call that deeper level 'God' is a matter of taste and definition. [I] believe that we human beings are built into the scheme of things in a very basic way".

Albert Einstein also marveled at the order and harmony he and his fellow scientists observed throughout the universe. In the book, *The Quotable Einstein* (Princeton University Press, 1996), he is noted to have commented on the religious feeling of the scientist:

"Everyone who is seriously involved in the pursuit of science becomes convinced that a spirit is manifest in the laws of the Universe—a spirit vastly superior to that of man".

"My religion consists of a humble admiration of the illimitable superior spirit who reveals Himself in the slight details we are able to perceive with our frail and feeble minds. That deeply emotional conviction of the presence of a superior reasoning power, which is revealed in the incomprehensible universe, forms my idea of God".

Phillip E. Johnson, a University of California law professor, writes in *Darwin on Trial* (InterVarsity Press, 1993):

"Just about everyone who took a college biology course during the last sixty years or so has been led to believe that the fossil record was a bulwark of support for the classic Darwinian thesis, not a liability that had to be explained away. . . The fossil record shows a consistent pattern of sudden appearance followed by a stasis, that life's history is more a story of variation around a set of basic designs than one of accumulating improvement, that extinction has been predominantly by catastrophe rather than gradual obsolescence, and that orthodox interpretation of the fossil record often owe more to Darwinist preconception than to the evidence itself. Paleontologists seem to have thought it their duty to protect the rest of us from the erroneous conclusions we might have drawn if we had known the actual state of the evidence".

Martin Rees, professor of astronomy at Cambridge University, and science writer John Gribbin, discuss how finely tuned scientists have found the universe to be in *Cosmic Coincidences: Dark Matter, Mankind, and Anthropic Cosmology* (Bantam Books; 1st Printing edition, 1989):

"...the conditions in our Universe really do seem to be uniquely suitable for life forms like ourselves, and perhaps even for any form of organic complexity ... Is the Universe tailor-made for man?"

Sir Fred Hoyle, English Astronomer, quoted from *Fred Heeren, Show Me God: What the Message From Space Is Telling Us About God* (Daystar Productions, 1996), said:

"A common sense interpretation of the facts suggests that a super intellect has monkeyed with the physics, as well as with chemistry and biology, and that there are no blind forces worth speaking about in nature. The numbers one calculates from the facts seem to me so overwhelming as to put this conclusion almost beyond question".

Robert Jastrow, Astronomer, Physicist, Cosmologist and NASA scientist writes in *The Enchanted Loom: Mind in the Universe* (Simon & Schuster, 1981):

"Scientists have no proof that life was not the result of an act of creation, but they are driven by the nature of their profession to seek explanations for the origin of life that lie within the boundaries of natural law. They ask themselves, 'How did life arise out of inanimate matter? And what is the probability of that happening?' And to their chagrin they have no clear-cut answer, because chemists have never succeeded in reproducing nature's experiments on the creation of life out of nonliving matter."

"Scientists do not know how that happened, and, furthermore, they do not know the chance of its happening. Perhaps the chance is very small, and the appearance of life on a planet is an event of miraculously low probability. Perhaps life on the earth is unique in this universe. No scientific evidence precludes that possibility".

Francis Hitching, member of the Royal Archaeological Institute, the Prehistoric Society and the Society for Physical Research, writes in *The Neck of the Giraffe: Darwin, Evolution and the New Biology* (New American Library, 1983):

"There are about two hundred and fifty thousand different species of fossil, plants and animals in the world's museums. This compares with about 1.5 million species known to be alive on earth today. Given the known rates of evolutionary turnover, it has been estimated that at least one hundred times more fossil species have lived than have been discovered... But the curious thing is that there is a consistency about the fossil gaps: the fossils go missing in all the important places. When you look for links between major groups of animals, they simply aren't there; at least, not in enough numbers to put their status beyond doubt. Either they don't exist at all, or they are so rare that endless argument goes on about whether a particular fossil is, or isn't, or might be, transitional between this group and that . . . there ought to be cabinets full of intermediates—indeed, one would expect the fossils to blend so gently into one another that it would be difficult to tell where the invertebrates ended and the vertebrates began. But this isn't the case. Instead, groups of well-defined, easily classifiable fish jump into the fossil record seemingly from nowhere: mysteriously, suddenly, full-formed, and in a most un-Darwinian way. And before them are maddening, illogical gaps where their ancestors should be".

Michael Denton, biologist and medical researcher, says in Evolution: A Theory in Crisis (Adler & Adler, 1986).

"When estimates are made of the percentage of [now] living forms found as fossils, the percentage turns out to be surprisingly high, suggesting that the fossil record may not be as bad as is often maintained".

"...of the three hundred and twenty-nine living families of terrestrial vertebrates [mammals, birds, reptiles and amphibians] two hundred and sixty-one or 79.1% have been found as fossils and, when birds (which are poorly fossilized) are excluded, the percentage rises to 87.8%"

H.J. Lipson, F.R.S. Professor of Physics, University of Manchester, UK, writes in *A physicist looks at evolution* (Physics Bulletin, 1980, volume 31):

"If living matter is not, then, caused by the interplay of atoms, natural forces and radiation, how has it come into being? I think, however, that we must go further than this and admit that the only acceptable explanation is creation. I know that this is anathema to physicists, as indeed it is to me, but we must not reject a theory that we do not like if the experimental evidence supports it".

In summary, these scientists conclude:

• The universe is finely tuned to operate under a set of predetermined rules of mathematics, physics, biology, chemistry and so forth.

• The structures that make up the base forms of biology are consistent with the assumption that they are made by a higher intelligence.

• Near enough to ninety percent of current species of amphibians, reptiles and mammals have been found in the fossil record but there is not one find of any creature that can be clearly identified as a transitional step.

- Paleontologists know that the fossil record does not prove the theory of evolution and that, in fact, the evidence is clearly of new creatures appearing fully made from nowhere. There is no evidence of any gradual changes from one form to another.

- Scientists are influenced by the nature of their professions to explain things based on natural laws rather than give conclusions based on the actual evidence.

The purpose of this book is not to dispute science. We need science. We need the men and women who dedicate their lives to exploring the intricacies of the universe. They provide us with information to help us live better, healthier and more fulfilled lives. They allow us greater understanding and to see things we wouldn't normally be exposed to. As we can see from these quotes above, there are many scientists who have concluded that the universe is designed by a super intellect. Although many steer clear of using the word, they mean God. They don't necessarily say it in such simple terms but the comments are clear in their intent. You can see their reluctance to use the word God, or even Creator, as it might drag them into a religious debate that is not their intent or domain of expertise.

It would be fair to the general public if science in general, and individuals within each scientific discipline, could acknowledge publicly that concepts proposed under the (unofficial) Theory of Evolution are not scientifically defensible and that the conclusion from the information they possess points to a higher intelligence at work. Further, that the existence of God doesn't contradict anything natural. Then, their future efforts could be focused on understanding more about how God constructed the universe so we might be able to have a more abundant life. In other words, find out how God did things and then understand why He did it that way and how we can benefit from it.

The Final Word For Evolution

As science has no official Theory of Evolution or any agreed detail about how such processes could have occurred, how do we actually work out where everything came from and how it was made? The only way to do this is by taking a selective range of laws, observations and measurements about the universe, the stars, our solar system, and the living things on our planet and trying to work backwards to determine what processes could have happened naturally by themselves in one long continual process. The tools we can use include laws based on mathematics, physics, chemistry, biology, geology, and other sciences.

In doing so, we discover that most past activities are unable to be explained as happening by themselves as they contradict natural laws. The result is that any Evolutionary Theory just becomes a concept that predominantly ignores known facts. Those proposing evolutionary theories often defend their concepts with diversions such as:

"It hasn't been found yet," or

"It used to happen that way but doesn't now," or

"There is no natural explanation possible," or

They fill in blanks with more made up theories.

If we simplify theories of evolution, we find that there are three principles on which they are based:

Principle 1:

Matter and energy self-created from nothing and arranged themselves into the finely tuned, structured, functioning layout of the universe.

Principle 2:

Life self-created from inert chemicals.

Principle 3:

The original life form, microbes (bacteria), mutated into every other life form, concluding with human beings.

If we stick with the proven observations and tests that science has presented, we can safely say:

Principle 1 is scientifically impossible:

Something cannot make itself out of nothing.

The only way for physical matter to exist is if someone (higher intelligence) made it.

Principle 2 is scientifically impossible:

We have discovered that life is made up as cells that are based on chemical components. We know that cells, and therefore life, have to be made. Life cannot make itself.

Principle 3 is scientifically impossible:

Life forms cannot change into other different life forms.

Factual scientific information allows us to confidently conclude that there is no realistic evidence to support Evolutionary Theory. The concept of evolution is based on the assumption that there is no Creator to make everything. Evolutionary explanations are merely ideas about what would have to happen for the universe to exist by itself without input from a higher intelligence. Evolutionary theories cannot use the scientific evidence that exists as it proves those theories to be impossible, irrelevant, misleading, and does not represent the facts that we know. That is why evolution is just a theory – because it isn't fact and this is known by its proponents.

This book has tried to present the known scientific evidence (facts) regarding the origins of life and the universe. It is now up to you to draw your conclusions of how everything came into existence.

It is up to you to follow the facts.

---

* * * * * *

THE MAJOR GAPS IN EVOLUTIONARY THEORY

There are many missing pieces in the story of evolution; many gaps for which science has contradictory evidence. These unexplained barriers prohibit evolution from being a valid account for the existence of the universe and life. In the following pages, there are two summaries that will assist in overviewing the claims of evolution:

THE GAP MAP

In this diagram, eight of the most significant events claimed by evolution are mapped out in visual form. Each of these has a brief explanation of the scientific reasons for the gap.

THE GAP TABLE

The Gap Table is a broader summary detailing the thirty seven most significant evolutionary gaps. This can be used as an easy reference guide as it includes the chapter location of the information for that topic within the book.

EVOLUTION EVENT 1
IMPOSSIBLE

Science proves matter and energy cannot self-create out of nothing.

EVOLUTION EVENT 2
IMPOSSIBLE

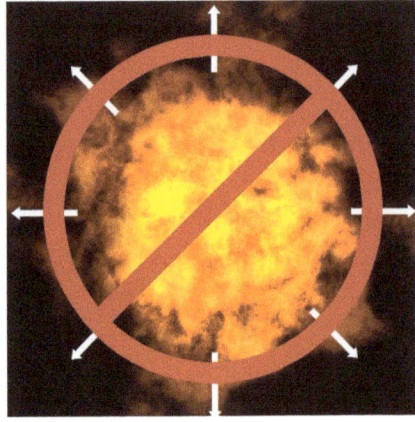

Science proves the big bang did not occur.

EVOLUTION EVENT 5
IMPOSSIBLE

Science proves oxygen and water did not appear on earth as described by Evolutionary Theory.

EVOLUTION EVENT 6
IMPOSSIBLE

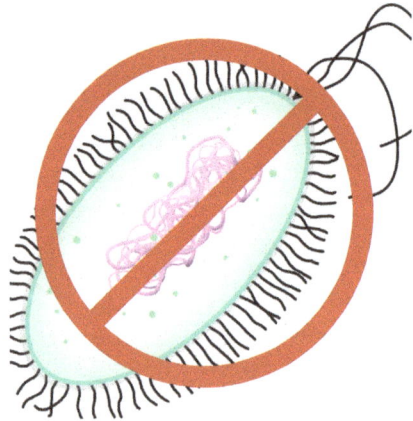

Scientist Louis Pasteur proved that life cannot and did not create itself.

EVOLUTION EVENT 3

IMPOSSIBLE

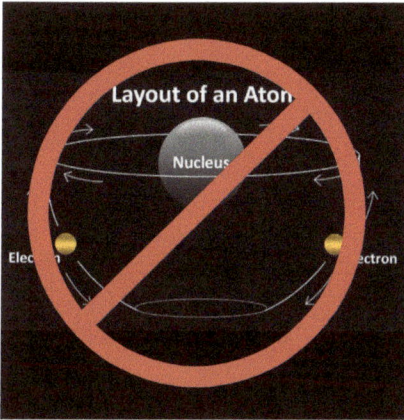

Science proves atoms, elements and stars cannot self-create.

EVOLUTION EVENT 4

IMPOSSIBLE

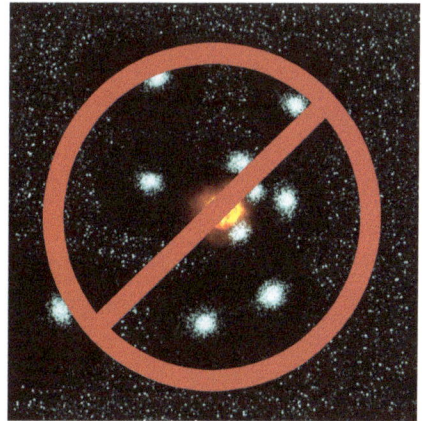

Science proves our solar system did not self-form.

EVOLUTION EVENT 7

IMPOSSIBLE

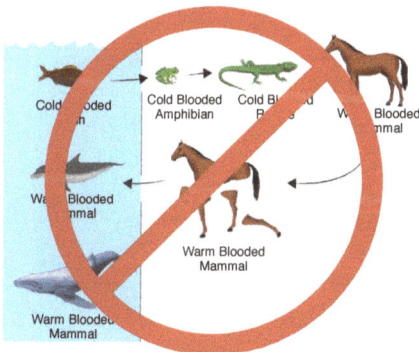

Science proves that ocean creatures did not move to land, that reptiles did not mutate into birds, and that mammals did not return to water.

EVOLUTION EVENT 8

IMPOSSIBLE

The scientific evidence in fossil records shows that apes and humans appeared fully formed from nowhere.

Gap	Evolutionary Event	Evolutionary Theory	Gaps Unraveled	Chapt. no.
	Before Time	Nothing exists		
Gap 1	Primal Matter – Something From Nothing	Matter and energy appear.	• Matter and energy cannot self-create from nothing – the First Law of Thermodynamics. • The alternate evolutionary proposal that matter has always existed contradicts The Second Law of Thermodynamics – all matter and energy would be infinitely old having already degenerated into inert particles rendering the universe as lifeless.	2
Gap 2	Primal Matter Compaction	This material compacted itself together into a tiny superheated area.	• There is no evidence for this compaction. • Science cannot explain the composition of this matter. • If atomic components did exist, they could not survive the intense heat of billions of degrees. • There is no explanation about where this heat resistant material is now.	3
Gap 3	Big Bang	The compacted energy exploded (the big bang), causing the following instantaneous events: 1. Quarks, leptons and Higgs boson (particles) self-created, gathering together vast amounts of energy into miniscule structures. 2. Positive and negative electrical charges	• Energy does not congregate together or compress itself into the miniscule components of atoms (e.g. quarks, leptons, protons, neutrons etc.) rather it does the opposite: spreading itself into its surroundings. • With such immense gravitational pull, the physics of this compacted material would most likely form a massive black hole rather than an explosion. The physics-based activities forecast in a big bang explosion contradict those that operate in the observable universe. • Science does not state the type of material involved in the compaction that would allow it	3

Gap 3 Cont.	Big Bang	self- invented, attaching themselves only to specific particles. 3. Various quarks made neutrons. Other quarks and Higgs particles combined to make protons. Some leptons became electrons.	to explode, rendering the idea of a big bang as speculation, not fact. • There are no known scientific processes that allow countless numbers of sub-atomic components to self-create, particularly amid the chaos of the largest ever explosion. • Further, all of the miniscule particles had to perfectly align with each other in less than a second in order for atomic matter and energy to later form and for life to self-create billions of years after.	3
Gap 4	Faster Than Light Expansion	This newly made matter burst outwards exceeding the speed of light, starting the universe and its expansion. Later, at unidentified times, the Higgs field, the photon field, dark matter, anti-matter and dark energy self-created from unidentified materials. All of these are invisible.	• There is no explanation regarding how these different components and fields made themselves or how they were instantaneously perfectly compatible in size, function, layout, interaction and electric charges by chance. • The claimed rate of expansion cannot have occurred as nothing can travel faster than light. • The unique and structured shape of the universe bears no resemblance to the effects of a big bang explosion.	3
Gap 5	Higgs Field & Particle	Higgs particles, which attached themselves to protons, rub against the omnipresent Higgs field creating friction, mass and substance in the universe.	• The composition of the Higgs field is unknown. • Its self-creation and continued expansion into the growing universe is not explained. • The Higgs particles and field have specifically aligned and matching synergies that defy scientific explanation, both in their existence and values.	3

Gap 6	Supernovas	Supernovas formed.	• Science cannot explain what supernovas were comprised of at this time as atoms are not thought to have existed.	4
Gap 7	First Atoms	1. Protons invented 'strong interaction' forces enabling them to group together in the nuclei of atoms that did not yet exist. 2. Atoms self-created by protons and electrons pairing off to make hydrogen atoms.	• Without strong interactions, protons would move apart because their positive charges repel each other, as do electrons with negative charges. • There are no explanations about how massive groups of these two components could congregate together in the same areas so they could pair off. • There is no explanation about how these two sub-atomic particles managed to have perfectly balanced electrical charges despite their enormous size differences.	4
Gap 8	First Stars	Billions of stars self-created by groups of hydrogen atoms locking themselves into pairs of hydrogen molecules (H_2) which congregated together. Immense compaction forces caused stars to start-up by fusing hydrogen into helium, resulting in violent explosions that occurred in a controlled manner.	• Hydrogen molecules (H_2) have limited charges of attraction with each other. They do not congregate together under gravitational attraction, making the formation of stars impossible by natural events. • Most of the hydrogen molecules in a forming star would have been expelled by the initial massive, uncontrolled fusion explosions, preventing stars from continuing for long periods, if at all. • Based on the balance point between compaction pressure and the number of atoms required to create that pressure, stars sizes should be consistent – but they vary enormously by thousands of times.	4

Gap 9	First Neutrons	Masses of neutrons, without charges of attraction, scattered them-selves among hydrogen mole-cules in order to attach to the hydrogen nuclei when fusion explosions sub-sequently began, forming helium in new stars.	• The existence of masses of neutrons, specifically in the billions of billions of locations where stars were forming, is unexplained by science.	4
Gap 10	First Elements	Ninety new elements (other than hydrogen and helium) self-created.	• There are no known methods that allow the ninety 'natural' elements to make themselves. • Atomic mass gaps 5 and 8 are natural barriers that prevent the self-formation of any elements heavier than hydrogen and helium.	4
Gap 11	First Galaxies	Stars congregated into groups forming billions of galaxies, which have black holes at their centers.	• The proposed cause of black holes (being collapsed stars) is not scientifically validated. • The outer stars in galaxies hold their relative positions while in motion despite enormously greater distances to travel than inner orbiting stars. • A powerful, invisible force maintains the formation of swirling galaxies in contradiction of the natural laws of physics.	4
Gap 12	The Universe Shape	The universe self-formed into hollow ball shapes.	• The unusual shape of the universe does not occur naturally. • New dark matter/energy is currently being inserted into the universe by an un-identified process, causing the	4

Gap 12 Cont.			universe to accelerate outwards from all points. • New material is therefore being created in the universe at the present time – an activity acknowledged by science.	4
Gap 13	Our Solar System	Our solar system self-created in one process from random materials drifting in space. Planets self-created around stars. Moons self-created around the planets. Other debris gathered around planets.	• Planets and moons could not self-create because the sun's start-up explosions would have blown away the material that would have made them. • The interactions between the bodies in our solar system defy the natural laws of physics because moons and debris should all have been swallowed up by the planets or the sun – their existence requires enormous, undetected, invisible and unnatural forces to hold them in place. • The ages of various planets and moons differ enormously, prohibiting their formation in a continuing evolutionary process.	5
Gap 14	The Earth Forms	The earth self-formed with a unique range of materials not known to exist together elsewhere.	• Earth is in the perfect position in orbit around the sun • It has the perfect size, crust thickness and a vast, unique range of elements and components necessary to support life. • There is no scientific explanation for the coincidental self-existence of all the earth's elements in one place. • Many aspects of earth's construction defy evolutionary timelines.	7
Gap 15	The Sun Rotates	Our sun self-started its own rotation in an anti-clockwise direction (from north).	• There is no reason why this would occur, nor any known method by which it could happen.	5

Gap 16	The Planets' Spins & Orbits	The sun's gravity and spin caused the planets to start orbiting around it and also rotating around their own axes in an anticlockwise direction.	• The forces between the sun and the planets are thirty-five thousand (35,000) times the wrong way based on angular momentum equations, making this idea scientifically invalid. • Venus and some moons spin in the opposite (clockwise) direction. • Neptune spins on its side. These anomalies are scientifically inexplicable, notwithstanding external influences.	5
Gap 17	Nitrogen Appears On Earth	Enormous volumes of nitrogen appeared in earth's atmosphere.	• It is theorized that substantial levels of ammonia existed and were broken down into water and nitrogen by oxidation. • However, there is no evidence of such enormous levels of ammonia, nitrogen or oxygen in the earth's history at that time. • Further, the chemical interaction of oxidation would make it impossible to produce the necessary molecules in the atmosphere. • There are no viable scientific explanations for the abundance of nitrogen in the current atmosphere.	10
Gap 18	Water Appears On Earth	Masses of water appeared on earth early in its formation with the levels unchanged ever since.	• The hydrogen and oxygen required to make water did not exist on earth. • There is no evidence of water arriving from space. • Because there was no atmosphere, water pouring out of the earth's crust after it cooled would have rapidly evaporated into space. • There is no viable explanation about how water created itself	10

Gap 18 Cont.			or how it arrived on earth in such immense volumes. • Water has to be made with bursts of energy forcing hydrogen and oxygen electrons to combine in linked orbits. This requires a contained space that resists the effects of the subsequent explosion. Science is unable to validate any evolutionary explanation for the existence of water on earth – none of the ideas proposed is supported by actual evidence. The method by which water was originally made in the universe is also unknown and is without any scientific evidence.	10
Gap 19	First Life	A series of components self-created in water, gathered together and jointly created a strong, flexible and permeable external covering and became the first ever life, being a bacterial (microbe) cell. The major components included ribosomes, proteins, acids, (including RNA and DNA), chromosomes and flagellum, which allowed movement.	• All life is made from cells – science has proven that life cannot make itself from non-life. • The base ingredients required to make life are not believed to have existed on earth at that time. • The existence of life requires the intervention of an undetermined external power.	8

Gap 20	Microbes Replicate	Microbes replicated by splitting into two, appearing in vast numbers and locations across the world. They mutated into new, different forms over 1.5 billion years, and then suddenly stop mutating.	• The actual evidence is that new, different microbe forms appeared over a specific period of time, not that they mutated then stopped. • There is no biological method whereby microbes can change into entirely different forms because they are limited by natural barriers. • There is no scientific or logical explanation why microbes ceased mutating after such rampant growth for over 1 billion years.	8 & 11
Gap 21	Oxygen Appears On Earth	Oxygen appeared in vast volumes in a relatively quick burst then settled at twenty-one percent by volume - the perfect level for later life forms. The level of oxygen gas (O_2) self-regulated in combination with ozone (O_3) and solar intensity.	• The theory that oxygen appeared as a result of the biological separation of water into its base components is not supported by any evidence of the masses of plant life required, nor their fate once the level had reached twenty-one percent (21%). • There is no evidence of the masses of hydrogen that would also have been made, nor of reduced water levels. The self-balancing of oxygen gas and ozone is fundamental to the systems that support life.	10
Gap 22	An Atmosphere Forms On Earth	The atmosphere self-created into a series of layers and barriers, performing different functions that collectively protect later life forms that did not yet exist.	• The processes and reasoning for the self-creation of such specific components in a clever sequence of densities, thicknesses, and depths is unexplained by science.	7

Gap 23	First Nucleated Cells	After 750 million years of inactivity, microbe cells suddenly joined together to make highly complex nucleated cells, then immediately stopped mutating for the second time in history.	• There are no known processes that allow single cells to make nucleated cells. • The actual evidence is that nucleated cell life appeared from nowhere in the form of living bodies, and not that microbes made nucleated cells. • Cell mutation is not a process that can be turned on or off – it is a perpetual process according to the tenets of evolution.	11
Gap 24	First Living Creature	Nucleated cells joined together to make a partial living creature in water.	• The existence and survival of independent nucleated cells is unexplained by science because the cells that make up creatures' bodies die once separated from that body – they do not live independently. • There are no known processes by which cells can make a living body independently of a mother who must pre-exist. • A body cannot live with only some of its systems developed. A body cannot self-create through the random grouping and mutation of cells.	12
Gap 25	The First Male & Female	A genderless, unidentified creature produced two creatures of differing genders by an unknown process. These were perfectly matched in shape, size, reproductive processes and genetics.	• The self-creation of two completely different but matching genders defies all known biological processes. • Nor is there any scientific explanation how these creatures produced matching chromosomes.	13

| Gap 26 | The First Baby | The new male and female genders invented reproduction through gamete/sex cell releases into water and produced the first baby creature in history.

This required new and complicated processes involving timed and coordinated cell duplication programs.

Later mutations resulted in a completely different reproduction process that required sexual contact and internal fertilization in a female. | • These two newly self-created creatures had to:
1. Both have the first-time knowledge that they had to release their sex cells into water so the male's sperm could fertilize the female ova.
2. Be in the same place at the same time.
3. Have matching chromosomes both in number and type.
4. Have complementary X and Y sex cells to make an embryo of either gender.
• A new cell development program starting with just one cell had to self-invent without knowing the end purpose or the sequence of millions of steps required to make a baby creature.
• This program had to include instructions for the life-long operation of the body. Such a program had to contain billions of sequential, timed instructions without fault first up.
• There are no known natural processes that allow such complex and perfect self-invention of so many components and programs even with foreknowledge of the outcomes – which did not exist at that time.
• There are no biological processes that allow changes to reproductive processes from external to internal fertilization. | 13 |

Gap 27	Different Species Exist On Earth	The embryos of these new creatures mutated to self-invent new body parts and processes. Over time, tens of thousands of different types of aquatic creatures self-created by gradual cell mutations.	• Permanent physical changes by mutation from one creature to a different type of creature are not known biological processes. • There is no evidence of such changes ever actually happening. • Scientists have already determined that species and nature do not have the capacity to rearrange their nucleotides to make different creatures or body parts and acknowledge that such changes require external intervention.	15
Gap 28	Sea Life Moves Onto Land	An ocean-based fish that breathed by separating dissolved oxygen from water, left the ocean and lived on land, changing into an amphibian and breathing air.	• There are no known biological processes that allow these fundamental physiological changes to occur. • The reality is that water-based creatures die quickly when out of water. • Amphibians as an evolutionary transition from water to land is a poor choice because amphibians are fresh water creatures.	16
Gap 29	First Shelled-Eggs	Later, land-based reproduction self-invented, changing from external eggs in water to eggs with coverings that could protect them in air.	• Complete changes to reproductive processes are not known to be biologically possible.	19
Gap 30	First Insects	Invertebrates (predominantly insects) appeared on land in a vast range of differing species that had two matching genders.	• There is no scientific explanation for this class of creatures having the same cell types, genders and reproductive processes as other creatures, particularly as they have no biological links.	17

Gap 31	First Plant Life	New and different types of nucleated cells self-created resulting in plants self-inventing.	• There is no scientific explanation regarding the self-creation of plants or how their cells managed to differ from those of creatures. • There are no biological processes allowing plants to change into other different types.	18
Gap 32	First Seeds	Seeds which were critical for plant life to continue, self-invented.	• The third development of genders in differing but unconnected life forms has no scientific explanation (i.e. fish, invertebrates and now plants).	18
Gap 33	Plants Develop Toxins	Toxins and defensive systems self-invented in plants.	• The self-invention of toxins and defense systems would require pre-knowledge of chemistry and biology – a scientific impossibility.	
Gap 34	Reptiles, Dinosaurs, Birds & Mammals Exist On Earth	As time progressed, tens of thousands of different species of reptiles, dinosaurs, birds and mammals self-created by mutating cells in matching genders with vastly different structures, processes and abilities. Sexual reproduction changed from internal to external and vice versa on several occasions.	• The fossil record evidence is that all of these types of creatures arrived on earth fully developed with no signs of mutating from other creatures. • There are no known biological processes that allow any creature (or plant) to change into another different creature. • There are biological barriers that include protoplasm and sexual security codes that forbid such changes. • Complete changes to reproductive processes are not known to be biologically possible.	19, 20, 21

Gap 35	Mammals Return To The Sea	Mammals returned to water reconfiguring their land-based bodies and became 'fish' again. This included some types losing all their legs while others changed them into swimming aids.	• There are no biological methods whereby a mammal can lose its legs and develop fins and become aquatic based bodies.	21
Gap 36	Humans Appear On Earth	Humans appeared as the last known new life form on earth.	• With humans' appearance, the claimed mutations in every life form on earth have coincidentally stopped. • As evolution's claim that mutations were rampant, random, unstoppable and fundamental to life development, and theoretically being at their peak rate, their cessation cannot be explained by science.	23
Gap 37	Missing Links	The earth should be awash with billions of partly mutated living life forms with few, if any, identifiable species. The fossil records should be full of examples of transitional forms of creatures mutating from one to another.	• There is no evidence of creatures mutating to other forms now, nor is there any evidence in the past. • All 'missing link' finds purporting the evolution of humans from apes have been debunked. • The evidence from the fossil record does not support the self-mutating development of anything living.	23

Other Things You Should Know

Section A

Which Came First – The Chicken Or The Egg?

The question of which came first – the chicken or the egg? – is meant to point out the simplicity of the observation that chickens only make chickens and nothing else. Chickens only come from chicken eggs. There is no other natural way of getting a new chicken.

If biological evolution is true, at some point in the past, a 'non-chicken' had to be the very first to lay an egg that produced an actual chicken. But where did the 'non-chicken' come from and where did it go to as there are no 'non-chickens' in evidence today or in the fossil record. Producing a chicken in a hard-shelled egg requires a partnering rooster, raising a further issue: where did the male chicken come from?

This question highlights the absurdity that the same situation applies to every creature born from a hard-shelled egg, not just chickens. Therefore, for egg laying reptiles and birds, which came first, the creature or the egg? The answer for one must be the same for all, and it has to be the same for plants that come from seeds.

Science has now delivered some astoundingly good news.

Research at the Sheffield and Warwick universities has discovered that the formation of chicken egg shells relies on a protein called ovocledidin-17 (OC-17), which is only found in chicken ovaries. This protein assists in the speed of the shell's development by converting calcium carbonate into calcite crystals. Therefore, without the chicken to provide the OC-17, there cannot be a chicken egg[105].

So which came first? The chicken or the egg?

The answer is now indisputable: The chicken – which had OC-17 in its ovaries. And at the same time, the rooster had to appear so more chickens could be made. The first chicken and the first rooster made the first baby chicken in a hard-shelled egg. As there is no evidence whatsoever that any creature can, or has ever changed into a chicken, what other answer could there possibly be?

Following this logic, we can now reasonably assume all of the other shelled-egg-based creatures, and all seeded plants, first appeared complete as their own 'species' and ready to reproduce their own kind. These self-

---

105. This information was released in the paper Structural Control Of Crystal Nuclei By An Eggshell Protein, first published online in June 2010.

contained external fertilized ovum processes appear to be clear evidence of a process beyond evolutionary accidents and luck.

Section B

The Truth About Charles Darwin

Charles Darwin is most famous for noting changes within a living class and species in his book: *On the Origin of Species by means of Natural Selection, or the Preservation of Favoured Races in the Struggle for Life*, (first published by John Murray, 24 November 1859).

In 1835, visiting the Galapagos Islands, Darwin captured twenty-six kinds of birds and all except one were peculiar to the islands. Among them, he found thirteen finches that all resembled each other in the shape of their bodies, plumage and beaks but still could be called different species[106]. But they were all still finches. Essentially, what he observed were minor changes within a group.

He later identified the influence of sexual reproduction in changing characteristics within a class in a second book: *The Descent of Man and Selection in Relation to Sex*, (first published by John Murray, 1871).

Most people know little of Charles Darwin's studies but principally what he observed was:

1. All living things vary.

2. All living groups tend to increase in geometric ratio.

3. The number of species tends to remain fairly constant.

From these, he came up with two conclusions:

1. There is a struggle for existence.

2. The struggle is survival of the fittest.

Darwin came to believe that man had risen by slow and interrupted steps from some primordial cell. In other words, he could see what he thought was progress but could not actually find any evidence of evolution (mutation) from one class to another. In a sense, Darwin's observations were correct, as the actual evidence is that increasingly complex forms arrived on earth. But his conclusion that they were somehow connected was unsupported with any evidence as he was never able to explain why there is no proof of the missing links, except to say that discovery of fossils

---

106. Notwithstanding that even today there is dispute over the definition of a species.

is a slow process and that the regions likely to give up these remains had not yet been searched. Considering the advancements in fossil research and the extent of man's use of land in the last century and a half since he lived, it is doubtful that his statement could still stand today.

To quote Darwin himself from the *Origin of the Species:* "Why, if species have descended from other species by fine graduations, do we not everywhere see innumerable transitional forms?"

Darwin openly queried why every layer in geological formations was not full of these examples. To quote: "Geology does not reveal any such finely graduated organic chain: and this, perhaps, is the most obvious and serious objection which can be urged against the theory".

In fact, if you think through the logic of Darwin's original proposal that there are intermediate steps but they are in places that haven't yet been found, then we have a most peculiar situation. What this means is that every fully complete creature lived in a place where other fully complete creatures lived and when they died their remains might be found as some sort of fossil. However, if any creature started to change its form into another different type of creature then it would have to go somewhere else, over possibly millions of generations, and only return to the 'normal' fully developed area when it had totally changed into a new complete creature. This 'exodus' and the later 'return' would have to be common with every creature across the entire planet in water, air and on land. The millions of places where they go to 'change' or develop would never be able to be found by anyone, ever. This theory would require a conscious act on the part of every creature, regardless of its type – and plants as well.

These types of theories not only fail to make any sense at all but they actually contradict the evidence that exists. As above, Darwin himself openly stated that the evidence contradicted his theory and yet still today he is referred to as the 'father of evolution'.

What we see today is that life did not take over the world locked into conflict, but that life exists by networking with differing forms cooperating together in symbiotic relationships. This fact actually conflicts with Charles Darwin's observations of 'survival of the fittest'.

Despite enormous leaps in science in the last century, we can observe that not one class of creature can sexually combine with another class to form an offspring. Each class of creature on earth is totally separate from every other class despite the theory that one came from the other.

Using this line of reasoning, surely even man and ape, being such close species in the same class (not relatives) would easily combine to make a hybrid – but even this is not possible. Their genetics do not match and the unique fertilization security codes do not allow it. Based on the known scientific evidence, the conclusion is that one did not come from the other – human from ape – and that they are totally separate and independent creatures, each with separate and independent functions on earth.

In conclusion, Darwin made some valid and some incorrect observations about the variations in living forms, but never provided proof of the ability of organisms to change from one species or class to another. In fact, by his own words, it is quite fair to say that Darwin acknowledged his theory was not supported by any evidence which, in itself, was sufficient to disprove it – but that his hope was that future finds would do so.

Section C

The Permian Extinction

Around 250 million years ago, paleontologists believe there was an event, or series of events, possibly lasting up to 10 million years, that killed the majority of species living on earth at that time. Estimates range from seventy to ninety-five percent of marine species and seventy percent of land-based vertebrate became extinct. There is much debate among scientists about the cause and exact dates and whether single or multiple events occurred. This mass extinction happened during the changeover from the Permian to Triassic geologic periods and is the only known mass extinction to include insects. There are also studies that identify extinctions occurring in different geographical locations but not affecting the same families in other locations.

The geologic records across the world suggest conflicting evidence about times and causes. The main theorized causes tend to center around gradual environmental changes, such as the release of methane from the ocean floors or massive volcanic eruptions or hydrogen sulfide excesses in oceans, to one-off catastrophes, such as a meteor or comet collision with earth. Various measurements show possible increases in carbon dioxide levels between eight- and ten-fold, raised levels of ultraviolet radiation permeating the atmosphere and a temperature elevation of fourteen degrees Fahrenheit, with oceans believed to have reached one hundred and four degrees Fahrenheit and, presumably, too hot for life. Other evidence points to oxygen deficiency in the oceans. Some claim a combination of several of these potential causes occurring simultaneously, adding to the mystery. Suffice to say, there is no definitive agreement on the cause or the severity of extinction or the exact timeline.

Of the vertebrates on land, there are claims that large numbers of species of amphibians, reptiles and herbivores disappeared with a broader variety of species appearing after the extinction. However, the volume of fossils across the Permian – Triassic boundary is low, not allowing any clear conclusions to be proven. Insect fossils after the Permian extinction are quite different from those before, with increased size being an observable factor[107].

Michael Denton, a British-Australian biologist and medical researcher, observed that, of the current living animal life forms, almost eighty percent of those families have been found in the fossil record (*Evolution: A Theory in Crisis*, Adler & Adler, 1986). This indicates that there may not be as many life forms lost in the 'Permian extinction' as thought.

There are various studies about different life forms before and after the Permian extinction, such as complex marine ecosystems becoming far more prevalent afterwards. The main consensus appears to be that life in the form of diversity took some time to recover, such as land vertebrates, estimated by M. J. Benton, a British paleontologist, to have taken up to 30 million years for the appearance of multiple diverse forms.

The end result of all of these conflicting ideas is that there is no definitive proof or even notable theory upon which to base any claims of evolutionary self-development around this assumed mass extinction. Yet many claims are made about fast recovery rates and new and broad diversity of life forms as proof of evolution in action, but these don't necessarily consider the fossil record findings or the extended timelines.

The 'Permian Extinction', perhaps even named beyond its real effect, gives us no clear direction in pursuit of the history of self-created life.

Section D

Reliability Of Fossil Dating Methods

While the previous sections have summarized a misleading history by some scientists about the origins of man, there is also a substantial question mark hanging over the methods used for dating fossils. We have learned that many of the actual dating results have been ignored or put aside in favor of matching the same timelines of the strata in which the fossils have been found. This means that scientists have ignored the test results that indicate the age of the fossils and have changed these to match the age of the material in which it was found. For example, if a 1 million year reading of a fossil was found in strata that was 3 million years old then the date of the fossil may

---

107. Refer to discussion in Chapter 10, *Where Did Oxygen Come From?*

have been changed to 3 million years. Consequently, either one, or both, dating methods for the fossils and the strata are giving the wrong results relegating the claimed dating results as neither reliable nor consistent.

It is important for us to understand the methods that are used to work out the dating of events because evolution assumes they are chronologically linked to each other in sequence; and if the methods or results of dating are wrong, then Evolutionary Theory is automatically also wrong.

If an event, such as the existence of ocean life, is shown to have occurred prior to land life, then evolution claims that ocean life must have been the cause of land life. Otherwise, life must have started independently more than once on earth – an event considered impossible in the first place. It is certainly not possible on multiple occasions in totally different environments (air and water) that living entities accidentally end up having the same basic components, structures and functions.

How scientists date things has always been a controversial issue. With any dating system, there cannot be any absolute proof that the methods used actually give accurate results. The reason is that we assume that the conditions, such as atmosphere, levels of gases, temperatures, pollutants, rates of decay and so forth were constant over billions of years of time, but there is no absolute way of knowing this.

We are now going to look at various dating methods to see how they work and how they compare with each other.

Radioactive Dating

Uranium (element 92 in the periodic table) naturally decays[108] into lead (element 82 – an isotope) by losing protons and electrons from its atom. By working out the levels of uranium and lead atoms in a rock sample, it is assumed one can calculate the theoretical age of the rock. However, this constant rate of decay is unprovable, particularly as it is over such a staggeringly long period of time. Uranium has several limitations when used in this measurement. For example, as it goes through the decay process there are several other isotopes that form. One of these is radon (element 86), which is a gas and can escape from the rock, thereby distorting the age.

Further, there are two different decay chains involved being:

(1) Uranium series with a half-life of about 4.5 billion years, which is the period most commonly referred to, and

(2) The actinium series with a 704 million year half-life.

---

108. Known as radioactive decay.

As there are two differing decay routes, there is more than one method used and therefore more than one timeline proposed. However, many dates that are regularly referred to are still based on rates of uranium decay, leading to doubt over their accuracy.

Science has moved forward in this area by using a number of other radioactive elements (for example, potassium) that are in greater supply and give more consistent results. The rate of decay of these radioactive elements is claimed to be unchanged regardless of the circumstances that the rock encounters. The only limitation is that the rock must be solidified. A melting rock will destroy the isotopes but not the original elements and 'reset' the clock back to zero. The benefit of radiometric dating is that it tells us how long a rock has been in solid form, however, its accuracy is apparently limited to being a minimum of one million years old – which, of course, is an arbitrary number, not one with scientific proof.

Among the evolutionary ideas to explain the existence of matter, there is a proposal that matter has existed forever – thereby avoiding any explanation for its creation from nothing. According to the rate of decay of elements such as those above, there had to be a moment when they first appeared, otherwise if they were infinitely old, elements such as uranium would have already turned to lead and there would be no uranium left in the entire universe. The existence of elements today, in their original form, is clear evidence that there was a beginning to the matter that makes up the universe. This means that matter had to appear from nothing, which we know is impossible by itself.

Radiocarbon Dating

This is known alternatively as Carbon 14 dating. Martin Kamen, a Canadian physicist involved in the Manhattan project, is thought to have discovered this method in 1940, though some books attribute it to Willard F. Libby, an American physical chemist, who won a Nobel Prize for his research development in 1949.

Carbon 14 (C14) is radioactive and has a claimed half-life of 5,700 years. There is a very small amount of C14 in the carbon dioxide in the atmosphere. It is regularly renewed when cosmic rays hit nitrogen in the atmosphere maintaining its presence. Organisms absorb carbon dioxide in photosynthesis and therefore will contain the same ratio of C14 to C12 as the atmosphere in which they live. When the organism dies it stops absorbing C14, which then begins a process of radioactive decay.

By working out the level of C14 in dead remains, an age can theoretically be calculated on the assumption that the level of C14 when the plant died was the same as today. However, this is specious because the rate of C14 renewal is thought to be about thirty percent more than the rate of decay. Therefore, the calculations will give a much higher reading than is actually the case, to the point where, in reality, all organic materials are possibly no more than 10,000 years old. This method has also been used for dating artefacts made from plant materials, for example, wood and textiles, and would, of course, give the same misleading results.

The basis on which the age measurements are calculated relate to the ratio of C14 to C12 in tree rings. It is assumed that the inner rings of trees are inert but that the outer ring acts with the atmosphere. As we can count the number of rings (years of life) of a recently dead tree we can test for the C14/C12 ratio for that year. This method has given information about carbon ratios back to about 5,000 B.C. However, for bodies that at some time have been in seawater, this method is inaccurate due to very low levels of C14. We know that in the early years of life for earth the atmosphere had a much higher level of carbon dioxide, so it cannot be assumed that dating will be accurate anyway.

Measuring C14 levels is a method limited for use in dating objects that were once alive, such as plants and animals and is assumed to be reasonably accurate. It cannot be used for dating inert matter such as rock. However, it is assumed that if a rock contains a previously living entity then it is assumed the age of the rock is the same.

Dating By Tree Rings

Trees appear to have a built-in calendar. Every growing season a tree will gain a new growth ring. This is assumed to occur every year as the warmer seasons arrive. These rings tend to be broader in hotter periods and narrower in colder years. The rings in trees that live in tropical climates are less evident as the weather is more constant and the rings are harder to identify. There are patterns in rings that show us how the climate changed over previous periods and we are able to match these with other wooden fossils and calculate the ages of a range of artefacts such as timbers from buildings and tools. The oldest tree cut down was in Nevada in 1964 and it displayed almost 5,000 rings but, generally, trees over 1,500 years are rare. This method limits dating by tree rings to approx. 5,000 years.

Curious Dating Methods

When looking at how science tries to put dates on some objects one can only be puzzled at the process. It seems amazing to see various discoveries given dates such as '5 million and 15 years' old. Naturally, most of us think there is some incredibly accurate method of identifying ages but this is not the case. This type of figure comes about, for example, when a dating is estimated as '5 million years old' in 1970 and then referred to in 1985 (15 years later) when the additional time is added on. This process is basically telling us is that 15 years ago, it was estimated at 5 million years old. If a better method of dating is discovered in 1986 and the time is adjusted to 4 million years, then the 'additional' years start from zero again! So, in 1986, our dating is 4 million years whereas in 1985 it was 5 million and 15 years.

Out Of Place Artefacts

To add further doubt to the validity of fossil dating, there has been a variety of archeological finds that cannot be explained by modern science. Thousands of these have been found, with most stored in museums waiting for an explanation. Referred to as Out of Place Artefacts (Ooparts), these are finds of relatively modern tools, jewelry and ornaments in soil, rocks or coal that are dated well before the appearance of mankind. In many cases, producing these artefacts would have required excavation, processing and smelting methods, as well as skills and tools, unknown at those times.

Some examples are:

- In the mid- to late-1780s at a quarry near Aixen in Provence, France, a stonemason's yard with tools used in the 18th century was unearthed around fifty feet below the surface under eleven layers of rock. The types of materials found were stone pillar stumps, petrified wooden tools, a long work-board with rounded, wavy edges, pieces of partially worked stone and some coins. The limestone rock at this site has since been dated at 300 million years old.

- In 1844 at a quarry in Tweed U.K., a gold thread was discovered embedded in stone that was aged around 320 to 360 million years old, being from the Early Carboniferous period when life on land had only recently appeared.

- In 1865 at the Abbey Mine, Nevada USA, a two-inch long metal screw was found in a piece of 21 million year old feldspar.

- In 1889 near Nampa, Idaho USA, at a depth of three hundred and twenty feet, drillers found a figurine made from baked clay. A fifteen feet thick layer of basalt lava that was drilled through nearer the surface was dated at more than 15 million years old, dating the find as likely over 100 million years old.

- In 1877, while tunneling near Table Mountain, Tuolumne County, California USA, J. H. Neale, a company superintendent, identified foot long spearheads and a mortar and pestle within a few feet of each other. These were found in bedrock gravel dated around 30-50 million years old.

• In 1891, near Cleveland, Tennessee USA, fifteen hundred feet into a tunnel, a sandstone wall about a thousand feet long was discovered in gravel deposits dated over 250,000 years old. The wall was two feet thick, eight feet high and made from sandstone blocks with clay mortar and showed many areas of plaster. Some of the blocks displayed hieroglyphs of an unknown language along with images of unidentified animals.

• In 1936 in London, Texas, USA, Max and Emma Hand found an ancient hand-made hammer embedded in rock. The hammer head was comprised of ninety-six percent iron, which would have required modern smelting methods, while part of the wooden handle had commenced transforming into coal. Archaeologists dated the rock at over 400 million years old while the hammer was dated at more than 500 million years old.

• In 1936 in Plateau Valley, Colorado, USA, a pavement made of five inch square tiles laid in mortar sourced from a different location was found in the same layer as a 30 million year old three-toed horse.

• The Klerksdorp spheres:

At the Wonderstone silver mine at Ottosdal in South Africa, miners uncovered about two hundred hard metal spheroids (some are spheres, other oblong) in pyrophyllite rock 2.8 - 4.5 billion years old. These measure between one and four inches in diameter. There are two types of spheres: solid metal and those with one quarter inch thick shells with a spongy material inside that disintegrates into dust when in contact with air. The metal, a nickel steel alloy, does not occur naturally. Some of the spheres have parallel grooves etched around their equators. It is claimed that the balance of these items is almost perfect, which, using today's technology, would require manufacture in zero gravity conditions.

One of these spheroids, locked in a vibration-free display case on exhibit at the South African Klerksdorp Museum, is claimed by curator Roelf Marx to be gradually rotating on its axis by its own power.

Finds In Coal

Coal is made from decaying vegetation and is claimed to take up to 400 million years to form under layers of sediment. Any object that is found in coal, by logic, must have been in the vegetation prior to the layers of sediment covering it. Archaeologists therefore date such artefacts with the same timing as the coal they are found in.

- In the 1800s a rancher in Colorado, USA found an iron thimble in a piece of coal dug from three hundred feet below the surface.

- In 1885, an Austrian foundry worker found a hand-made forged iron cube inside a piece of coal claimed to be millions of years old.

- In 1891, Mrs. S. W. Culp of Morrisonville, Illinois USA, is reported to have found an eight-carat gold chain in a lump of 230 million year old coal (estimated). The ten inch long chain was considered to have required skilled workmanship.

- In 1944, a brass alloy bell made from a mix of metals that differs from modern alloys was found in a piece of coal by a young boy, Newton Anderson. The bell appeared to be crafted by hand, having an iron clapper and a sculptured handle. The age of the coal seam where the bell was found was around 300 million years old.

- In 1928, in a coal mine near Heavener, Oklahoma USA, Atlas Mathis found a stone wall in blast residues two miles below the surface. The wall comprised of one foot cubic blocks made from concrete, the surfaces of which were so smoothly polished that they looked like mirrors. Other parts of the wall were located up to 150 yards away. The coal was deemed to be about 285 million years old.

Other Unexplained Finds

- Human footprints have been identified in the same areas as dinosaur prints in multiple non-related locations on earth. The most controversial of these was discovered in 1968 by paleontologist Stan Taylor who found human footprints next to fossilized dinosaur footprints in the Paluxy river bed in Texas, USA. These prints also extended under other rocky material that was later excavated. All prints had to be made while the river bed was in a soft, malleable state. There are claimed to be over 60 million years between the extinction of dinosaurs and the arrival of man.

- In the mid-1800s in a gold mine dated 12 million years old at Table Mountain, California USA, various animal bones were found along with a human skull, a stone bowl and mortar and a grinding disc. The animals included mammoths, hippopotami, rhinoceros, bison, horses, and camels.

- In 1927 at Fisher Canyon, Nevada USA, an imprint from the heel of a shoe displaying double-row stitched leather and clear impressions of the twists of the threads was found in rock dated at 225 million years old. Assumedly, the heel print occurred when the rock was still forming in its "soft mud" state.

- In 1973 near Moab, Utah, USA, two human skeletons were found in strata dated at over 100 million years old.

- In the 1980s an American landowner found a fossilized human finger among road gravel that was being quarried from 100 million year old cretaceous limestone.

Summary Of Dating Methods

- The above research indicates that there are no dating methods currently used about which we can be absolutely certain provide consistent, accurate or reliable dates, particularly as we don't know the history of the item being dated.

- The existence of Ooparts in rocks, coal and soils that are hundreds of millions and billions of years old undermines the reliability of present dating methods. There are no scientific explanations for these finds; their existence is impossible during those ancient times, based on the methods used for dating materials. This is particularly highlighted by the Klerksdorp spheroids which are dated almost as old as the earth, billions of years prior to any life existing.

Being aware of the variables, the enormous time frames associated with dating methods and our inability to prove the methods are valid, we should be cautious when given dates that are assumed to be correct. It is impossible to guarantee that any dating method is correct or even really represents a proper 'ball park' time frame.

Section E

How Scientists Make Their Conclusions

As with dating methods, we should be careful about what we accept as 'fact', so we are going to look at an example from physics using the theories relating to the attraction of physical bodies. We will see from this exercise how science flexes and adapts through creative thinking, not facts, in order to account for the unaccountable.

Scientists have discovered that there are four main forces that influence the universe. These are: -

- Gravity – which governs the movements of the galaxies, stars, planets, and moons.

- Electromagnetism – which governs the movement of atoms.

- Strong interactions – are the forces that hold protons together in an atom's nucleus. Without this force the positively charged neutrons would repel each other.

- Weak interactions – are caused by virtually undetectable neutrinos that carry away a very small amount of energy in some processes, for example, the transformation of hydrogen into helium in the sun.

In looking at the 'law' of gravity, we find that there are three separate levels to its application:

- Quantum mechanics only applies on a microscopic scale.

- Newton's Law of Gravity will only work if gravitational forces are small.

- Einstein's Theory of General Relativity applies only in strong gravitational fields.

As a result, to describe the effect of the attraction between bodies, we have three different laws of physics. Essentially, the problem that this creates for the Theory of Evolution is that these laws cannot have been in operation at the proposed moments of the beginning of the universe during the theoretical big bang because they contradict what was meant to be happening at that time. This means that the 'evolving' or expanding universe operates differently now from how it behaved when it first erupted. Even though these laws contradict the theory of the big bang, science doesn't admit the big bang can't have occurred without, by default, acknowledging the only other option – a Creator. This places science in a position where creative thinking is the answer and that is exactly what has happened.

Some scientists, in this case Stephen Hawking and Jim Hartle, made up concepts of 'Real and Imaginary Time', 'Quantum Cosmology' and a 'No Boundary Universe'. In other words, they thought up some amazing new ideas, that can't be proven and for which there is no evidence, to skirt around the problem of why the theory of one single initial state of things, where everything contracted together to start the big bang, doesn't match the laws of physics[109].

---

109. This seems to be a case of one inexplicable theory explaining another inexplicable theory.

As brilliant as these ideas may be, they still overlook the obvious conclusion that fits all the known information that we have already reviewed: there was no 'big bang'. The latest scientific data indicates that matter has been, and is being, made in a continuing process; it was not made in one moment of time, which is why the big bang theory doesn't comply with the laws of physics – because it can't have happened.

This process above, of thinking up new concepts when a current theory doesn't match all the current data, is a good example that illustrates how some scientists will either change a theory or invent new ones to try to fit known or new data.

In other words, they ignore the obvious conclusion from the actual data, in this case that some other process is required to explain the existence of matter that has been made and is still being made, and they invent unprovable ideas to avoid acknowledging the evidence that clearly points to a logical conclusion – a Creator.

Section F

False Interpretation Of The March Of Progress Illustration

The image above is referred to as *The March of Progress* illustration, which was included in the 1965 *Early Man* volume of *Time-Life Books, Life Nature Library*. It displays fifteen different types of creatures walking in a line with progressive developments showing an ape gradually changing into a human. The image (drawn by Austria-Russian natural history painter, Rudolph Zallinger, 1919–1995) was a visual interpretation of the text (by American anthropologist, Francis Clark Howell, 1925–2007) that challenged the analysis by some scientists that various partial fossil finds did actually represent the steps of human history on earth.

The image, sometimes referred to as *The Road to Homo Sapiens* has become popular with many different versions being drawn over the past fifty years. However, author Howell commented that, "The artist didn't intend to reduce the evolution of man to a linear sequence, but it was read that way by viewers... The graphic overwhelmed the text. It was so powerful and emotional".

508

When viewed in context of the article, the image clearly supports the tone of the text that such progress, when seen visually, was obviously too linear, and too focused on humans as the final end point, to make it a realistic proposition.

Science itself does not support any sequential linear progression from ape to human, for example, now having dismissed the first five and the seventh of Zallinger's figures. Of his other figures, those having colored stripes across their tops represent a lack of continuity between extinct and current ancestries, yet others are known to have more than one possible lineage[110].

In essence, science knows that the assumed evolution of life will branch out and be affected by external events such as natural extinctions, rather than a predictable direct linear sequence. Yet, despite science knowing the truth about this image and its supporting text, many people have this concept embedded in their thinking as an actual representation of human evolutionary history, whereas the opposite is true: The March Of Progress is a challenge to such thinking.

Section G

DNA Similarities In Apes And Humans

There is growing information that apes have very similar DNA to humans, leading to the proposal that this is proof that one evolved from another. For example, chimpanzees and humans share ninety-eight percent of their DNA code. Is there an alternative explanation?

According to Wikipedia, "Deoxyribonucleic acid, or DNA, is a nucleic acid that contains the genetic instructions used in the development and functioning of all known living organisms (with the exception of RNA viruses). The main role of DNA molecules is the long-term storage of information. DNA is often compared to a set of blueprints, like a recipe or a code, since it contains the instructions needed to construct other components of cells, such as proteins and RNA molecules. The DNA segments that carry this genetic information are called genes, but other DNA sequences have structural purposes, or are involved in regulating the use of this genetic information".

DNA is the basic construction module of life. DNA defines the blueprints for the type of being and for its makeup. It makes sense that if we were

---

110. The reality regarding many of these 'finds' are detailed in Figure 189: Table Of Ape-To-Human Disproven Finds, Chapter 23.

able to play God and make similar creatures we would use similar DNA because that is the base component. Such a conclusion is therefore not exclusive to evolution.

For example, if we make a cake we typically use flour, water, baking soda, salt, butter, eggs, and sugar. We have lots of different cake recipes which use all, or some, of these and other ingredients. Because we use base ingredients for our cake, we could use our imagination to make a different type of cake from them. This doesn't mean that our vanilla cake changed or evolved into a strawberry cake; it means that we, the maker, simply used certain base ingredients and changed them a little to get different cakes.

Similarly, if we had the power to create all the creatures on earth, then we would most probably use similar DNA structures as the base ingredient for our animals and tweak it depending on the type of creature we want. For similar creatures, such as apes and humans, we would use much of the same DNA, perhaps even as much as ninety-eight percent the same.

Section H

First Sex Claim

In October 2014, a group of Australian scientists announced that they believed the first creatures that were linked to humans and had performed sexual intercourse were the *microbrachius dicki* from 385 million years ago. This species were Paloderms found in Scotland (and were also found in China and Estonia), being the earliest known vertebrates which, by default, Evolutionary Theory considers to be humans' primary ancestors.

The male's anatomy includes bony L-shaped genital limbs believed to have been used for holding on to the female who had small paired bones which may have been used to lock in the male sex organs. Effectively, the claim is that the anatomy and processes we use in sexual intercourse started with these creatures hundreds of millions of years ago, making this a substantial scientific breakthrough.

Two surprising aspects to this claim that seem at odds to our own anatomy are that:

1. These creatures were under three inches (eight centimeters) long.

2. Intercourse had to be performed side by side, as if sitting or floating next to each other.

Among the unexplained issues are:

1. There doesn't appear to be any suggestion of other uses for these body parts.

2. There is no explanation regarding how the male and female anatomy were perfectly matched, as it was only coincidence that they would independently develop sex organs and the necessary attaching limbs.

3. How these creatures found themselves in the same location and with the same intent by chance.

4. How or why sexual pleasure and desire invented themselves separately in both genders, coincidentally at the same time and for no specific reason.

5. How they invented the first, complex, timed cell development programs to make the first baby in history perfectly from just one fertilised cell.

6. By what biological processes the future locations of sex organs were constantly changed in the millions of their male and female descendants - who still managed to match each other by pure chance.

This watershed claim raises further issues that show why there is no officially endorsed Theory of Evolution.

Professor John Long of Flinders University South Australia, who is the lead author of the study, pointed out that prior to this find evolutionists believed that internal fertilization such as this could only 'evolve' from creatures that were originally spawning externally, but not the reverse.

To quote Professor Long from an interview with ABC Science:

"This discovery implies something that was thought to be impossible in biology, [that is] fishes went from internal fertilisation back to external spawning... Placoderms lost their claspers when they began evolving into modern bony fishes, at which point they go back to fertilising in water again".

The previously held view that the reproductive process of external spawning was able to change to internal fertilization, but not the other way around, is now redundant. If this new claim is to be accepted, reproduction needs to be able change both ways, which was previously considered impossible.

Professor Long also said:

"So they were highly successful little beasties, switching to internal fertilisation as a strategy worked for them better than external fertilisation".

These comments about strategy imply that these primal creatures, being among the first on the planet, had some knowledge of reproductive processes that were not yet invented and also the ability to physically change their own fertilization processes. Both of these concepts oppose the foundation of Evolutionary Theory in that everything is by random chance with no foreknowledge of events and processes that had not yet appeared.

Of note here is that:

1. This new 'discovery' cannot be included in the official Theory of Evolution, as it does not exist.

2. The general public is not specifically made aware of this significant change of evolutionary thinking because nothing official has changed.

New concepts such as this one cast further doubt over our acceptance of the validity of any evolutionary theories - as they can, and do, change in an instant.

Bibliography

(Not previously referenced)

P.M Dauber and R.A. Muller, *The Three Big Bangs (Addison-Wesley Publishing Company, 1995)*

J.P. McEvoy and Oscar Zarate, *Introducing Stephen Hawking: A Graphic Guide (Icon books Ltd, 1999)*

Paul Davies, *The Last Three Minutes (Orion Books Ltd. 1995)*

David Lindley, *Where does the Weirdness Go?: Why Quantum Mechanics Is Strange, But Not As Strange A You Think (Basic Books, A Division of HarperCollins Publishers Inc, 1996)*

Evry Schatzman, *L'expansion de l'univers (Hachette Littérature, 1989)*

Paul Davies, *More Big Questions (ABC Books, 1998)*

David McNab and James Younger, *The Planets (Yale University Press, 1999)*

Michael E. Bakich, *The Cambridge Planetary Handbook (Cambridge University Press, 2000)*

Lynn Margulis and Dorion Sagan, *MICROCOSMOS (Summit Books New York, 1986)*

John Reader, *The Rise of Life: The First 3.5 Billion Years (William Collins Sons and Co. Ltd. London, 1986)*

Robert A Heinlein, *Time Enough for Love (Ace Books, 1973)*

Internet reference materials

Marshall Brain

http://www.howstuffworks.com/

http://www.enchantedlearning.com/

http://www.solbaram.org/articles/humind.html

http://atschool.eduweb.co.uk/sirrobhitch.suffolk/invert/inverteb.htm

http://www.aloha.net/~smgon/eyes.htm

http://www.seps.org/oracle/oracle.archive/
LifeScienceEvolution/2001.11/001004995595.18266.html

http://www.oup.co.uk/oxed/children/yoes/pictures/

http://uk2planets.org.uk/common-questions/what-different-elements-are-the-planets-made-of/

http://www.letusfindout.com/what-are-planets-made-of/

http://science.howstuffworks.com/life/human-biology/babies-kneecaps1.htm

http://www.geneagenetics.com.au/Resources/What-are-chromosomes

http://www.genome.gov/26524120

http://en.wikipedia.org/wiki/Chromosome

http://ghr.nlm.nih.gov/handbook/basics/chromosome

http://insects.about.com/od/antsbeeswasps/a/10-cool-facts-about-ants.htm

http://www.insecta-inspecta.com/termites/macrotermes/index.html

http://www.desertmuseum.org/books/nhsd_termites.html

http://en.wikipedia.org/wiki/Mutation

http://learn.genetics.utah.edu/archive/mutations/

http://www.pawnation.com/2013/04/22/12-oldest-animal-species-on-earth

http://listverse.com/2010/05/14/top-10-prehistoric-fish-alive-today/

http://phys.org/news/2012-05-shift-shore-extinct-tetrapod-ichthyostega.html#jCp

http://www.answers.com/topic/amphibian

http://www.starfish.ch/reef/echinoderms.html

http://www.heifer.org/blog/2012/03/an-eggs-periment-to-test-the-strength-of-a-shell.html

http://www.plosone.org/article/info%3Adoi%2F10.1371%2Fjournal.pone.0039056

http://io9.com/how-spontaneous-combustion-really-happens-698848769

http://www.ecology.com/2011/09/12/important-organism/

http://diet.yukozimo.com/category/reptiles/

http://en.wikipedia.org/wiki/Human_tooth

http://www.abc.net.au/news/2014-10-20/ancient-fish-to-thank-for-sex/5825594

http://www.ukapologetics.net/08/50reasons.htm

http://www.theguardian.com/science

INDEX OF FIGURES

INDEX

microbes xxii, 86, 87, 113, 114, 118, 120, 122, 124, 127, 129, 130, 131, 133, 148, 149, 150, 153, 154, 163, 221, 222, 235, 437, 440, 455, 456, 461, 478, 489, 490

micro-evolution 7

Milford H. Wolpoff 324

Milky Way 30, 33, 57, 58, 65, 75, 76

Miranda 92

missing link 119, 326, 494

mitochondria 127, 151, 154, 323, 428

molybdenum 125, 455, 464

monophosphate (AMP) 153

monophosphate (TMP) 153

mosses 249, 254

Mount Olympus 86

Mount Stromlo Observatory 33

mouth 171, 226, 229, 237, 263, 270, 299, 344, 345, 347, 348, 351, 359, 364, 366, 408, 409, 410, 423, 432

mucous slime 175

Multiregional Hypothesis 323, 324

mutation xxii, 5, 124, 129, 133, 160, 166, 167, 174, 181, 182, 184, 193, 194, 195, 197, 198, 199, 223, 224, 225, 227, 229, 232, 239, 240, 257, 267, 270, 271, 281, 283, 295, 310, 315, 320, 329, 331, 332, 335, 336, 343, 351, 376, 377, 378, 379, 381, 383, 384, 388, 389, 414, 437, 490, 492, 496

mycorrhia 253

Myrmecologists 242

N

NADPH 258

natural selection 4, 187, 384

neocortex 307

Neptune viii, 74, 75, 83, 84, 93, 94, 95, 97, 453, 487, 516

nerves 166, 172, 173, 176, 213, 228, 229, 231, 346, 347, 348, 357, 360, 364, 391, 409

nervous x, xiii, 172, 364, 365, 522

neurons 173, 347, 391, 392, 393, 394, 396, 398, 400, 404, 406, 407, 419, 431, 434

Neuropathies 424

neurotransmitter 311

neutrons 38, 44, 46, 49, 50, 450, 451, 482, 483, 485, 507

Newton's Law 69, 71, 80, 507

New World monkeys 320

niche fulfillment 156

night star 85

nitrogen 65, 72, 92, 93, 114, 120, 125, 130, 139, 143, 144, 145, 146, 455, 456, 462, 487, 501

Nix 94

nocireceptors 433

non-living matter 6, 8, 122, 203, 206, 207, 455, 472

non-ruminants 299

North Atlantic Ocean 225

nuclear explosion 51

nuclear fusion 51, 54, 63, 136, 451

nucleated cells 16, 120, 154, 155, 158, 203, 207, 458, 490

nucleoid 121

nucleotides ix, 153

nurseries 243, 244

nuts 256, 259

O

Old World monkeys 320

olfaction 407

olfactory bulb 213, 346, 392

omnivores 146

On the Origin of Species 4, 496

optical 176

optic nerve 177, 404, 406

ovum 182, 186, 194, 195, 198, 204, 207, 269, 301, 339, 340, 341, 342, 344, 358, 368, 373, 377, 378, 388, 496

oxidization 138

oxygenated 171, 362

ozone 102, 131, 132, 222, 489

P

Pacinian 433

Pain xiv, 424, 425, 432

Palomar Observatory 94

Pasteur 122, 123, 128, 445, 455, 480

Paul Davies 473, 513

penis 187, 268, 270, 271, 340, 347

perception 173, 404

RANDALL HARRIS
BIOGRAPHY

Randall Harris was born in Adelaide, South Australia in 1953 and has tertiary qualifications in Business Management. He has worked in International Banking and several family owned companies while most of his working career was managing a range of business units for a large international conglomerate. He has travelled extensively throughout Europe, Asia and America giving him exposure to many differing views about life and its origins.

In 1999, Randall began researching the universe and was surprised to learn that the knowledge held by science on most topics under the banner of evolutionary theory was very different from normal public understanding. The disparities were so significant that he decided to compile a comprehensive picture of the relevant scientific information that existed so that he could share it with others. Over a period of fifteen years, he uncovered a staggering array of wonderful and amazing revelations that are presented in this book, Evolution Unraveled.

Randall hopes that you will be enlightened by exposure to these scientific facts rather than accepting as truth many of the current evolutionary ideas and concepts that are known to be unproven. The vast range of information presented will help you to draw your own conclusions and will challenge your current understanding on accepted science. This is a comprehensive analysis of evolutionary theory starting from the beginning of time, using the facts known by science that are not generally presented to the public. It is a journey of wonder, reflection and revelation that will challenge many of the fundamental beliefs on which our lives are based.

www.ingramcontent.com/pod-product-compliance
Lightning Source LLC
Chambersburg PA
CBHW042313210326
41599CB00038B/7110